Local and Urban Governance

Series Editor

Carlos Nunes Silva, Institute of Geography and Spatial Planning,
University of Lisbon, Lisbon, Portugal

This series contains research studies with policy relevance in the field of sub-national territorial governance, at the micro, local and regional levels, as well as on its connections with national and supranational tiers. The series is multidisciplinary and brings together innovative research from different areas within the Social Sciences and Humanities. The series is open for theoretical, methodological and empirical ground breaking contributions. Books included in this series explore the new modes of territorial governance, new perspectives and new research methodologies. The aim is to present advances in Governance Studies to scholars and researchers in universities and research organizations, and to policy makers worldwide. The series includes monographs, edited volumes and textbooks. Book proposals and final manuscripts are peer-reviewed.

The areas covered in the series include but are not limited to the following subjects:

- Local and regional government
- Urban and metropolitan governance
- Multi-level territorial governance
- Post-colonial local governance
- Municipal merger reforms
- Inter-municipal cooperation
- Decentralized cooperation
- Governance of spatial planning
- Strategic spatial planning
- Citizen participation in local policies
- Local governance, spatial justice and the right to the city
- Local public services
- Local economic development policies
- Entrepreneurialism and municipal public enterprises
- Local government finance
- Local government and sustainable development
- Anthropocene and green local governance
- Climate change and local governance
- Smart local governance

The series is intended for geographers, planners, political scientists, sociologists, lawyers, historians, urban anthropologists and economists. **Local and Urban Governance—now indexed in Scopus**

Daniel du Plessis
Editor

Territorial Governance and Spatial Transformation in Post-apartheid South Africa

Springer

Editor
Daniel du Plessis
Geography & Environmental Studies
Stellenbosch University
Stellenbosch, South Africa

ISSN 2524-5449 ISSN 2524-5457 (electronic)
Local and Urban Governance
ISBN 978-3-031-85488-0 ISBN 978-3-031-85489-7 (eBook)
https://doi.org/10.1007/978-3-031-85489-7

This Springer imprint is published by the registered company Springer Nature Switzerland AG
The registered company address is: Gewerbestrasse 11, 6330 Cham, Switzerland

This book is dedicated to all South African urban and regional planners and other built-environment professionals who tirelessly dedicated their time and effort over the last three decades to create improved human settlements.

Preface

This book provides a reflective account of the evolution of territorial governance in South Africa over a period of 30 years after the transition to a post-apartheid society in 1994. It considers the role and influence of evolving territorial governance institutions and processes and their associated outputs (in the form of policies and plans) on different dimensions and at different scales of spatial transformation. Although it is way beyond the scope of a single book or publication to adequately address all dimensions of a complex phenomenon such as spatial transformation and associated institutional processes, the key findings documented in this book provide a contribution to contemplating important future focus areas for territorial governance and spatial transformation. A valuable contribution of this book is the insights provided through the extensive practical experience of all authors in territorial governance institutions and processes dealing with various elements of spatial transformation in post-1994 South Africa. The contributors are thus well-positioned to provide insightful longer-term reflections informed by personal real-world engagement in these processes and their outcomes. This ensures that the content is not only based on academic debates and theoretical concepts but also informed by practical experience gained from diverse backgrounds and contexts. These lessons and experiences that focus on the interplay between evolving territorial governance and urban spatial transformation are potentially not only of significance in the South African context but also globally in countries and regions faced with territorial governance and spatial transformation challenges. This compilation will be a useful resource for researchers, academics, students, government officials, practitioners in private practice, businesses, and community leaders involved in any of the numerous dimensions of spatial transformation. This book contributes to new insights on the future of territorial governance under conditions of societal change and transformation.

Stellenbosch, South Africa

Daniel du Plessis

Acknowledgements

The editor would like to acknowledge:

- All the authors who contributed chapters to this book and their efforts devoted to this project.
- All the peer reviewers who volunteered their time and energy to ensure the quality and academic rigour of this work through their valuable comments and inputs.
- Dr. Ruth Massey for her editorial assistance with the compilation of the draft combined manuscript.
- Jessica Eichhoff of the Centre for Geographic Analysis for her assistance in the compilation of some of the maps in this book.

Contents

Editor and Contributors

About the Editor

Daniel du Plessis is an urban and regional planner with more than 30 years of experience in the fields of urban management, policy development, and spatial analysis and planning of cities and urban systems. He spent the first half of his career as a town and regional planning consultant, during which time he successfully completed more than 200 research and contract assignments, both locally and internationally. These include spatial development plans and frameworks, housing plans and strategies at local, regional, and national level, long-term city development strategies, regional growth and development strategies, integrated development plans, and environmental management frameworks. He joined the Centre for Regional and Urban Innovation and Statistical Exploration (CRUISE) at Stellenbosch University in 2009 and has served as its director since November 2019. His research focusses on urban spatial policy and planning, innovation in urban planning systems, urban transformation, and spatial statistical analysis of cities and urban systems.

Contributors

Reece Alberts Unit for Environmental Sciences and Management, North-West University, Potchefstroom, South Africa

Andrè Brand Statistics South Africa, Pretoria, South Africa

Dirk Cilliers Unit for Environmental Sciences and Management, North-West University, Potchefstroom, South Africa

Antony K Cooper Council for Scientific and Industrial Research, Pretoria, South Africa

Department of Geography, Geoinformatics and Meteorology, University of Pretoria, Pretoria, South Africa

Ryan Daniels Centre for Regional and Urban Innovation and Statistical Exploration, Department of Geography and Environmental Studies, Stellenbosch University, Stellenbosch, South Africa

J. Ernst Drewes Unit for Environmental Science and Management, North-West University, Potchefstroom, South Africa

Daniel du Plessis Centre for Regional and Urban Innovation and Statistical Exploration, Department of Geography and Environmental Studies, Stellenbosch University, Stellenbosch, South Africa

Herman Geyer Centre for Regional and Urban Innovation and Statistical Exploration, Department of Geography and Environmental Studies, Stellenbosch University, Stellenbosch, South Africa

Stephan Gildenhuys Department of Urban and Regional Planning, University of the Free State, Bloemfontein, South Africa

Trynos Gumbo Urban and Regional Planning Department, University of Johannesburg, Johannesburg, South Africa

Anele Horn Department of Geography and Environmental Studies, Stellenbosch University, Stellenbosch, South Africa

Retsepile Kalaoane Urban and Regional Planning Department, University of Johannesburg, Johannesburg, South Africa

Tinus Kruger Council for Scientific and Industrial Research, Pretoria, South Africa

Karina Landman Department Town and Regional Planning, University of Pretoria, Pretoria, South Africa

Kiara Lawrence Town and Regional Planning, Durban University of Technology, Durban, South Africa

Martin Lewis Department of Urban and Regional Planning, University of the Free State, Bloemfontein, South Africa

Cecil Madell City and Regional Planning, University of Cape Town, Cape Town, South Africa

Masilonyane Mokhele Faculty of Informatics and Design, Cape Peninsula University of Technology, Cape Town, South Africa

Tshepiso Monnakgotla Department Town and Regional Planning, University of Pretoria, Pretoria, South Africa

Mark Napier Council for Scientific and Industrial Research, Pretoria, South Africa

Verna Nel Department of Urban and Regional Planning, University of the Free State, Bloemfontein, South Africa

Alexander Paterson Institute of Marine and Environmental Law, University of Cape Town, Cape Town, South Africa

Engela Petzer Council for Scientific and Industrial Research, Pretoria, South Africa

Claus Rabe Palmer Development Group, Cape Town, South Africa

Matšubu Ragoasha Council for Scientific and Industrial Research, Pretoria, South Africa

Francois Retief Unit for Environmental Sciences and Management, North-West University, Potchefstroom, South Africa

Claudine Roos Unit for Environmental Sciences and Management, North-West University, Potchefstroom, South Africa

Nosizo Sebake Council for Scientific and Industrial Research, Pretoria, South Africa

Mfaniseni Fana Sihlongonyane School of Architecture and Planning, University of the Witwatersrand, Johannesburg, South Africa

Avhatakali Sithagu School of Architecture and Planning, University of the Witwatersrand, Johannesburg, South Africa

Catherine Stone Independent Urban and Regional Planning Consultant, Cape Town, South Africa

Mariske van Aswegen Unit for Environmental Science and Management, North-West University, Potchefstroom, South Africa

Acronyms

ACSA	Airports Company South Africa
AI	Artificial intelligence
ANC	African National Congress
BEPP	Built Environment Performance Plan
BNG	Breaking new ground
BRT	Bus rapid transit
CBD	Central business district
CEF	Capital expenditure framework
CHE	Council for Higher Education
CKRSDF	Central Karoo Regional Spatial Development Framework
CLMS	Customary Land Management Systems
CMP	Coastal management plan
CPD	Continuing professional development
CSIR	Council for Scientific and Industrial Research
CSP	Cities Support Programme
DALRRD	Department of Agriculture, Land Reform and Rural Development
DCOGTA	Department of Cooperative Governance and Traditional Affairs
DDM	District Development Model
DFA	Development Facilitation Act
DPME	Department of Planning, Monitoring and Evaluation
DPWI	Department of Public Works and Infrastructure
EIA	Environmental impact assessment
EMF	Environmental management framework
EO	Earth observation
EPWP	Extended Public Works Programme
GDP	Gross domestic product
GEAR	Growth, Employment and Redistribution Strategy
GHSL	Global Human Settlement Layer
GIS	Geographic information system
GP	Growth potential
GPS	Global positioning system

HOA	Homeowners Association
ICM	Integrated Coastal Management
ICT	Information and communication technology
IDP	Integrated development plan
IDZ	Industrial development zone
IRPTN	Integrated public transport network
IUDF	Integrated urban development framework
KPI	Key performance indicator
LCPC	Land consumption per capita
LCR	Land consumption rate
LDO	Land development objectives
LED	Local economic development
LFTEA	Less Formal Township Establishment Act
LUE	Land use efficiency
LUMS	Land Use Management System
LUS	Land use scheme
MFMA	Municipal Finance Management Act
MISA	Municipal Infrastructure Support Agent
MLG	Multi-level governance
MRT	Mass rapid transit
MSA	Municipal Systems Act
MSDF	Municipal Spatial Development Framework
NDP	National Development Plan
NEMA	National Environmental Management Act
NEMBA	National Environmental Management: Biodiversity Act
NEMPAA	National Environmental Management: Protected Areas Act
NGO	Non-governmental organisation
NIPF	National Industrial Policy Framework
NPC	National Planning Commission
NPDP	National Physical Development Plan
NSDF	National Spatial Development Framework
NSDP	National Spatial Development Perspective
NUA	New Urban Agenda
OECD	Organisation for Economic Co-operation and Development
PCA	Public coastal access
PGR	Population growth rate
PHSHDA	Priority Human Settlement and Housing Development Area
PPA	Planning Profession Act
PRASA	Passenger Rail Agency of South Africa
PSDF	Provincial Spatial Development Framework
RAI	Regional Authority Index
RDA	Regional Development Agency
RDP	Reconstruction and Development Programme
REDZ	Renewable Energy Development Zone
RET	Radical economic transformation

RPL	Recognition of prior learning
RSA	Republic of South Africa
RSDF	Regional Spatial Development Framework
SACN	South African Cities Network
SACPLAN	South African Council for Planners
SADC	Southern African Development Community
SAITRP	South African Institute for Town and Regional Planners
SALGA	South African Local Government Association
SANCO	South African National Civic Organisation
SAPI	South African Planning Institute
SAQA	South African Qualifications Authority
SDF	Spatial development framework
SDG	Sustainable Development Goals
SES	Socio-ecological system
SEZ	Special economic zone
SOE	State-owned enterprise
SONA	State of the Nation Address
SPLUMA	Spatial Planning and Land Use Management Act
TOD	Transit-oriented development
UN	United Nations
UNA	Urban network analysis
UNDP	United Nations Development Programme
UNECA	United Nations Economic Commission for Africa
URP	National Urban Renewal Programme
USA	United States of America
WCPSDF	Western Cape Provincial Spatial Development Framework

Chapter 1
Territorial Governance and Spatial Transformation in South Africa: An Introduction

Daniel du Plessis

Abstract This chapter summarises the influence of apartheid planning on the spatial characteristics of South African cities and towns and highlights that spatial transformation in the post-1994 era has been more challenging than initially anticipated. The five key dimensions underpinning the notion of territorial governance in policy discourse are described as reference framework applied in the various individual chapters. Different definitions and approaches to spatial transformation are reviewed. The final section describes the overall objectives and general approach and provides a summary and roadmap of the individual chapters in the book.

Keywords Territorial governance · Spatial transformation · South African cities · Post-apartheid · Spatial planning

1.1 Introduction

South Africa's national spatial development pattern, as well as the spatial structure of individual cities and towns has been strongly influenced by both colonial and apartheid spatial development policies. The terms 'apartheid planning' and 'apartheid cities' thus became uniquely associated with the spatial structure of South African cities and are widely used concepts in local and international literature. The concept of apartheid cities can be interpreted in two related but distinct contexts. On the one hand, it refers to a historical phenomenon of the spatial patterns of cities and towns that were shaped during the period when enforced racially segregated development was pursued as a national policy after 1948 until the late 1980s. On the other hand, it also refers to contemporary post-1994 South African regions and cities that,

D. du Plessis (✉)
Centre for Regional and Urban Innovation and Statistical Exploration, Department of Geography and Environmental Studies, Stellenbosch University, Stellenbosch, South Africa
e-mail: ddp@sun.ac.za

D. du Plessis (ed.), *Territorial Governance and Spatial Transformation in Post-apartheid South Africa*, Local and Urban Governance,
https://doi.org/10.1007/978-3-031-85489-7_1

despite 30 years of democratic government rule, still reflect many of the pre-1994 physical and socio-economic characteristics. The spatial characteristics of apartheid cities are widely recognized and well-documented and include aspects such as segregated and fragmented spatial form, urban sprawl and comparatively low population densities compared with cities in other countries with similar incomes, peripheral location of low-income settlements, inverted density profiles increasing with distance from urban core areas, and unequal access to economic and social opportunities (Lemon 1991; Landman 2006; Todes 2006; Turok 2011; Parnell and Crankshaw 2013; SACN 2016). As a result, the notion of 'spatial transformation' emerged as a key element of territorial governance and spatial planning in South Africa after 1994 and became entrenched as an overarching national priority and policy objective. This national focus on spatial transformation required an alignment of territorial governance processes across all geographic scales as a precondition for successful implementation.

However, opinions on the effectiveness of territorial governance and the degree of success in achieving the transformation of national and urban space however remains divergent and it is increasingly recognised that spatial transformation in the post-1994 democratic era has been more challenging than initially anticipated. South Africa reached an important milestone in 2024 when it celebrated 30 years of democracy. This landmark provides an ideal opportunity to reflect on the successes and shortcomings of the territorial governance framework since 1994 and its influence on achieving the national goal of spatial transformation. This book applies the concept of territorial governance to reflect on the vision of spatial transformation pursued as a national priority in South Africa since 1994. As implied by the title the book fundamentally deals with two related concepts: territorial governance and spatial transformation. This introductory chapter starts by first reviewing the concepts of 'territorial governance' and 'spatial transformation' as applied in this book. Thereafter it provides a summary of the overall objectives and general approach of the book, and in the concluding section provides a summary and roadmap of the individual chapters.

1.2 Territorial Governance as a Concept

Defining the concept of territorial governance implies an understanding of the terms 'territorial' and 'governance'. The concept 'territorial' fundamentally implies a geographical area defined by aspects such as authority, legality, or administrative purpose (or combination thereof). Governance on the other hand implies institutional structures responsible for implementing processes (such as rules and procedures) to provide direction and with some form of authority within a given geographic territory (Davoudi et al. 2008).

The five separate but interrelated dimensions underpinning the notion of territorial governance in policy discourse as defined by Stead (2013) provide a useful reference framework for territorial governance as applied in this book. The first dimension is the vertical or horizontal coordination of policies and/or actors (including transnational and cross-border institutional cooperation). The different scales at which territorial governance can manifest range from national and supra-national levels right down to functional urban areas at the micro level (Davoudi et al. 2008). The second dimension of territorial governance is the promotion of participation and consensus-building among public and/or private actors. While the different interests within a defined territory may share certain objectives, it invariably also must deal with conflicting goals and aspirations. The nature of territorial governance thus requires a participative process in which various stakeholders cooperate to achieve convergent objectives and negotiate compromises between the conflicting goals and aspirations (Davoudi et al. 2008).

The third dimension is the devolution of powers and/or resources to lower levels of decision-making (Stead 2013). This implies a transition from hierarchical top-down government towards more relational and participative forms of governance involving the various manifestations of the state, local communities, and the market (Storper 2004). It involves defining territorial capital assets that serve the common good of the region and the socio-political valorisation of territorial capital (Storper 2004). Dimensions 4 and 5 are defined as the delivery of territorial cohesion by jointly devising territorial visions and strategies and assessing the territorial impacts of these visions and strategies.

In summary it can thus be stated that territorial governance as a process is concerned with managing territorial dynamics by organising and coordinating actors to develop territorial capital to enhance territorial cohesion (Davoudi et al. 2008). This concept of territorial governance thus implies two important principles: authority and influence shared among players at multiple levels of governance, and a focus on place-specific long-term strategies (Böhme et al. 2015). A key component of the broader concept of territorial governance that gives effect to these two principles is territorial planning. UN-Habitat (2015: 2) defines territorial planning as a "decision-making process aimed at realizing economic, social, cultural and environmental goals through the development of spatial visions, strategies and plans and the application of a set of policy principles, tools, institutional and participatory mechanisms and regulatory procedures." Spatial governance can be regarded as a specific subset of territorial governance which specifically deals with the management of land and space within a defined geographic area (territory). This also implies a paradigm shift in which "space" gave way to the "territory" as the basis of planning and development (Dias et al. 2022). Essential elements of spatial governance are achieving spatial cohesion through aspects such as connectivity improvements and addressing spatial disparities and imbalances (Zaucha and Böhme 2020).

Despite the widespread use and application of territorial governance as a concept it however not without criticism and challenges. Faludi (2012:198) questions the focus on multi-level governance and argues that "governments are in the business of governing" which implicitly includes governing at various levels. He also questions the view of the world as a collection of fixed territories (or territorial units) and points out an alternative view of territory as a flexible social construct (Faludi 2012). A recurring challenge in territorial governance associated with dimensions two and three outlined above is the risk of diluting the accountability and competency of decision-making associated with the challenge of lower-scale decentralisation risking the hijacking of public policy by special interest groups (Fishell 2002).

1.3 Spatial Transformation in a South African Context

As outlined in the previous section territorial governance is focused on the overall management of a defined territory and includes various dimensions such as land (where spatial governance plays an important role) and the associated social, economic, and environmental resources/systems. Spatial governance and spatial transformation are thus inextricably linked in the South African context. Although widely used in both academic literature and in policy and legislative documents, spatial transformation is a fluid term, and definitions vary in their focus and nuance depending on the scale and theme under consideration. The term is also widely used in the vocabulary of professionals, officials, and politicians alike, often without any clear understanding of the practical meaning and implications thereof. Spatial transformation as a concept is also not unique to the South African context and has been used in plans and policies in many parts of the world where it is generally associated with some form of major urban restructuring or change (Turok 2014), and often associated with the physical conversion at various scales (from neighbourhoods to city level) through processes such as urban regeneration and renewal (Harrison and Todes 2015).

Oranje and Van Huyssteen (2011) argued that the restructuring of the spatial economy at various spatial levels aimed at sustainable, equitable and inclusive growth should be a key component of spatial transformation. Harrison and Todes (2015) further also emphasized the importance of aspects such as improved access to urban opportunities and greater efficiencies, as well as reducing the economic marginality of townships as a key element of spatial transformation and emphasizes the role of more compact and integrated urban forms to achieve these desired changes. The outcome of this transformation should ultimately facilitate access to a range of opportunities and contribute to a more inclusive and cohesive society (Gasa 2016). A further critical dimension of spatial transformation not often considered is preventing spatial segregation from re-establishing itself in new forms of post-apartheid spatial segregation and spatial inequality based on class and income (SALGA 2012).

The overall prominence of spatial transformation as a national priority is reflected in the National Development Plan (NDP) that identified the need for a national focus on spatial transformation supported by policies, plans and instruments across all geographic scales. It proposes that a national focus on spatial transformation should result in tangible outcomes such as reduced travel distances for poor households, a larger proportion of the population living in closer proximity to their places of work, and the availability of safe, reliable and energy efficient transport (NPC 2012). The Spatial Planning and Land Use Management Act of 2013 further expects spatial planning and spatial plans to be transformative and conforming to the principles of spatial justice to redress past spatial and other development imbalances. Building on the recommendations of the NDP, the Integrated Urban Development Framework (IUDF) provides more detail on the intentions of Chapter 8 of the NDP dealing with the transformation of human settlements and the national space economy (DCOGTA 2016). The IUDF describes spatial transformation as changing existing city footprints to produce compact, coordinated and connected cities, using transit-oriented development (TOD) and other urban planning strategies. Its aim is to reduce urban sprawl and the encroachment of valuable agricultural and natural land by encouraging higher-density development, mixed land use, and the revitalization of existing urban areas (DCOGTA 2016). It was however only with the finalization of South Africa's National Spatial Development Framework (NSDF) in 2022 that an official and unifying definition for the concept of spatial transformation was established for the first time (DALRRD and DPME 2022: 19–20): "The carefully and collaboratively planned and well-managed process of (1) locating infrastructure, (2) providing social services, and (3) allowing, placing and incentivising economic activities in settlements in such a way that the segregated spatial patterns inherited from colonial and Apartheid times are broken down, and the inefficiencies, injustices and inequalities in access to economic and other opportunities resulting from these past patterns are corrected."

However, despite this focus on spatial transformation, the NDP recognised that spatial transformation in the post-1994 democratic era has been more challenging, and the rate of actual change slower than initially anticipated (NPC 2012). It also identified the need to establish a firm understanding of why expectations have not been met and identify appropriate responses to accelerate further future spatial transformation towards 2030 and beyond. The emphasis of spatial transformation debates thus shifted from a focus on 'how' to achieve transformed sustainable human settlements during the first decade post-1994, to the more recent focus of why spatial transformation has progressed slower than anticipated and in certain instances not been achieved (SALGA 2012).

As indicated earlier, spatial governance is a key component of broader territorial governance. Spatial Development Frameworks at various spatial scales as envisaged in the national Spatial Planning and Land Use Management Act (Act No. 16 of 2013) (SPLUMA) are thus a central instrument in territorial governance to achieve spatial transformation and is closely associated with dimensions 4 and 5 of territorial governance as described in Sect. 1.2 (the delivery of territorial cohesion by jointly devising territorial visions and strategies, and assessing territorial impacts). The role of Spatial Development Frameworks (SDFs) in territorial governance and its influence on spatial transformation is thus an essential element considered in more detail in several of the subsequent chapters. Municipal SDFs is of particular relevance as local government is arguably best placed to drive change in the built environment that leads to spatial transformation (SACN 2016).

1.4 Purpose and Objectives of This Book

This book uses the different dimensions of territorial governance as described in Sect. 1.2 as the lens for reflecting on progress with transforming territorial governance institutions and processes and its influence on spatial transformation in post-1994 South Africa. These five different dimensions clearly permeate the themes covered in the various subsequent chapters. The official NSDF definition of spatial transformation as outlined in Sect. 1.4 provides the overall reference framework for understanding spatial transformation, and more context specific and nuanced interpretations are also considered in the various individual chapters. Based on these two central principles of territorial governance the purpose of the book is thus twofold. It firstly provides an overview of the evolution of the territorial governance framework in South Africa after 1994 in various sectors, contexts and localities. Secondly, it considers the role and influence of these evolving governance structures and their associated outputs (in the form of policies and plans) on different dimensions and at different scales of spatial transformation. This is pursued through three related objectives:

- To provide a reflective account of the post-1994 period focusing on experiences and lessons learnt from the transformation of territorial governance institutions and processes.
- To consider the role and influence of the changing governance framework and its resulting 'products' in the form of plans/policies on different dimensions and scales of spatial transformation over the last three decades.
- To synthesise the lessons learnt and the potential implications for future territorial governance and spatial planning approaches.

The approaches and methods applied by the authors of the various chapters include a combination of historical and contextual analysis, policy content analysis, policy evaluation, case studies, and primary quantitative and qualitative analysis. All authors have extensive practical experience of territorial governance institutions and processes dealing with various elements of spatial transformation in post-1994 South Africa and hence well-positioned to provide insightful longer-term reflections informed by personal real-world engagement in these processes and their outcomes.

1.5 Structure and Content

The rest of the book consists of three parts that are aligned with the objectives outlined above:

- The chapters in Part One focus on the post-1994 transformation of several key components of territorial governance including regional governance (Chap. 2), governance and spatial transformation at municipal level (Chap. 3), evolving

approaches to participatory planning in territorial governance (Chap. 4), and the transformation of the planning profession in South Africa post-1994 (Chap. 5).
- Part Two consisting of Chaps. 6, 7, 8, 9, 10, 11, 12 and 13 focusses on the influence of territorial governance and its resulting governance instruments on various dimensions of spatial transformation. These include aspects such as land use efficiency patterns and trends of the South African urban system (Chap. 6), territorial governance and spatial transformation within the conservation and protected areas context (Chap. 7), the role of territorial governance in South Africa's urban spatial economy (Chap. 8), the influence of coastal gated developments on spatial and territorial control (Chap. 9), the role of public transport in urban spatial transformation (Chap. 10), the influence of territorial institutional arrangements on airport-centric developments (Chap. 11), customary land management systems as a form of territorial governance in informal peri-urban settlements (Chap. 12), and considering the implications of territorial governance under Traditional Authority on urban spatial transformation (Chap. 13).
- The chapters in Part 3 contemplate several potential considerations and interventions for future territorial governance and spatial transformation. Chapter 14 considers the vexing question whether spatial development frameworks are effective as strategic plans and their continued role and relevance in spatial transformation, Chap. 15 considers whether smart city concepts can contribute to spatial transformation, and Chap. 16 concludes with some sobering thoughts regarding a potential more pragmatic approach to territorial governance and planning.

In Chap. 2 Ernst Drewes and colleagues reflect on the changing regional planning landscape in South Africa over the last three decades and evaluate the existing structures and instruments focused on spatial transformation. It explores various forms of multi-level governance (MLG) as a potential tool for effective regional governance towards a more transformed national space in several case studies. They identified regional government as the critical factor in the implementation of regional policy and regional frameworks, and decentralised or deconcentrated authority (or a combination thereof) to strengthen the role of this level of government. The vital role of establishing locally informed and nationally supported regional strategies as enablers of national objectives and respondents to local needs in the regional milieu is highlighted. Both regionalisation (nationally downwards) and regionalism (locally upwards) are suggested as tools for more robust regional tier support and to reinforce regionally focused governance.

Chapter 3 reviews territorial governance and its influence on spatial transformation at municipal level. In this chapter Verna Nel and Stephan Gildenhuys point out that although South Africa has made significant strides in reshaping its urban landscape and transitioning from apartheid-era policies to more inclusive and sustainable approaches, challenges still persist. They identify the need for territorial governance that promote justice and inclusivity, specifically in Land-use schemes (LUS) and SDFs as key instruments of spatial transformation. The chapter indicates that current approaches lack the required impact due to a reliance on anecdotal evidence instead of empirical assessment, resulting in irrelevant plans and ineffective monitoring and evaluation. It suggests some potential alternative solutions such as configurational approaches in spatial planning that integrate network science and complexity theory, and the transect concept that aims to promote compact, pedestrian-friendly, and sustainable urban forms that harmoniously integrate with natural ecosystems. It concludes that merging the configurational approaches with the transect concept might bridge this knowledge gap and promote successful spatial transformation in South Africa as both a spatial planning and land use management tool.

In Chap. 4 Anele Horn reflects on one of the most profound changes to territorial governance processes in South Africa after 1994—the recognition of the importance of public participation in various planning instruments. Since 1994, there has been deliberate efforts towards stakeholder participation emphasised in post-apartheid legislation and guidelines for Integrated Development Plans (IDPs) and SDFs. These frameworks prescribe in detail how such engagements can and should take place, however most stakeholder engagement processes still seem unable to produce plans that appreciate and reflect community dynamics that can lead to community social and spatial transformation. This chapter focusses on civil society organisations such as NGOs as emerging critical role players in spatial and social transformation. Their proximity to communities often affords them a better understanding of local opportunities and challenges than municipal authorities. This chapter reviews the lessons learnt from traditional consultative approaches applied in territorial governance processes in South Africa since 1994, considers the impact of these processes on spatial transformation in general, and reflects on the potential value of novel approaches such as community-based networks that have emerged.

Urban and regional planners should by implication be central role players in the process of driving spatial transformation. This however requires a transformed planning profession with the knowledge, skills, and desire to improve the existing situation. Martin Lewis and Verna Nel addresses three areas of change in Chap. 5: the planning institutions, the content of planning programmes, and the knowledge and abilities planners need. It first describes the reform of planning institutions and their role in transitioning to a multiracial profession post-1994. Thereafter, it identifies the core knowledge and skills required by planners and its influence on planning qualifications at universities to ensure that graduates can contribute to transformation. It also explains the 'internship' period which ensures that planners can apply the skills in practice, and the continuing professional development requirements emphasising the need to remain up to date with theory and practice.

The chapters in Part Two consider the influence of evolving governance structures and their associated outputs (in the form of policies and plans) on different dimensions and at different scales of spatial transformation. In Chap. 6 the authors provide a detailed analysis of the Land Use Efficiency (LUE) patterns and trends of the South African human settlement system. It analyses and compares the LUE of different categories of cities and towns within the South African urban system over the period 1990 to 2020 by applying the United Nations SDG Indicators as a global comparative measure to assess LUE. The results indicate that territorial governance instruments introduced since 1994 have contributed to an overall improvement in LUE, albeit highly uneven amongst different settlement categories and parts of the country. These results may be sobering for the future of spatial planning and policy direction in South Africa and may indicate the need to further refine local spatial plans and policies taking into consideration the full range of perspectives of LUE within varying local contexts.

South Africa is well-known for its biodiversity and national protected areas as iconic tourist destinations. Dirk Cilliers and colleagues focus on the country's national parks, which are directly impacted by surrounding activities with a specific focus on buffer zones as territorial governance instrument to mitigate these impacts in Chap. 7. They point out that within the context of spatial transformation at a national scale it is essential to integrate these buffer zones into spatial planning processes. This chapter examines how municipal spatial development frameworks (MSDF) incorporate national park buffer zones, by analysing ten MSDF case studies from nine national parks. The results show that many MSDFs don't adequately recognise or define these buffer zones, and they lack specific land use guidelines for them. The majority of MSDFs evaluated failed to integrate the applicable national park buffer zones into their development visions and land use guidelines. The authors argue that the incorporation of national park buffer zones into MSDFs symbolises more than mere environmental conservation but represents an embodiment of South Africa's broader aspirations for spatial transformation and territorial governance.

Chapter 8 explores and reflects on the role of territorial governance and public policies in transforming the spatial economy of urban areas in South Africa during the post-apartheid era. In this chapter Cecil Madell argues that contemporary South African cities and towns are reproducing the legacy of colonial and apartheid patterns of economic exclusion and inequality, perpetuating the entrapment of the majority of urban inhabitants in dormitory townships, devoid of meaningful economic opportunities, underserved in terms of basic utility services, remotely located from established urban centres and poorly served with public transport and other critical social services. The chapter reflects on why state spatial and economic policy interventions inadvertently obstruct the poorest residents of towns and cities from accessing economic and other opportunities, even though South Africa's constitution and plethora of legislation are geared at building a socially, economically, and spatially just and inclusive society. This critique brings into focus the implications and contradictions in the state's policy and legislation over the last 30-years and reflects on the required changes in the approach to urban development and territorial governance to advance economic transformation of South Africa's urban spatial economy.

Gated developments have proliferated in South Africa as a response to security and local management challenges. However, these manifestations of territorialisation have numerous implications for the country's aspirations towards spatial transformation and justice. Tshepiso Monnakgotla and Karina Landman in Chap. 9 focus on gated developments within the context of public coastal access in South Africa.

It deals with negotiated privilege by exploring the relationship between luxury gated estates, municipalities, and fishermen communities on the coast. Public coastal access is often restricted through gated developments, which has implications for spatial transformation related to coastal access for poor communities. The chapter examines how municipalities deal (or fail to deal) with public coastal access restrictions in the broader context of territorial governance and spatial transformation. The case studies considered in this chapter reveal how some gated developments negatively impact spatial transformation and highlights the role of local and district governments in relation to territorial governance. It also highlights the fact that coastal estates also reflect notions of spatial or territorial control and private micro-governance that in some cases may constrain democratic accountability.

Trynos Gumbo and colleagues in Chap. 10 discuss the role, influences, and outcomes of urban public transport systems in South Africa focusing on the achievements that have been recorded in the Gauteng and KwaZulu Natal provinces in formulating and implementing public transport policy and planning. The approach adopted includes a systematic review of literature and a bibliometric analysis. The results reveal meaningful progress in formulating plans and strategies at these provincial levels which have significant potential for transforming urban spaces through large-scale urban public transport programme implementation. Spatial inequality however remains a significant issue, with many communities still facing long-standing structural and systemic barriers to accessing economic opportunities and essential services.

Large-scale infrastructure development projects such as the 2010 soccer world cup stadiums, the Gautrain express commuter rail system, and various airport related developments had significant impacts on the spatial development of South African cities since 1994. In Chap. 11 Masilonyane Mokhele analyses the institutional arrangements that affect the planning of development that hinges on airports. Framed under institutional economic geography and collaborative planning, the chapter focuses on the stakeholders' roles, interdependencies, synergies and contentions in the Ekurhuleni Aerotropolis planning. It established that although there was ultimately consensus among the stakeholders, there was no territorial cohesion and a shared vision for the Ekurhuleni Aerotropolis. There were strong disagreements on the extent of the territory, with some stakeholders arguing the territory should be the municipality as defined by the administrative boundaries. In contrast, others contended it should be the Gauteng City-Region defined by the functional linkages across municipalities. It is recommended that in South Africa, the focus could be on linking airport-centric development planning to regional spatial development frameworks instead of limiting them to municipal spatial development frameworks.

Chapter 12 analyses the territorial governance of informal development in developing regions within a deregulated policy framework and a hybridisation of regulatory authorities. In this chapter Herman Geyer focuses on the co-federal collaboration between informal horizontal territorial regimes operating through customary land management systems (CLMS) in informal settlements, and the formal vertical governance structure of municipal and provincial planning and policies. In this mode of territorial governance, the CLMS provides residents in micro-regions with accessible and transparent mechanisms to participate in the region's governance through the South African National Civic Organisation (SANCO), an organisation representing the interests of thousands of street committees within local and national government. The chapter established that formal regulatory authorities in various municipal and provincial agencies and specialised units take on a new role in selectively enforcing policy regulations based on the normative principles of good governance by negotiating compliance and compromises with community representatives to effectively manage certain challenging aspects of informal development. The CLMS thus becomes an extension of the state by ensuring partial compliance and self-regulation of policy and practices.

Mfaniseni Sihlongonyane and Avhatakali Sithagu in Chap. 13 deal with transformation of territorial governance under Traditional Authority in peri-urban areas and the associated spatial implications for urban change. The Constitution of the new South Africa (1996) recognised traditional authorities and afforded them opportunities for a political resurgence. This is occurring in a period of rapid urban growth especially in peri-urban areas within the sphere of extended urbanisation. This chapter shows that while municipal governance attempts to control urban growth using legal and planning instruments, traditional authorities are shaping the location, direction, and expansion of urban growth through customary systems of tenure rights and responsibilities. It uses the case of Thembisile Hani Local Municipality (THLM), a peri-urban area in the Mpumalanga Region, South Africa, to discuss the factors shaping the transformation of territorial governance under traditional authority in peri-urban areas and the associated spatial implications for urban change. The central argument is that the changing status of traditional authorities and their integration into global/economic dynamics of urban change have made them counterparts in shaping urban transformation, especially in peri-urban areas.

In Chap. 14 Catherine Stone explores the effectiveness of Spatial Development Frameworks (SDF) as key strategic planning tools within the territorial governance system in South Africa and a primary policy instrument for driving spatial transformation within this system. As frustration mounts with the slow pace of spatial transformation of cities and towns, many conclude that spatial plans, and spatial planners, notwithstanding the policy commitments made in SDFs, have failed. In this chapter the author shows that there have been significant steps taken after the introduction of SPLUMA to embed the role of the plan more rigorously in governance systems and hold decision-making practices accountable to implementation of the spatial plan. Taking cognisance of the systemic challenges in the current territorial governance framework and because transformation of the built environment is inherently slow and incremental, the role of SDFs in influencing decision-making is not always visible and perhaps under-estimated. This chapter firstly explores how the effectiveness of a spatial plan or SDF should be assessed, drawing on suggestion that a strategic plan is a tool for supporting and influencing decision-making and cannot dictate outputs and outcomes. As such its effectiveness should be assessed based on its role as a 'learning framework' or decision support tool. Secondly, the chapter explores contextual and structural factors that influence the implementation of SDFs and argues that scrutiny of these factors is important to more accurately understand the causes of the inertia that hinder significant change, despite the best laid out spatial plans.

Municipalities, as the territorial governance structures closest to citizens and mandated to lead local development, have employed various planning concepts with the aim of effecting spatial transformation. In Chap. 15 Engela Petzer and colleagues reflect on the concept of smart cities, including its international origins in the early 1990s and the lack of a collective understanding and varied definitions of the term. The chapter explores what a smart approach to spatial transformation could look like and discusses lessons that can be learned from local South African smart city initiatives. The chapter concludes that the impact of current South African municipal smart city initiatives on spatial transformation is not immediately apparent; however, when integrated into the system of territorial governance—which aims to unify a place-based approach and multi-level governance—the smart city concept may contribute to spatial transformation over the long term, given an appropriate implementation approach.

Chapter 16 contrasts the situational merits of normative planning in the nation-building period (1994–2008) with its role in the subsequent economic malaise gripping South African cities. Drawing on the life and death of three urban policy concepts, ‘Local Economic Development’, ‘Transit-oriented Development’ and ‘Smart Cities’, Claus Rabe argues that out of necessity the South African body politic is implicitly reassessing the role of the State in the realisation of its economic objectives. It further argues that urban practitioners face a choice either eschewing or embracing pragmatism in practice and discourse. Rabe argues that the one option—the continued elevation of abstract planning ideals over the messy reality on the ground—will lead to the deepening irrelevance of the planning profession in broader policy debates and co-option by short-term political interests. In contrast as an alternative option, he suggests the pragmatic planner as absorbed by identifying and understanding interventions that work—that impact citizens’ material wellbeing in an immediate and measurable way. The chapter argues that within this alternative approach the pragmatic planner appreciates the role of functional theory as an essential precondition for effective policymaking. By augmenting each stage of the policy process with empirical evidence and applied theories on proximate causes and effects, the pragmatic planner is better positioned to regain a foothold at the policy table.

References

Böhme K, Zillmer S, Toptsidou M, Holstein F (2015) Territorial governance and cohesion policy. European Union

Davoudi S, Evans E, Governa F, Santangelo M (2008) Territorial governance in the making. Approaches, methodologies, practices. Boletin de la AGE 46:33–52

Department: Agriculture, Land Reform and Rural Development (DALRRD) & Department: Planning, Monitoring and Evaluation (DPME) (2022) National spatial development framework, Pretoria

Department: Cooperative Governance and Traditional Affairs (DCOGTA) (2016) Integrated urban development framework. A new deal for South African cities and towns. DCOGTA, Pretoria
Dias RC, Seixas PC, Lobner N (2022) State transformations through public, multilevel, and territorial governance in European union: towards a metagovernance paradigm. Eur Policy Anal 8:467–483. https://doi.org/10.1002/epa2.1163
Faludi A (2012) Multi-level (territorial) governance: three criticisms. Plan Theory Pract 13(2):197–211. https://doi.org/10.1080/14649357.2012.6775
Fishell W (2002) The home voter hypothesis: how home values influence local government taxation, school finance, and land-use policies. Harvard University Press, Cambridge, MA
Gasa B (2016) In pursuit of a comprehensive economic transformation agenda. In: Vision 2030: our future our plan. Topco Media, Cape Town, pp 247–251. https://issuu.com/topcomedia/docs/vision_2030_publication_2015
Harrison P, Todes A (2015) Spatial transformations in a "loosening state": South Africa in a comparative perspective. Geoforum 61:148–162
Landman K (2006) Privatising public space in post-apartheid South African cities through neighbourhood enclosures. GeoJournal 66:133–146
Lemon A (1991) The apartheid city. In: Lemon A (ed) Homes apart. South Africa's segregated cities. Paul Chapman, London, pp 1–25
National Planning Commission (NPC) (2012) National Development Plan: vision for 2030. National Planning Commission, Pretoria, p 6
Oranje M, Van Huyssteen E (2011) Nestling national 'transformation' imperatives in local 'servicing' space: critical reflections on an intergovernmental planning and implementation project. Town Reg Plan 58. http://www.ajol.info/index.php/trp/article/viewFile/77204/67655
Parnell S, Crankshaw O (2013) The politics of 'race' and the transformation of the post-apartheid space economy. J Housing Built Environ 28:589–603. https://doi.org/10.1007/s10901-013-9345-6
SACN (South African Cities Network) (2016) State of the cities report 2016. SACN, Johannesburg
SALGA (2012) Spatial transformation barometer: towards refinement of indicators for municipalities. SALGA, Pretoria
Stead D (2013) Dimensions of territorial governance. Plan Theory Pract 14(1):142–147. https://doi.org/10.1080/14649357.2012.758494
Storper M (2004) Territories, flows, and hierarchies in the global economy. Reading Econ Geogr 3:271
Todes A (2006) Urban spatial policy. In: Pillay U, Tomlinson R, du Toit J (eds) Democracy and delivery. Urban policy in South Africa. HSRC Press, Cape Town, pp 50–74
Turok I (2011) Deconstructing density: strategic dilemmas confronting the post-apartheid city. Cities 28:470–477
Turok I (2014) Settlement planning and urban transformation. South African Cities Network (SACN), Johannesburg, pp 74–99
UN Habitat (2015) International guidelines on urban and territorial planning. UN Habitat, Nairobi
Zaucha J, Böhme K (2020) Measuring territorial cohesion is not a mission impossible. Eur Plan Stud 28(3):627–649. https://doi.org/10.1080/09654313.2019.1607827

Part I
Transformation of Territorial Governance

Chapter 2
Regional Governance and Spatial Transformation in South Africa Post-1994

J. Ernst Drewes, Mariske van Aswegen, and Andrè Brand

Abstract The governance institutions in South Africa saw immense change in structure and application after 1994, based on an expanded number of provinces and determining district- and metropolitan government municipalities. The post-apartheid instruction of spatial transformation was provided through various pieces of legislation and policy, but the application thereof differs significantly when applied to this intermediate level of regional governance. The Constitution of South Africa clearly spells out the statutory role local government must play as a decision-maker on spatial planning matters—the role of regional governance is much more an advisory and compliance role in facilitating spatial planning and the resultant focus of transformation.

This chapter aims to reflect on the changed regional planning landscape in South Africa over the last three decades and evaluate the existing structures and instruments focused on spatial transformation. The policy and legislation available to regional governments will be analysed against the backdrop of effective spatial transformation. Furthermore, this chapter will explore case studies as a form of multi-level governance (MLG) and a potential tool for more effective regional governance towards a more transformed national space.

Keywords Regional government · Adaptive governance · Spatial transformation · Regional Spatial Development Framework · Multi-level governance

J. E. Drewes (✉) · M. van Aswegen
Unit for Environmental Science and Management, North-West University, Potchefstroom, South Africa
e-mail: ernst.drewes@nwu.ac.za

A. Brand
Statistics South Africa, Pretoria, South Africa

D. du Plessis (ed.), *Territorial Governance and Spatial Transformation in Post-apartheid South Africa*, Local and Urban Governance,
https://doi.org/10.1007/978-3-031-85489-7_2

2.1 Introduction

According to Walker et al. (2004), the concept of transformation relates to the deep-seated renewal of a system when the existing conditions, be it social, political, economic, or ecological, are not sustainable or do not reach their core goals (Scheffer et al. 2003). Transformation is often the only resolution to a state of "lock-in" or stagnation within a particular system or region (Allison and Hobbs 2004). This chapter will focus on spatial transformation from the regional government's perspective in South Africa. It firstly focuses on the types of regional government from a functional and international perspective, including the importance of decentralisation of government. The subsequent focus in Sect. 2.3 will be on the evolution of regional government in South Africa through three main eras in South Africa's recent history. Spatial transformation using regional plans will be analysed in more detail through case studies in two regions, utilising their Spatial Development Frameworks (SDFs) in Sect. 2.4.

2.2 The Role of Regional Government in Regional Planning

Regional planning, often denoted as regional or local planning, intra-regional planning, or regional strategic planning, is a multifaceted discipline that encompasses various approaches to spatial and economic development (Glasson and Marshall 2007). For this chapter, the term "regional" is defined as a tier of planning that exists between national planning and local (settlement-level) planning. This intermediate level serves as a critical nexus, effectively linking national economic strategies with local physical planning initiatives (Keeble 1969).

The significance of regional planning lies in its ability to address issues that transcend local boundaries while still being grounded in the specific needs and characteristics of individual communities.

Regional planning constitutes an encompassing approach to socio-economic, political, environmental and infrastructural elements within a specific region. Regional planning assists in shaping the pattern of human activity within a specific region (Chaudhuri 2001). This form of spatial planning is often characterised by friction as it attempts to simultaneously take on practical problems and abstract and theoretical concepts. This confrontation of abstract issues such as equity and efficiency, stand directly opposed to the more practical issues of disequilibrium or underdevelopment (Cumberland 1971).

The responsible authority for regional planning and development is crucial in the process of policy formulation, which in turn impacts on spatial frameworks and development projects, as well as the allocation of development funding. From an international perspective, the responsible authority for regional planning and development plays a crucial role in policy formulation and the assignment of funds for development projects. The function of regional development is often found within the central government, or that of an interministerial committee. The literature highlights that this function mostly resides under another ministry, and rarely constitutes a ministry of its own. The task of regional development usually falls within the management of an urban or rural development ministry and could also be part of an agricultural department. These national ministries are more concerned with a national spatial vision, international linkage, and infrastructure provision, which could lead to the neglect of regional-level spatial planning and development. On the regional level, the implementation of national plans and the coordination of the local level of government are the focus. An emphasis on regional spatial policy, rather than a national level focus, strengthens endogenous development strategies (Angell et al. 2016). This enables a localised focus on the affected communities` prosperity, rather than a common focus across the entire country.

Conventional approaches to vertical top-down governance approaches are dwindling in popularity as their failures and lack of adaptability come under scrutiny (Folke et al. 2005; Lebel et al. 2006) during an era of a growing understanding of the complexities associated with socio-ecological systems (SES). The old paradigm of policymaking from the national government in a top-down approach, often led to segregation between central and regional government and other institutions (OECD 2010; Van Aswegen 2018). This urges a more adaptive approach to governance able to deal with the continuous uncertainties and changes of a fundamentally complex world.

2.2.1 Multi-Level Governance (MLG)

It is acknowledged that the challenges faced by the developed and developing world are unique (Pahl-Wostl 2009) and therefore the approaches should inform this uniqueness. Struggles with ineffective governance, exploitation, and inefficient service delivery are most often associated with developing tendencies. On the other hand, developed countries also face various challenges including dominance of economic growth over environmental development, sectoral fragmentation, overregulation and sustainable development. It is postulated that progressive polycentric and multi-layered government, coupled with efficient leadership will enhance institutional resilience through its adaptive capacity. Such a government will enable participation, and be accountable, deliberate and just in its actions. Such an evolutionary change from bureaucratic governance will furthermore stimulate critical thinking in the policy-making process (Pahl-Wostl 2009) to include various actors and institutions through a multi-layered approach. Pahl-Wostl (2009) indicates that governance structures exhibiting greater polycentricity and a balance between top-down and bottom-up approaches tend to possess higher adaptive capacities, implying improved sustainability and a more diverse governance landscape. Within such systems, top-down and bottom-up networks coexist alongside each other or run in parallel within a framework of intermediate complexity. Internationally governments have increasingly been focused on dispersing authority away from the central state concept to that of a multilevel approach with the term "multilevel governance" (MLG) (Marks 1993; Hooghe and Marks 2001) growing in popularity (Pahl-Wostl 2009). Through the MLG approach autonomy is focused at the regional level and in support of local development, reminding of the fundamentals of regionalism, This bottom-up approach proposes community participation and engagement to address specific problems (Campbell 1996), often fixated on localised social and environmental concerns. Regionalism from the local level encourages local socio-economic development, whereas MLG recognises the need for a twofold focus including a national-to-local input supported by a local-to-national approach, in essence combining the concepts of regionalisation and regionalism.

Governments are faced with challenges of ineffective coordination between various spheres, as well as untangling the roles and responsibilities of each sphere while respecting and maintaining their mutual dependence (Allain-Duprè 2020). The OECD (2017, 2019) highlights the importance of MLG in the context of growing regional inequality, policy complexities, and pressures from increasing urbanisation and climate change, advocating for "robust multi-level arrangements among national, regional and local government, as well as across jurisdictions". In a similar fashion and expressing this urgent need for the decentralisation of powers, the UN (UN-Habitat 2017) in the New Urban Agenda (NUA) stresses the importance of "inter-municipal cooperation on a global, regional, national, subnational and local scale", with a focus on adapting the more traditional inter-governmental roles, the various frameworks among these levels and newly defined relationships of mutual dependence among spheres of government (Bache and Flinders 2004).

MLG also requires a multi-level policy approach, especially considering that spatial concerns will differ across territorial boundaries (Hooghe and Marks 2001). The way MLG is designed directly impacts the quality and results of development policy (Allain-Duprè 2020). The OECD (as a leading organisation in MLG policy-making) upholds that more integrated policy across various levels of government, in a vertically and horizontally integrated manner, will result in a more successful policy response across all levels of social, economic, and environmental development.

MLG is often mistaken for the reduction of responsibility and loss of control at the central/national government level. This is, however, not the case, given that MLG has a twofold focus: (i) the transformation or reform of the role of the central government and (ii) an asymmetric regional government, translating to multiple types of MLG systems coexisting at the same level of government (OECD 2019). This will be elaborated on in Sect. 2.2.2.

As with any concept, MLG is often criticised for its lack of prescriptive measures, unclear scope, and difficulty to measure. Various obstacles could arise with the implementation of a form of MLG, which include the following: (i) Initial costs of formal procedures associated with coordination; (ii) Cross-sectoral coordination challenges within the same level of government; (iii) Capacity challenges on the various levels of government; (iv) failure to successfully coordinate vertically between levels of government, leading to gaps in policy design and implementation.

2.2.2 *Devolving Power*

Internationally, sub-national/regional levels of government have attained power through devolvement from the national government, This is especially true for strategic planning, and the allocation of fiscal resources to projects as identified from the national level. This devolvement of power to a deconcentrated, or decentralised regional authority armoured with more power and responsibility, is recognised as an inevitable tool for policy creation and enactment of national objectives.

The United Nations defined decentralisation and deconcentration as the:

> "...process of reorganisation of the State that involves a gradual transfer of responsibilities originally concentrated in central government towards other levels of government (regional, provincial governments, or municipalities) accompanied by the necessary resources to fulfil such responsibilities. The principle behind these processes is the belief that decision-making and implementation are more efficient if taken as close to citizens as possible (subsidiarity principle)" (UN-Habitat 2015: 1).

The OECD (2017), through various case studies and practical applications, notes that there is no single optimal design for MLG through the devolution of power and that the success of the process will depend on its unique design for each case (Allain-Duprè 2020).

Ribot (2003) attests that the decentralisation of authorities is strengthened by heightened public participation, leading to an increase in downward responsibility and a more efficient local development focus. Whereas Bergvall et al. (2006) indicate the twofold impact of the decentralisation of government, on the one hand, the increase in the responsibility of the regional authority, but on the other hand, a continued and often increased dependence on the national government for support and resources.

Therefore, decentralisation not only affects the subnational government but also the central government. Decentralisation implies that the central government will have a more strategic role, complimented by a coordination and alignment role pertaining to policy, monitoring and alignment across all regions in the national space (Allain-Duprè 2020). Decentralisation reforms involve a shift from a direct role in service delivery to one of enabling, advising, assisting, ensuring consistency, and facilitating the work of regional governments. This requires building new capacity at the central government level to manage these new functions (OECD 2019) effectively. Often, this changing role of the central government is underestimated by countries undertaking decentralisation reforms, taking into consideration that for the same level of government, diverse types of responsibilities or MLG arrangements could exist (Allain-Duprè 2020). This could furthermore lead to the duplication of certain functions or responsibilities resulting in additional costs. This transfer of power and certain functions to the regional level through decentralisation of powers is guided by two mechanisms to govern this process (Oates 2005), Firstly regional strategies are viewed as supplementary to national policies, aiming to localise the national initiatives from the regional level. Hudson (2006), however, argues that for regional authorities to become more accountable and active in the future, they should be responsible for developing and enforcing unique policies and utilising unique resources. Secondly, regionalisation as a mechanism towards decentralisation is viewed as an approach by national powers to enhance the efficiency and effectiveness of their programs on a subnational level (Campbell 1996), while retaining control of regional growth through monetary measures.

As stated earlier, regionalisation in most EU countries occurred as a natural process due to EU cohesion policy and funding. Allain-Duprè (2020) confirms that decentralisation dynamics (as measured by the Regional Authority Index (RAI) for certain OECD countries), within various regions have led to the strengthening of regions, and an increase in functional and effectual subnational government, strengthened by municipal cooperation and metropolitan governance.

Deconcentrated authorities in turn, refer to "the process by which the agents of central government control are relocated and geographically dispersed" (Sayer et al. 2005), or similarly "a transfer to lower-level central government authorities, or to other local authorities who are upwardly accountable to the central government" (Ribot 2003: 56). Internationally, deconcentrated authorities are often observed in the manifestation of Sectoral Agencies or Regional Development Agencies (RDAs). RDAs are assigned various responsibilities ranging from environmental management, environmental monitoring, rural development, and infrastructure development (including regional transport) as well as social matters of education, employment, immigration etc. In this instance, the sectoral agency takes over the role of the regional/subnational authority and contributes to the central government's programs and aims through regional development. Internationally, various countries make use of the RD/Sectoral Agency concept for targeted investment as identified in their national policies. In this instance, the regional government will hold the RD/Sectoral Agency liable as the latter acts separately from the traditional levels of government (OECD 2016). These RD/Sectoral Agencies are recognised for their place-based focus on specialised and localised development initiatives which evolved from the lack of skill and competence on the regional level.

2.2.3 Adaptive Governance

The decentralisation modifications are often designed and executed in a symmetric approach, i.e., all subnational governments, even though vastly different, are treated similarly. In a constantly changing environment, a more adaptive approach is needed, two forms of adaptive MLG are subsequently discussed, i.e., asymmetric decentralisation and polycentric decentralisation. *Asymmetric decentralisation* is a more flexible and efficient alternative, growing in popularity (Allain-Dupré 2018; Hooghe et al. 2016). When the aforementioned MLG tools (deconcentrated and decentralised governance) are combined, it is referred to as asymmetric decentralisation. Asymmetric decentralisation is employed for political, efficiency and administrative reasons. Politically, the diffusion of regional or ethnic tensions could be more effectively addressed asymmetrically. Similarly, the achievement of national economic development and growth goals is addressed more effectively with this approach. Lastly, capacity challenges and efficient implementation of policy are administratively more readily engaged with through the use of asymmetric decentralisation (OECD 2010; Allain-Duprè 2020). Asymmetric and polycentric institutions provide a balanced and more flexible approach between decentralised and centralised control (Imperial 1999) as they traverse various levels and sectors of government.

Another approach to adaptive MLG is that of *polycentric* governance (Ostrom 2001, 2005), which is grounded in the theory of complex adaptive systems. Ostrom initially focused on polycentric political systems, referring to it as a system 'of many centres of decision making which are formally independent of each other' (Ostrom et al. 1961). More recently, polycentric governance systems have been defined as "complex, modular systems where differently sized governance units with different purpose, organization, spatial location interact to form together a largely self-organised governance regime". These systems are not hierarchical in the traditional sense, but rather multilayered with multiple authorities (McGinnis 1999; Cash 2000) to address the various and complex concerns regions grapple with. The polycentric governance system is regarded as a more adaptive form of governance, taking into consideration the constant changes governments and regions are subject to (Ostrom 2001, 2005; Pahl-Wostl 1995, 2007) and allowing more freedom at the various levels of governance. It is noted that a polycentric adaptive system could (and should) rely on a network approach, not only between levels and sectors of government but also other agencies, institutions, organisations, and individuals (Folke et al. 2005). This allows for continuous adaptation through flexible and collaborative learning (adaptive co-management) (Folke et al. 2003).

The critique against adaptive MLG, and especially asymmetric decentralisation, revolves around finding a balance between equity and heterogeneity, and ensuring that the processes and implementation are transparent, and governance is held accountable (Allain-Duprè 2020). Further regional imbalance could be the result of unequal spending between regions, and lead to a form of physical migration between regions. Complexity regarding administrative systems, lack of accountability and inefficiency of governance (Bird 2003; Martinez-Vazquez 2007; Allain-Duprè 2020) also arise if the roles and responsibilities of all levels of governance are not clearly defined and monitored.

Against this background of territorial governance in a regional context, the subsequent section will focus on the evolution of subnational/regional governments in South Africa.

2.3 Regional Government in South Africa

This section compares the three pertinent approaches and timeframes to regional governance in South Africa, as they can be dissected in terms of spatial planning. To analyse regional (such as provincial) governance as a form of MLG, it is necessary to reflect on the origin of regional planning in SA. The evolution of regional government in SA shows similar trends to the international environment, moving from a state of strong national (central) governance to the devolution of powers and attempts to various forms of MLG.

2.3.1 Spatial Planning at the Provincial Level in South Africa Before 1994

As a result of South Africa's colonial history, South Africa implemented adapted British town and regional planning laws and replicated various regional planning initiatives. The concept of RDAs (also refer to Sect. 2.2.2) emerged as a solution to address the basic service needs and social and economic circumstances prevalent during and after World War II. Membership was voluntary, and the associations' constitutions emphasised the importance of integrating, coordinating, and communicating between relevant communities to advance the region's interests. The Social and Economic Planning Council was the first organisation to seriously address regional planning in South Africa in 1944 (Van Wyk 2012). They recommended stricter land-use controls, the creation of a department of physical planning and regional development, and the establishment of regional development agencies.

What made the South African approach different from its European counterpart was a complex and divisive approach to spatial planning based on a racially discriminatory framework. This policy of racial segregation (*apartheid*) defined South Africa's political landscape for over four decades, and spatial planning as a demarcation instrument was a crucial tool employed by the government. At the provincial level, the only regional platform at this time, spatial planning was implemented through a combination of laws and regulations, including the Land Acts of 1913 and 1936 (SA 1936) and the Physical Planning Act 88 of 1967 (SA 1967). In terms of later revisions of the Physical Planning Act, so-called *Guide Plans* could be compiled for strategic areas focusing on transportation, land use, and infrastructure planning for a 25-year timeframe. These plans were compiled for numerous city-regions but lacked implementation as they were not sufficiently binding on the affected constituent authorities (local government). They were, however, fundamental in structuring urban space up to the 1990s (Van Wyk 2012). The regional government, at this time, was based on the four provinces of the Republic (see Fig. 2.1).

By 1968, proposals were being put forward for a strategy for regional development in South Africa (Van Wyk 2012). What was envisaged was the country's division into socio-economic development regions for the purposes of survey, analysis, and planning. In 1975, the National Physical Development Plan (NPDP) was introduced as South Africa's national physical plan (SA 1975). The NPDP was novel in this context, as it demarcated 38 planning regions and four metropolitan regions and was, in essence, a decentralisation strategy (Drewes and Bos 1995) in terms of physical and governance decentralisation. Each development region was meant to have a nodal core enclosed within an area of interrelated and integrated economic

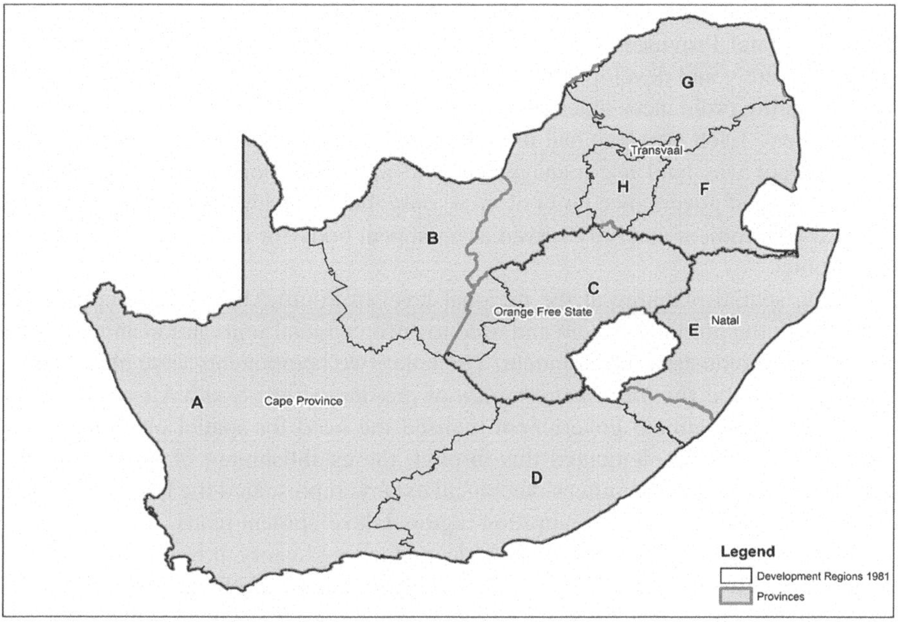

Fig. 2.1 Units of regional planning: pre-1994 (original to the author)

activity, bounded by physical boundaries and metropolitan areas. The NPDP was aimed at eradicating the growing concentration of people and economic activities in the relatively small metropolitan areas, as well as the increasing concern that the spatial development patterns of the Republic did not reflect the potential needs of its inhabitants. However, regional planning in the technical sense was still derived from the National Department of Planning and the Environment, even though five regional development advisory committees were established (Van Wyk 2012).

Regional planning evolved further with the Good Hope Conference and the subsequent Good Hope Plan in 1981, albeit with a strong focus on industrial development (SA 1981). The 38 NPDP (1975) regions were grouped into eight regions, each represented by a Regional Development Advisory Committee (see Fig. 2.1). These were technical, non-statutory committees, as regional planning was still a function of national government.

The Physical Planning Act (125 of 1991) was promulgated in 1991 and replaced the abovementioned 1967 act. At this time, the national government (top-down) still hosted physical planning encompassing regional planning through the Department of Planning, Provincial Affairs and National Housing (SA 1991). It made provision for appointing "*Administrators*" to facilitate and oversee the compilation of regional development- and regional structure plans in an attempt at MLG. These plans could only be approved by said national minister or the appointed administrator. Interestingly, no regional development- or regional structure plans were formally compiled in terms of this Act (Van Wyk 2012).

The four provinces that existed before the restructuring in the 1990s, i.e., Cape Province, Natal Province, Transvaal Province, and Orange Free State, were governed separately and developed their own spatial policies by adopting the so-called town-planning ordinances that were provincial legislation for dealing with land use management. Each province had its ordinance (devolution of power), but the legal focus thereof was land use management and administration (statutory planning). From a regional governance point of view, only the Townships Boards[1] reflected on regional government and only served as an appeal body for local land use and zoning disputes.

Lastly, spatial planning at the regional level in South Africa was deeply intertwined with the political system and used to enforce racial segregation and promote decentralised industrial development. The latter two components were also the relevant focus of the *spatial transformation* of pre-democratic South Africa. It is also evident that the national government realised the need for spatial planning at the regional level and implemented this through the establishment of non-regulatory advisory boards and committees. Technical experts represented the latter, with little or no public participation in compiling regional development plans.

Based on this timeframe in South African planning history, it is apparent that the intermediate governance level was acknowledged, but mostly in an advisory capacity. Moreover, the demarcation of four provincial boundaries was considered less effective, which resulted in the delineation of smaller planning regions such as the Good Hope Plan. This demarcation was the foundation for restructuring provincial boundaries into nine provinces following the 1994 government elections.

2.3.2 Regional Governance After Democratisation

Regional governance and planning underwent a significant transformation after the democratic elections in 1994. This ushered in a new era of democratic governance, bringing about a shift in spatial planning practices at the provincial and/or regional levels. This section reflects on the transitional period after democratisation (1994) and the formal adoption in 2013 of the existing normative and integrated approach to spatial planning in all three spheres of government.

This period in South Africa focused on dismantling the previous planning system and on spatial inequalities, with the overall goals of spatial integration and -transformation. In 1996, the adoption of a democratic constitution in South Africa had a profound impact on spatial planning at the regional level. The constitution has enshrined vital principles of equality, non-discrimination, and the right to adequate housing (SA 1996), which have been integral in shaping the country's policies and

[1] The functions of a Township Board included the exercise of competencies and responsibilities in terms of the relevant ordinance. They reported to the premier on hearing appeals (Van Wyk 2012).

strategies in this area. In terms of spatial planning, all three spheres of government were deemed to have legislative powers from a spatial planning perspective (MLG).

One of the first pieces of legislation promulgated was the Development Facilitation Act (67 of 1995), which included the so-called Land Development Objectives (LDO) aimed to speed up the land development processes to increase access to formal land. Although enacted by the national government, some delegations were made to the provincial government to "fast-track" land development and the delivery of housing (SA 1995). The creation of provincial planning tribunals to facilitate these applications was more effective than previous-era legislation. Still, sections of the act were later declared unconstitutional as it encroached on municipal powers and delegations (Van Wyk 2012). The Municipal Systems Act (32 of 2000) replaced other sections of the DFA (SA 1995). The latter act then introduced a vital feature of the existing spatial planning system in South Africa, namely Integrated Development Planning (SA 2000).

Integrated Development Plans (IDPs) reflect long-term planning documents incorporating social, economic, and environmental considerations at the municipal level. They aim to promote spatial transformation and address historical imbalances, becoming the primary instrument for guiding spatial development. This could be linked to the discussion above on deconcentrated authorities (refer to Sect. 2.2) with an "upward-accountability" of local levels to that of the central authority. Another significant contribution of the Municipal Systems Act was the introduction of Spatial Development Frameworks (SDF) as an integral component of IDPs (SA 2000). The SDF is regarded as a "regional strategy" and a tool to decentralise authority from the national to an intermediate and local level of governance (refer to Sect. 2.2). In essence, the latter document is a strategic spatial plan for the relevant geographical area implied in terms of this legislation.

What is essential in terms of this chapter is the onus on the provincial government, as part of their legislative power, to also compile SDFs. This was an essential addition to regional government, as previous spatial planning policies and legislation often led to segregated urban and rural areas and unbalanced development. Provinces started to focus on holistic development strategies that connected urban centres with surrounding rural areas, recognising the interdependence between the two. In terms of this legislation (SA 2000), and supporting legislation from an environmental perspective (SA 1998), environmental considerations gained its first prominence in spatial planning. Sustainable development principles were incorporated into provincial planning processes, highlighting responsible land use, conservation of natural resources, and climate resilience. Provincial SDFs heralded a new era for regional planning and governance in South Africa, and most provinces adopted these strategic regional plans during this interim phase. However, their local or municipal enforcement was voluntary, and no higher tier or national spatial plan was provided by the Government (Drewes 2015).

The transition from previous-era policies to democratic governance represented a meaningful change in planning paradigms, focusing on integration, sustainability, and social justice. It also reflected a timeframe with an extremely limited spatial planning focus at the regional governance level, and no national spatial planning initiatives or guidance (Drewes 2015). This was clearly an emphasis shift from top-down national planning to a grass-root spatial planning focus entrenched in local government.

2.3.3 *Regional Governance Post-2013*

Since its promulgation (2013) and implementation, the Spatial Planning and Land Use Management Act (SPLUMA) in 2015 has created significant changes in the realm of spatial planning in South Africa. Based on various draft policies and bills since 2001 (SA 2006; Van Wyk 2012), the eventual aim of SPLUMA was to streamline spatial planning and land use management throughout the country while granting provincial governments greater authority to shape their respective landscapes. The SPLUMA also prescribed, for the first time since 1994, the development of a national spatial development framework (NSDF). This section delves into the post-SPLUMA era of spatial planning at the regional- or intermediate governance level in South Africa, examining pivotal advancements and endeavours due to this legislation.

The publication of SPLUMA saw South Africa make a noteworthy shift in its spatial planning approach. The recognition that the rural and urban components of the country are interlinked (Todes and Turok 2018) and should not be treated as two separate entities was fundamentally crucial at the time. This act was designed to create a cohesive and organised framework for managing land use and spatial planning by replacing numerous earlier laws and regulations. One of the implicit goals of the act was to restore the emphasis on national spatial planning while also transferring the responsibility for spatial planning to regional and local governments. It also provided a normative planning approach based on similar international planning approaches (Laubscher et al. 2016).

SPLUMA (SA 2013b), subsequently, made it compulsory for national and regional governments to compile national- and provincial/regional spatial development plans, or “Spatial Development Frameworks” (SDFs), on the appropriate geographical scale (see Fig. 2.2). The drafting of said frameworks is compulsory not only on a national level but also on the provincial level and includes specific guidelines for community participation at appropriate geographical levels. Since the implementation of SPLUMA in 2015, all local-, district-, and metropolitan municipalities, as well as all the provinces, are supposed to compile SDFs based on the set requirements in the Act. This initial round of compilation of SDFs was concluded in 2023 with the publication of the NSDF, based, amongst others, on input from the lower levels of government. It is also compulsory that these spatial development frameworks be revised at least every 5 years (SA 2013a, b, 2023; Laubscher et al. 2016).

A novel contribution of SPLUMA is the creation of an additional level of regional planning and, per implication, governance. This approach touches on some of the characteristics of an adaptive governance tactic (refer to Sect. 2.3) in a polycentric

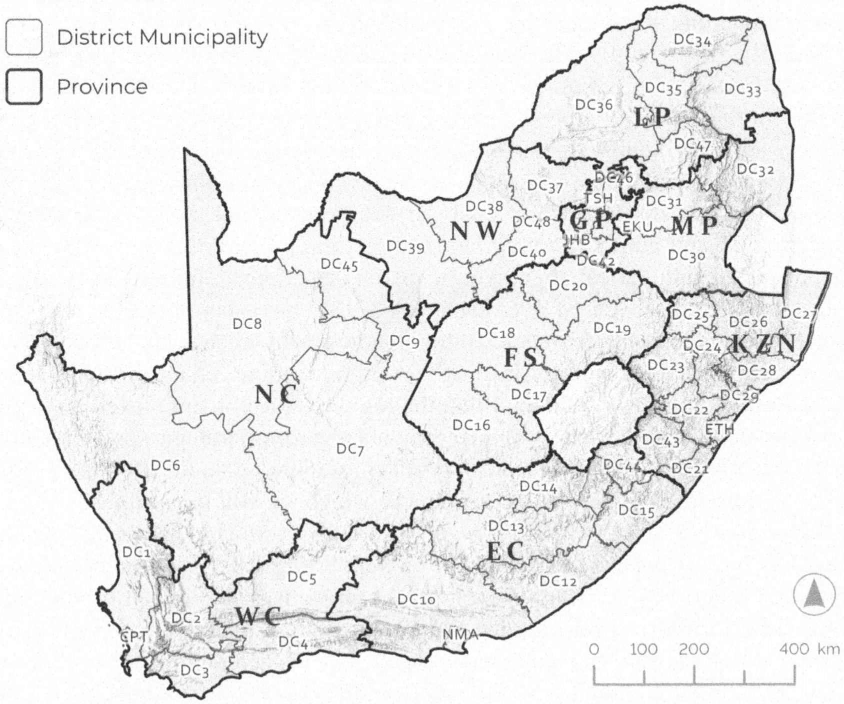

Fig. 2.2 Units of regional planning: post-1994 (original to the author)

manner. The relevant national ministry can facilitate the determination and compilation of a "Regional" SDF through any combination of cadastral boundaries, not necessarily linked to the boundaries of municipalities and provinces. It is assumed that the management of a promulgated Regional SDF lies with numerous government levels and entities, depending on the combination of administrative boundaries. The inclusion of this level of planning and governance supports global policy initiatives like the United Nations Sustainable Development Goals (UN 2015), which emphasises the spatial planning links between urban, peri-urban, and rural areas (Laubscher et al. 2016).

The adoption of SPLUMA (SA 2013b) changes the provincial government's role threefold. Firstly, all nine regional governments are legally required to compile a provincial SDF that *must* be consistent with the national-level strategic plan. Note the peremptory provision of Section 15 of this Act, in contrast to previous legislation (SA 2000). Secondly, in terms of Section 10 of said act, the provincial government should play a supportive and monitoring role for local government in terms of strategic plans (SDFs) and statutory processes (land use management). Lastly, the regional government must compile provincial legislation (see Schedule 1 of Act 16 of 2013) based on the national spatial planning legislation to support strategic- and statutory planning at the local and municipal levels.

Recently, frustrated by a lack of service delivery from local government in general, the President of South Africa proposed the District Development Model (DDM) for the country to improve cooperative governance (SA 2019). Although not a legally structured entity, the DDM is based on existing district municipality borders (between local municipal and provincial levels) and certainly supports the principle of MLG, as discussed previously. It is an operational model supporting the National SDF and numerous other national sectoral policies to enable more effective horizontal and vertical alignment of spatial development initiatives.

In summary, the advent of SPLUMA (SA 2013b) has ushered in a fresh epoch of spatial planning in South Africa, resulting in noteworthy impacts on regional governance and spatial planning, leaning towards a more adaptive approach to MLG. Before democratisation in 1994, the regional government played an explicit advisory role to national and local governments regarding strategic spatial planning. On the other hand, they played a primary role in statutory planning through promulgating land use planning ordinances, some of which are still in use today.

Between 1995 and 2013, the role of regional government in spatial planning terms has been eroded, with the focus on supporting local government strategic planning initiatives (e.g., Land Development Objectives, Integrated Development Plans, etc.). In terms of relevant legislation, they could undertake strategic spatial planning (SDF) at the said governance level, but no peremptory instruction prevailed. The land use planning ordinances from the previous era of governance were still used in most provinces, and the determination of Planning Tribunals (SA 1995) formed the focus of this governance level between 1995 and 2013.

This changed after the introduction of SPLUMA, and the nine provinces, as well as any regionally delineated initiative, can now assume a more active role in moulding their spatial domains with the aid of principles such as efficiency, inclusivity, sustainability, and good governance. They play a legislative role in both strategic- and statutory planning and advisory and monitoring roles for local and national government.

2.4 Regional Government and Transformation Initiatives: A Case Study of the Western Cape Provincial and Central Karoo Regional Development Frameworks

The previous section elaborated on regional governance before and after democratisation. To evaluate spatial planning (SDFs) as the collective vision of government, businesses, and civil society to promote social and economic inclusion and spatial transformation against the backdrop of effective regional governance and spatial transformation in practice, this section mainly focuses on two case studies namely: (1) the Western Cape Provincial Spatial Development Framework (WCPSDF); and

(2) Central Karoo Regional Spatial Development Framework (CKRSDF). The case studies will accentuate the period 1994 to 2013 and the period after 2013 with the enactment of SPLUMA, respectively. Both frameworks represented the prevailing spatial planning policy within the Western Cape Province and were prepared and adopted as an integral component of spatial transformation. Lastly, to further demonstrate integrated governance, focus is also placed on the wider Karoo accentuating how the Karoo Regional Spatial Development Framework (KRSDF) as a national initiative, contribute towards spatial transformation.

2.4.1 Western Cape Provincial Spatial Development Framework (WCPSDF)

The Constitution assigns provincial and regional planning as exclusive responsibilities of the Provincial Government (SA 1996). With reference to SPLUMA (2013b), provinces are required to compile and publish PSDFs. In essence, a PSDF must systematically align: (1) provincial and development strategies with policies of National significance; (2) plans and development strategies under provincial authority; and (3) plans and development strategies under regional and local dominion. Although the Constitution assigns shared responsibilities to each sphere of governance, a PSDF mandate necessitate coordinated and integrated alignment between all spheres of governance. Furthermore, although provincial governments cannot dictate where institutions or stakeholders should invest, PSDFs can influence investment decisions by establishing a coherent and logical spatial investment framework (South Africa 2013a, b; WCPSDF 2014). Accordingly, a PSDF guides how development will be communicated, coordinated, and delivered.

In 2005, the Western Cape Provincial Government introduced the first PSDF, followed by a second version in 2007 (WCPSDF 2005, 2007). Both PSDFs (pre-2013) were guided by the National Spatial Development Perspective (NSDP), later replaced by the National Development Plan (NDP). The 2005 and 2007 WCPSDFs were mainly guided by the following NSDP principles: (1) investments should focus on areas of economic potential; (2) inequalities should focus more on people; (3) economic development opportunities should be channelled into activity corridors and nodes; (4) rectify space economy; and (5) development should alternate current patterns of resource utilisation. Although these principles are still considered prevalent, in 2014 (post-2013) the Western Cape Provincial Government enacted its most recent PSDF (WCPSDF 2014) in response to a number of factors: (1) greater clarity that emerged regarding the spatial planning responsibilities of the three spheres of government emanating from the drafting of SPLUMA promulgated in 2013, and needed to be brought into full effect; (2) the release of the results of the 2011 Census that provided up to date information; and (3) the ONECAPE2040 and the NDP initiatives which were introduced in 2012 and 2013 respectively, that established new complementary development agendas within which to reframe the

Western Cape. The response placed the WC Province on a more focused regional governance and transformation path.

Although the mentioned planning principles that have underpinned the WCPSDF remain relevant, events since 2014 such as COVID-19 pandemic, gender-based violence, the fees must fall protests, violent demands for access to land, on-going load shedding, ailing conditions of State-Owned Enterprises, low to stagnant economic growth, and declining revenue and fiscally vulnerable local government have shaped new priorities. An amendment to the PSDF implementation chapter was thus required to articulate a path towards more impactful spatial governance building on progress that has been made and, shifting emphasis where needed to maximise effectiveness and impact of government programmes in space (WCPSDF 2020).

The 2014 WCPSDF focuses primarily on the Western Cape Province, which makes up 10.6% of South Africa's land surface and encompasses an area of 129,462 km^2, and according to Statistics South Africa (2022) generating more than 14% of South Africa's gross domestic product (GDP). The area of jurisdiction not only includes the five regional Districts and 24 Local Municipalities, but also the City of Cape Town categorised as a Metropolitan Municipality (see Fig. 2.3). The WCPSDF (2014) overall seeks to improve the effectiveness of investments in the Western Cape's built and natural environments by (1) spatially targeting and aligning the various investment programmes; (2) adopting credible spatial planning principles to underpin investment programmes; and (3) opening-up opportunities for community and business development in targeted locations.

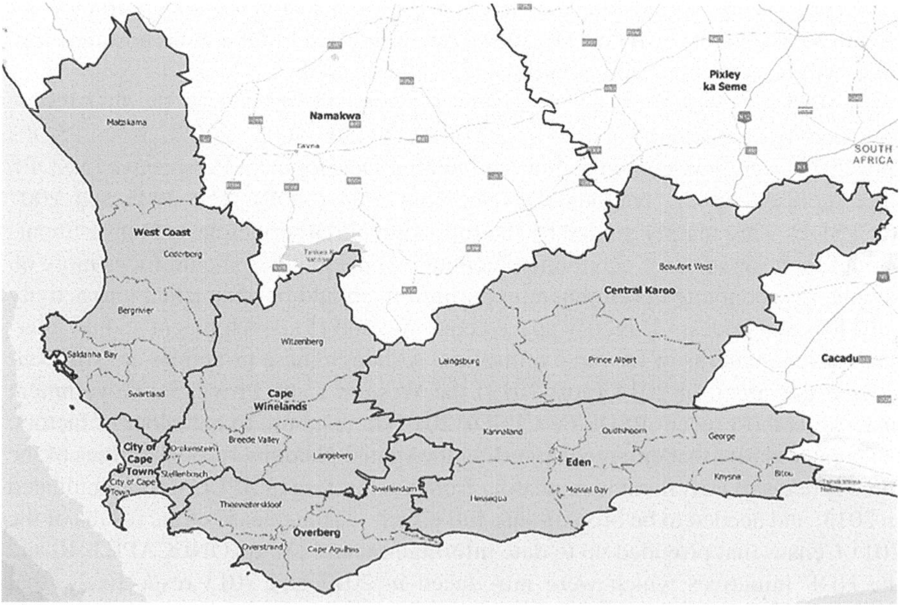

Fig. 2.3 Western Cape Provincial District and Local Municipalities (WCPSDF 2014)

The Cape Metro region dominates the Provincial economy generating more than 85% of GDP, making it the leading economic hub of the province. Although the Cape Metro is morphologically the smallest region, functionally it dominates the overall economic performance of the province. This translates to the interpretation that the Western Cape Province exhibits a strong monocentric spatial configuration. A monocentric configuration focuses on a group of connected distinct cities with obvious leading cities that don't share the same level of equity meaning, a country or province can be considered monocentric when a relatively large number of regions would not have such a leading city within its borders (Kraus 2006; You 2017). Such a spatial configuration challenges regional governance which can potentially compromise spatial transformation. Under such circumstances, the PSDF not only re-introduced the growth potential (GP), but also introduced the ONCECAPE2040 as a strategic spatial intervention to overcome these challenges. The GP as with the 2005 and 2007 PSDFs, played an important role in compiling the 2014 PSDF.

The primary objective of the GP (WCPSDF 2005, 2007, 2014; Stellenbosch University 2013) was to determine the growth potential of settlements located outside the City of Cape Town (Cape Metro) in relation to potential future economic, population and physical growth. The 2005 and 2007 PSDFs (see Fig. 2.4) showed that the GP in the Province is clustering around the City of Cape Town and the western part of the Cape Winelands, Saldanha Bay, the coastal part of the Overberg, and

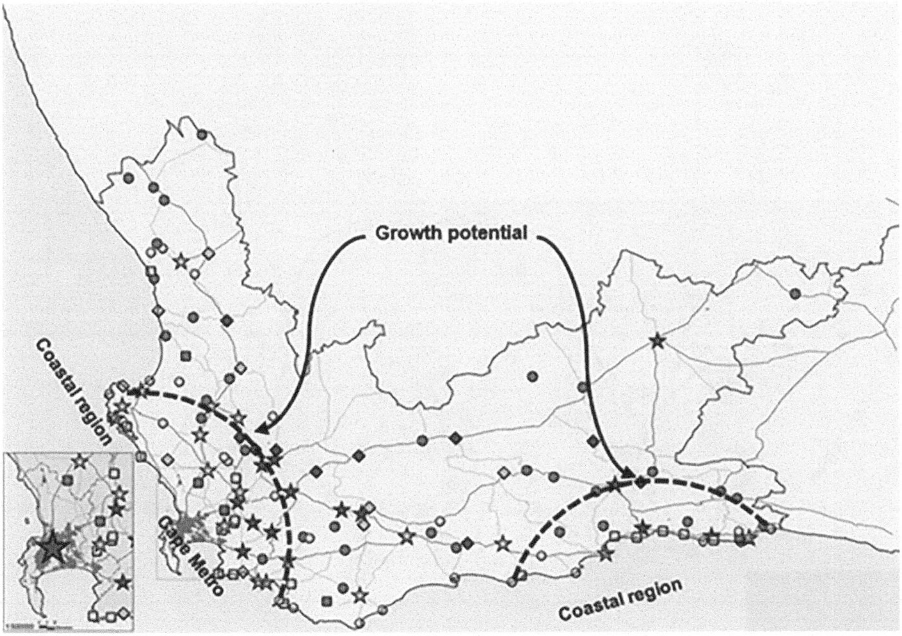

Fig. 2.4 Growth Potential of the Western Cape Province as depicted by the 2005 and 2007 PSDFs (WCPSDF 2005, 2007)

the Eden coast. With the introduction of the GP for the 2014 PSDF, the analysis hinged on two fundamental and related concepts: (1) growth preconditions; and (2) innovation potential. For the purposes of analysing the GP, thematic indexes were developed for each concept that included an economic, physical and infrastructure index for growth preconditions, and institutional and human capital index for innovation potential. The outcomes from the 2005, 2007 and 2014 PSDFs indicate a very similar pattern of clustering and promoting the same economic potential opportunity (see Figs. 2.4 and 2.5). In contrast, the northern West Coast and the Central Karoo hinterland regions did not form part of these growth potential areas and prompted the question of what potential opportunity can a lagging region such as the Central Karoo contribute to economic and physical growth that can enhance regional governance and transformation? In response, this aspect is discussed in more detail in the Central Karoo's RSDF.

ONECAPE2040 (2012) was initiated as an attempt to transition towards a more inclusive and flexible economic future for the Western Cape region. It articulates a vision about how the people, also prevalent in the previous PSDFs, can collectively develop an economic future. It seeks to set a common direction to promote a joint commitment and accountability to sustained long-term progress. ONECAPE2040's spatial dimension revolves around connectivity and concentration. Connectivity demand that cities/towns and hinterlands are connected to localities of opportunity (local and international), meaning connectivity will be accompanied by the

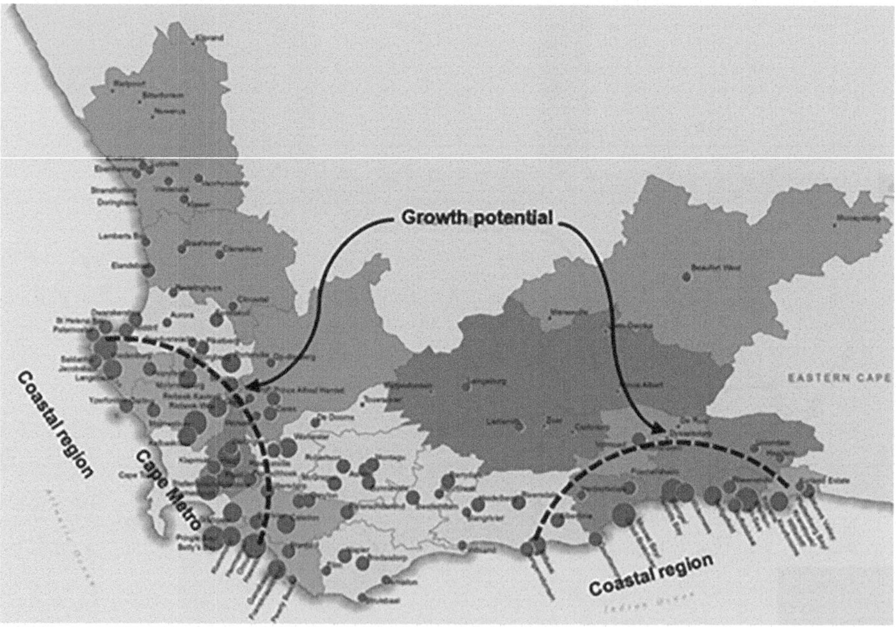

Fig. 2.5 Growth Potential of the Western Cape Province as depicted by the 2014 PSDF (WCPSDF 2014)

concentration of activity in key nodes promoting spatial targeting and making spatial transformation more explicit (see Fig. 2.6).

It is clear from Figs. 2.4, 2.5, and 2.6 that the places with a high to very high growth potential are in proximity of the City of Cape Town (Cape Metro), the western part and the eastern coastal regions of the province. These places emphasise the concentration of activity in key urban areas. Furthermore, viable connectivity is established through the promotion of regional development and transport corridors such as the Oliphant and Breede River Valley and the National Road (N2 and N7) corridor initiatives. Although the PSDF coordinates, integrates, and aligns plans, policies, and development strategies of national, provincial, district and local significance, spatial transformation on a regional level, was confined due to a lack of regional equity.

The regional planning response to this lack of regional equity, in the more enabling governance environment of SPLUMA (SA 2013b), was to (1) capitalise and build on the region's comparative strengths; (2) consolidate existing and emerging regional economic nodes on the regional spectrum (urban-rural); (3) connect and transform urban and rural markets and consumers, fragmented settlements and

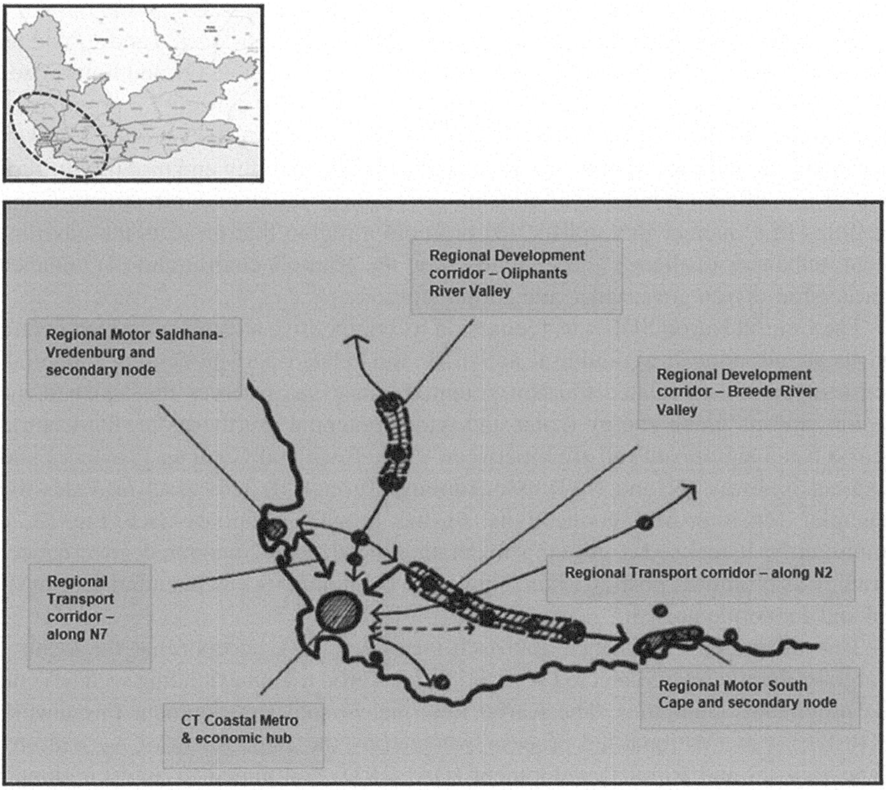

Fig. 2.6 Concentration and connectivity (ONECAPE2040 2012)

critical biodiversity areas; and (4) cluster economic infrastructure and facilities along important transport routes to maximise the coverage of potential investments (SA 2021). The overall strategy of the PSDF was to accelerate and give momentum to spatial transformation at provincial and regional scale by embarking on this series of sequential, targeted planning actions.

2.4.2 Central Karoo Regional Spatial Development Framework (CKRSDF)

This section deals with both the SDF of the Central Karoo District municipality area of jurisdiction, as well as on the SDF of the wider Karoo region extending beyond the boundaries of the Western Cape that also straddles the Northern-, and Eastern Cape, as well as the Free State Provinces.

The Central Karoo District is morphologically the largest district in the Western Cape Province but is sparsely populated and economically it contributes 0.5% of the provincial economy (CKRSDF 2020). According to the GP the district portrays a very low potential for economic and physical growth (see Figs. 2.4 and 2.5).

It was recognised that these regional inequalities are multi-faceted and should be addressed across all spheres of governance, also including State Owned Enterprises (SOEs) as well as key private sector role players. The spatial concept for progressive development in the Central Karoo SDF (CKRSDF 2020) revolves around four key strategies, namely: (1) improve regional and rural accessibility and mobility in support of a resilient economy; (2) allocate government resources, infrastructure and facilities in a manner that uplifts; (3) promote a region that protects the environment, enhances resilience and capitalises on the Karoo's charm; and (4) enhance partnership-driven governance and administration.

The Central Karoo SDF thus focuses on its competitive advantages which centre on its scenic appeal, agricultural activities and related agri-processing potential, sense of place and related tourism potential, shale gas reserves that exist in the region, and its green energy (solar and wind) potential. Furthermore, the Central Karoo has a significant rail (link between Cape Town and Gauteng Province) and national highway (N1and N12) asset running through it. This asset provides the potential for improved connectivity to the broader economy (see Fig. 2.7). Importantly, it realised a growing understanding that an 'integrated governance' strategy as promoted post-2013, is required to overcome its low potential for physical and economic growth.

This integrated governance approach (see Sect. 2.2), accepts that the Central Karoo is directly interconnected with the wider Karoo region extending well beyond the provincial boundaries. The Karoo Regional Spatial Development Framework (KRSDF) was a national-led process overseen by the Department of Agriculture, Land Reform and Rural Development (DALRRD) and aimed to apply regional-scale governance, conceptualisation and planning to: (1) harness the full

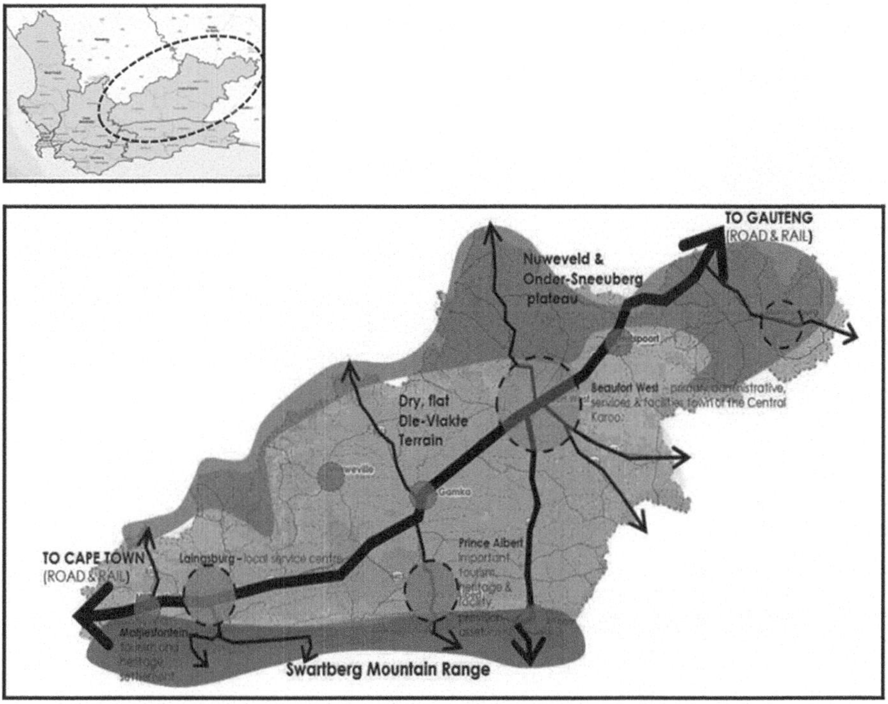

Fig. 2.7 Spatial concept for the Central Karoo (CKRSDF 2020)

development potential of the region in a sustainable and transformative way, and (2) address the shared, large-scale regional challenges, such as climate change, poverty, inequality, and unequal access to, and participation in the economy (SA 2021).

The Central Karoo's affiliation with the KRSDF rests upon three key spatial initiatives (see Fig. 2.8). These include establishing a Karoo Regional Airport in Beaufort West as a gateway to unlock economic development potential, Renewable Energy Development Zones (REDZs) for the development of large-scale wind and solar photovoltaic facilities, and strategic transmission corridors which support areas where long-term electricity grid infrastructure will be developed. It is envisaged that these initiatives would improve the Central Karoo District's low potential for economic and physical growth. These national initiatives underpin the realisation that planning- and transformation solutions must be approached from an integrated governance perspective to overcome the lack of regional equity. The main consideration is demonstrating integrated governance with provinces and its districts assuming a more active role in moulding their spatial domains and how national spatial planning initiatives such as the wider Karoo are integrated as a component of spatial transformation.

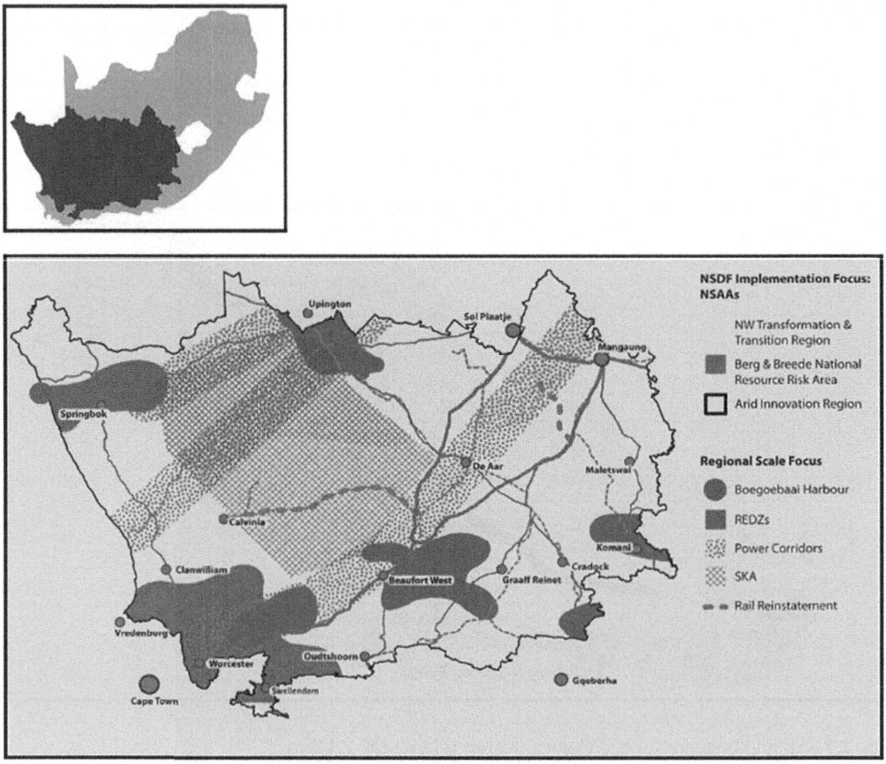

Fig. 2.8 Spatial targeting for potential investment in the wider Karoo region (SA 2021)

In conclusion, in contrast to pre-2013 conditions, various levels of regional planning and subsequent governance were established after 2013, as is evident in the above case studies.

The determination and compilation of these SDFs were facilitated through a combination of municipalities, provinces, national government, as well as non-governmental agencies. Furthermore, since there is no uniform distribution of economic activities across the Western Cape Province, sensitive areas, or areas with very low potential growth, such as the Central Karoo, require diligent governance to support future equitable growth and expansion. This is illustrated by the Karoo RSDF, initiated by the National Government, and overseen by a multi-level combination of governance in terms of compilation and application. Such a level of integration not only supports national policy initiatives but also a normative-oriented approach as envisaged by global planning initiatives (see Sect. 2.3.3). The case studies investigated show an adaptive governance designed to create a cohesive and organized framework by granting provincial government greater authority to shape their spatial landscape. This is facilitated by transferring the responsibility for spatial planning through any combination of provincial, regional, and local government. It further illustrates that the management of spatial planning lies with various government levels depending on the combination of administrative boundaries.

2.5 Conclusion and Recommendations

SDFs provide an overview of the economic, social, physical, and natural landscape of different spatial units of analysis and establish the groundwork for policies and action plans to promote integration and spatial transformation. The findings indicate that growth and development are based on economic agents distributed among various sectors of activity, with economic agents located in specific locations which leads to areas of productive specialisation and spatial transformation. This is in turn supported by the preconditions and innovation potential, not only with a focus on the inherent economic endowment but also on institutional capacity, socioeconomic conditions, and human capacity.

The successful implementation of regional policy and frameworks is only possible through a strong and devoted regional government, be it in a deconcentrated, decentralised or asymmetrical approach. It was established (Sect. 2.1) that regional strategies should be both locally informed and reinforced at the national level. This approach establishes both role-players as enablers of spatial objectives on a local as well as local level and reinforces the mutually dependent relationship across all spheres of government. Furthermore, tools for a more robust regional government were identified as a combination of top-down (regionalisation) and bottom-up approaches (regionalism).

Adaptive governance for complex systems should focus on the regional authority and its various linkages with national and local government in a polycentric/asymmetric manner. This asymmetric approach to decentralised governance in terms of regional planning has advanced the country into a new era of regional planning, albeit often lacking the capacity to implement these well-intentioned plans. Figure 2.9 elucidates the proposed implications from the preceding discussions and case studies for future regional governance in South Africa. Many practices and entities are already in place within the country; however, the lacking component relates to the interaction and integration among the various entities and the follow-through at the grassroots level.

Lessons learned through the above regional governance perspective and case studies point to necessary processes that should connect governance levels and improve vertical and horizontal coordination. It is critical to understand that agents from one level (e.g., the national level) should participate in dialogues at another level (e.g., the regional level or local level) in a polycentric manner to ensure strategic horizontal and vertical alignment in a shared vision approach to policy and frameworks. These agents may thus become actively involved in producing the rules that influence them. Secondly, it is important to realise that institutions (formal and informal) provided for at one level influence processes at another level. The

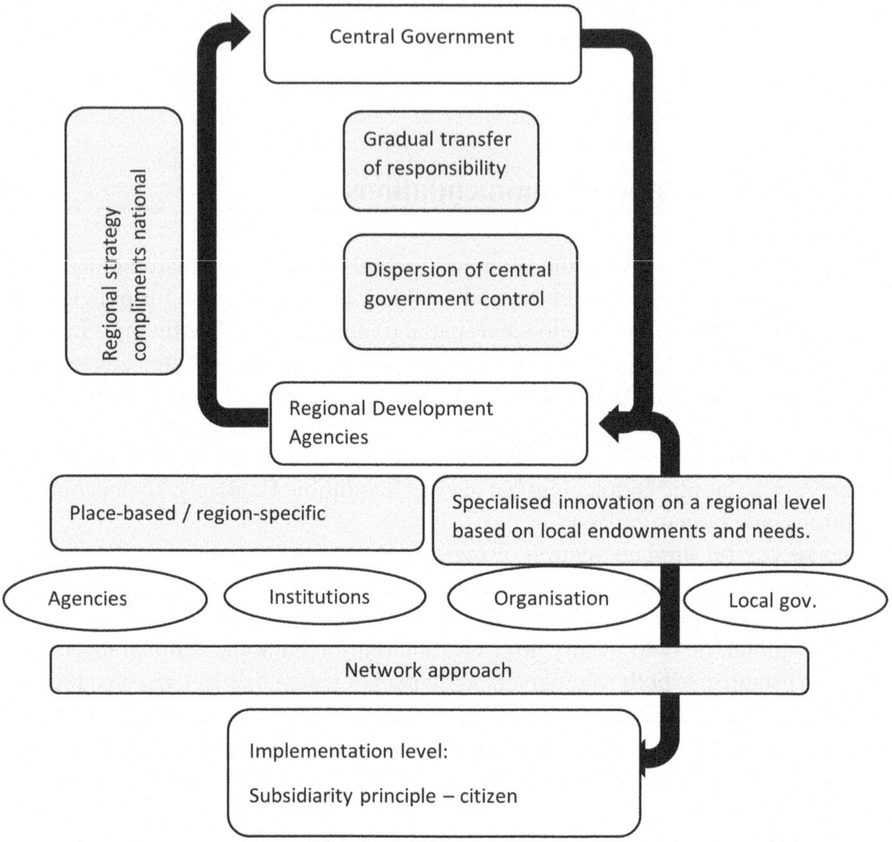

Fig. 2.9 Implications for future regional governance (original to the author)

flow of authority may then be both top-down as well as bottom-up. Thirdly, knowledge produced at one level influences processes at another level. Strategic spatial planning and relevant regional governance must take note of this and act proactively in support thereof in terms of policy, legislation, and implementation. The identification of Regional Development Agencies will be a step in the direction of bridging the gap between policymaking and practical implementation on a grassroots level. Lastly, it is proposed that the inclusion of the private sector in consensus building, together with community participation, should form a solid foundation of not only regional governance but all appropriate planning scales in the shape of a MLG approach. The dynamic implementation of these approaches, accompanied by actual adaptive governance is a prerequisite for effective spatial transformation.

References

Allain-Dupré D (2018) Assigning responsibilities across levels of government: Trends, challenges and guidelines for policy-makers. OECD Working Papers on Fiscal Federalism no. 24, 27 September. OECD Publishing, Paris
Allain-Duprè D (2020) The multi-level governance imperative. Br J Polit Int Rel:1–9. https://doi.org/10.1177/1369148120937984
Allison HE, Hobbs RJ (2004) Resilience, adaptive capacity, and the "Lock-in Trap" of the Western Australian agricultural region. Ecology and society, 9(1)
Angell E, Flo Y, Grimsrud GM (2016) The rural and regional policy of Norway. Uni Research, Bergen
Bache I, Flinders M (2004) Multi-level governance. Oxford University Press, Oxford
Bergvall D, Charbit C, Kraan D, Merk O (2006) Intergovernmental grants and decentralized public spending. OECD J Budgeting 5(4):111–158
Bird R (2003) Asymmetric fiscal decentralization: glue or solvent? Working Paper, No. 03-09. Andrew Young School of Policy Studies, Atlanta
Campbell A (1996) Regionalism, regionalisation, and natural resource management. Centre for Resource and Environmental Studies (CRES), Canberra
Cash DW (2000) Distributed assessment systems: an emerging paradigm of research, assessment and decision-making for environmental change. Glob Environ Chang 10:241–244
Chaudhuri JR (2001) An introduction to development and regional planning. Orient Blackswan, Hyderabad
CKRSDF (2020) Central Karoo District Municipal Spatial Development Framework. [Online] Available at: https://www.skdm.co.za/spatial-development-framework
Cumberland MH (1971) Regional development. Mouton, Den Haag
Drewes JE (2015) More explicit regional policy for South Africa, please Mr President. Town Reg Plan, No 67. ISSN 1012-280
Drewes JE, Bos DJ (1995) The regional industrial development programme: an evaluation. S Afr J Econ 63(2):247–270
Folke C, Colding J, Berkes F (2003) Synthesis: building resilience and adaptive capacity in social-ecological systems. In: Berkes F, Colding J, Folke C (eds) Navigating social-ecological systems: building resilience for complexity and change. Cambridge University Press, Cambridge, pp 352–387
Folke C, Hahn T, Olsson P, Norberg J (2005) Adaptive governance of social-ecological knowledge. Annu Rev Environ Resour 30:441–473
Glasson J, Marshall T (2007) Regional planning. Routledge, Oxfordshire
Hooghe L, Marks G (2001) Multi-level governance and European integration. Rowman & Littlefield, Boulder
Hooghe L, Marks G, Schakel AH, Osterkatz SC, Niedzwiecki S, Shair-Rosenfield S (2016) Measuring regional authority, a postfunctionalist theory of governance, vol I. Oxford University Press, Oxford
Hudson C (2006) Regional development partnerships in Sweden: a way for higher education institutions to develop their role in the processes of regional governance? High Educ 51(3):387–410
Imperial MT (1999) Institutional analysis and ecosystem-based management: the institutional analysis and development framework. Environ Manag 24:449–465
Keeble L (1969) Principles and practice of town and country planning, 4th edn. Estates Gazette, London
Kraus M (2006) Monocentric cities. In: Richard JA, McMillen DP (eds) A companion to urban economics. Blackwell Publishing Ltd
Laubscher N, Hoffman L, Drewes JE, Nysschen JJ (2016) SPLUMA: a practical guide. Lexis Nexis, Durban
Lebel L, Anderies JM, Cambell B, Folke C, Hatfield-Dodds S, Hughes TP, Wilson J (2006) Governance and the capacity to manage resilience in regional social-ecological systems. Ecol Soc 11(1):19. [Online]. http://www.ecologyandsociety.org/vol11/iss1/art19/

Marks G (1993) Structural policy and multilevel governance in the EC. In: Cafruny A, Rosenthal G (eds) The state of the European Community. Lynne Rienner, New York
Martinez-Vazquez J (2007) Asymmetric federalism in Russia: cure or poison?. http://scholarworks.gsu.edu/econ_facpub
McGinnis M (1999) Polycentric governance and development: readings from the Workshop in Political Theory and Policy Analysis. University of Michigan Press, Lansing
Oates WE (2005) Toward a second-generation theory of fiscal federalism. Int Tax Public Financ 12(4):349–373. https://doi.org/10.1007/s10797-005-1619-9
OECD (2010) Regional development policies in OECD countries. OECD Publishing, Paris
OECD (2016) OECD regional outlook 2016: productive regions for inclusive societies. OECD Publishing, Paris
OECD (2017) Multi-level governance reforms in OECD countries. OECD Publishing, Paris
OECD (2019) Making decentralisation work: a handbook for policy-makers. OECD Publishing, Paris
Ostrom E (2001) Vulnerability and polycentric governance systems. IHDP (International Human Dimensions Programme on Global Environmental Change) Newsletter UPDATE no. 3, 1, 3–4
Ostrom E (2005) Understanding institutional diversity. Princeton University Press, Princeton
Ostrom E, Tiebout CH, Warren R (1961) The organization of government in metropolitan areas: a theoretical inquiry. Am Polit Sci Rev 55:831–842
Pahl-Wostl C (1995) The dynamic nature of ecosystems: chaos and order entwined. Wiley, Chichester
Pahl-Wostl C (2007) Transition towards adaptive management of water facing climate and global change. Water Resour Manag 21(1):49–62
Pahl-Wostl C (2009) A conceptual framework for analysing adaptive capacity and multi-level learning processes in resource governance regimes. Glob Environ Chang 19:354–365. https://doi.org/10.1016/j.gloenvcha.2009.06.001
Ribot JC (2003) Democratic decentralisation of natural resources: institutional choice and discretionary power transfers in Sub-Saharan Africa. Public Adm Dev 23(1):53–65
Sayer J, Elliott C, Barrow E, Gretzinger S, Maginnis S, McShane T, Shepherd G (2005) The implications for biodiversity conservation of decentralized forest resources management. In: Colfer CJ, Capistrano D (eds) The politics of decentralization. Earthscan, London, pp 121–137
Scheffer M, Westley F, Brock W (2003) Slow responses of societies to new problems: causes and costs. Ecosystems, Volume 6: 493–502
South Africa (1936) State President's Office. Native Trust and Land Act, 18 of 1936. Government Printer, Pretoria
South Africa (1967) State President's Office. Physical Planning Act, no. 88 of 1967. Government Printer, Pretoria
South Africa (1975) Department of Planning and the Environment. National physical development plan. Government Printer, Pretoria
South Africa (1981) Good hope plan. Department of Foreign Affairs. Government Printer, Pretoria
South Africa (1991) State President's Office. Physical Planning Act, no. 125 of 1991. Government Printer, Pretoria
South Africa (1995) Department of Land Affairs. Development Facilitation Act, no. 67 of 1995. Government Printer, Pretoria
South Africa (1996) Constitution of the Republic of South Africa. Government Printer, Pretoria
South Africa (1998) The Presidency. National Environmental Management Act, no. 107 of 1998. Government Printer, Pretoria
South Africa (2000) The Presidency. Municipal Systems Act, no. 32 of 2000. Government Printer, Pretoria
South Africa (2006) The Presidency. National Spatial Development Perspective. Government Printer, Pretoria
South Africa (2013a) The National Development Plan 2030. Our future—make it work. Government Printer, Pretoria

South Africa (2013b) The Spatial Planning and Land Use Management Act. (Act 16 of 2013). Government Printer, Pretoria
South Africa (2019) Department of Cooperative Governance (DCoG). District Development Model: One Plan. Government Printer, Pretoria
South Africa (2021) Department of Agriculture, Land Reform and Rural Development. Karoo Regional Spatial Development Framework. [Online]. Available at: https://www.karoorsdf.co.za/
South Africa (2023) Department of Agriculture, Land Reform and Rural Development. National Spatial Development Framework. Government Printer, Pretoria
Statistics South Africa (2022) Census 2022: Facts about Provincial economies. [Online]. Available at: https://www.statssa.gov.za/?p=12056
Stellenbosch University. (2013). Growth potential of towns in the Western Cape—quantitative analysis of growth potential at settlement and municipal level. Interim Report
Todes A, Turok I (2018) Spatial inequalities and policies in South Africa: place-based or people-centred? Prog Plan 123:1–31
UN-Habitat (2015) Habitat III issue paper on urban governance. http://unhabitat.org/wpcontent/uploads/2015/04/Habitat-III-Issue-Paper-6_Urban-Governance-2.0.pdf
UN-Habitat (2017) New urban agenda: Habitat III. https://data.unhabitat.org/pages/new-urban-agenda
United Nations (2015) Transforming our world: the 2030 agenda for sustainable development. United Nations, New York
Van Aswegen M (2018) Planning for a resilient peripheral region. North-West University, Potchefstroom Campus (Unpublished Thesis (PhD))
Van Wyk J (2012) Planning law. Juta & Co. Ltd., Cape Town
Walker B, Holling C, Carpenter S, Kinzig A (2004) Resilience, adaptability and transformability in socialecological systems. Ecology and Society, 9(2), p. 5
Western Cape Government (2005) Western Cape Provincial Spatial Development Framework. Provincial Government of the Western Cape, Cape Town
Western Cape Government (2007) Western Cape Provincial Spatial Development Framework. Provincial Government of the Western Cape, Cape Town
Western Cape Government (2012) ONECAPE2040: The Western Cape agenda for joint action on economic development. [Online]. Available at: https://www.westerncape.gov.za/sites/www.westerncape.gov.za/files/one-cape-2040-narrative-4th-draft-19-october-2012_0.pdf
Western Cape Government (2014) Western Cape Provincial Spatial Development Framework. Provincial Government of the Western Cape, Cape Town
Western Cape Government (2020) Western Cape Provincial Spatial Development Framework: chapter 4 Amendment. Provincial Government of the Western Cape, Cape Town
You Y (2017) The classification of urban systems: a review from monocentric to polycentric. Adv Econ Bus Manag Res 42. https://doi.org/10.2991/isbcd-17.2017.1

Chapter 3
Governance and Spatial Transformation at a Municipal Level

Verna Nel and Stephan Gildenhuys

Abstract Territorial governance comprises the legislative, policy and planning frameworks that guide planning decisions and the land use management system where many of the decisions are taken. The policy has seen significant change over the past two decades, including the shift from master planning to a normative approach. However, the land use management component appears to be lagging, stuck in a modernist format that emphasises separation, despite the noble principles in the legislation to the contrary. This chapter outlines the evolution of territorial governance with a particular focus on the municipal level. The achievements and failures of the legislation, spatial planning policy and land use management components are discussed. The chapter considers proposals to change the governance systems and address the problems at local municipalities that hinder transformation.

Keywords Spatial transformation · Spatial governance · Spatial planning · Land use management · Municipal planning · Empirical assessment · Configurational approaches

3.1 Introduction

Territorial or spatial governance comprises the policy, legislative and planning frameworks that guide decisions regarding the planning of space and the present and future development of land. It can be likened to the spatial planning and land-use management system described in Section 4 of the Spatial Planning and Land Use Management Act (RSA 2013) (SPLUMA) that consists of spatial planning, the development principles, norms and standards, land-use schemes and land

V. Nel (✉) · S. Gildenhuys
Department of Urban and Regional Planning, University of the Free State, Bloemfontein, South Africa
e-mail: nelvj@ufs.ac.za

D. du Plessis (ed.), *Territorial Governance and Spatial Transformation in Post-apartheid South Africa*, Local and Urban Governance,
https://doi.org/10.1007/978-3-031-85489-7_3

development applications. It occurs on supranational scales (such as the European Union), and at national, provincial and municipal levels, including local precincts and should inform investments in infrastructure and promote sustainable development. Implementation of spatial plans occurs by direct government or private sector action, for instance, through infrastructure development or indirect means, such as land development actions.

Local and precinct plans usually have a more substantial design element, while plans at a national and provincial level tend to be more policy-oriented. Political will, influence and a clear agenda are essential to implement plans at the national or provincial level as it may take several years to realise the projects stemming from these plans. Many will involve land development applications when physical development occurs. It is during the implementation phase, when broad aims are translated into development on physical sites, that usually provokes the most contestation.

Since 2001, the Municipal Systems Act (RSA 2000) has required local governments to prepare and implement integrated development plans (IDPs). These plans are intended to be the municipality's strategic plan for a Council's term of office and inform the activities and budget of the municipality. IDPs must be reviewed every 5 years. Critically, they must also include a spatial plan.

In addition to spatial development frameworks (SDFs) at a municipal level, spatial planning is also mandatory at national and provincial government levels in South Africa, according to SPLUMA. Besides the projects identified in the SDFs that are carried over to the IDP and the municipal budget, there are regulatory tools to implement these municipal plans, such as land use schemes and decisions on development applications. Nonetheless, undertaking planning seems easier than implementing those plans. Impediments include a lack of funds to undertake preparatory work for infrastructure projects such as registering servitudes, obtaining environmental authorisations, and preparing designs and tender documents prior to the capital funding becoming available. Feasibility studies may not be conducted for proposed projects in shrinking economies (Denoon-Stevens et al. 2018) with various dependencies such as path dependencies; interdependencies and goal dependencies that prevent realistic planning (Marais et al. 2021; Matebesi et al. 2024). Consequently, territorial governance has had little impact on the spatial structure of the area or its economy (Du Plessis and Boonzaaier 2015).

This chapter traces the evolution of territorial governance in South Africa. Our focus is on the municipal level, which impacts citizens more directly than provincial or national planning. We briefly describe the situation prior to 1994 and then consider both elements of territorial governance, firstly, spatial planning and secondly, land use management. In the next section, we describe how the spatial planning component of territorial governance has changed from rational-comprehensive processes and master planning to more normative approaches which generally occurred post-1994. The sluggish transformation of the land use management component to date is also outlined in this section. Thereafter, we provide a summary of the aims, policy and legislation of the democratic government and the challenges in implementing them in municipal spatial planning processes and land use management of the governance system. We conclude the chapter with proposals to improve the governance system and address the issues that hinder transformation.

3.2 The Evolution of Territorial Governance

3.2.1 Early Origins

Towns and cities are closely associated with civilisation. The earliest cities known are those along the Tigris and Euphrates Rivers (Reader 2005). However, many earlier settlements could have been flooded by sea level rise with the melting of glaciers at the end of the last ice age. Other early settlements were along the Nile, followed by other civilisations on Mediterranean islands and shores (e.g. Malta and Crete). The Indus Valley settlements reveal evidence of urban planning and the provision of essential services such as public toilets and sewers. The Greeks largely followed a strategy of colonisation of new areas rather than enlarging existing cities. Hippodamus's ideal city was laid out in a grid with sacred, public and private areas. The acropolis was the fortified area, while the agora and the main temples were in the central public area and private residences on the periphery (Toynbee 1967). A grid layout and locating important public functions in the centre of a settlement have been common design principles practised by the Indus Valley settlements, Roman colonial towns and Chinese cities (Kotkin 2005; Reader 2005).

Most towns and cities were established for administrative functions, fortifications or demonstrations of power, as marketplaces, for religious activities, or combinations of these purposes. From the Medieval period to the Industrial Revolution, the organic growth of urban areas was common in Europe, with a few examples of planned development by rulers of territories or nations. Settler states such as Australia, New Zealand, Canada and the United States of America (USA) generally used grid layouts with generous plot sizes.[1]

Land use management practices also date back to ancient times, as evidenced in Harappa and Mohenjo-Dara in the Indus Valley. The Emperor Qin of China set out rules for planning cities that were captured in the *Choi Lui* (Book of Rituals) before 722 BC (Whitehand and Gu 2006). Treatises by Julian of Ascalon (531–533 AD) and Ibn al-Imam (940–996 AD) contained principles for health, safety and privacy and were applicable to Mediterranean and Islamic settlements, respectively (Hakim 2001, 2008a) while similar principles were contained in the 'Law of the Indies' (Rodriguez 2009) that applied to Spanish colonies. Hakim (2008b) refers to these principles as 'generative codes'.

In the 1800s, zoning was introduced in Europe to separate noxious industries from residential uses. However, a measure of overlap between industrial, business and residential uses was possible. New York's first zoning ordinance had similar

[1] Most of the USA is laid out on a grid with only car-oriented areas designed with cul-de-sac.

provisions (Talen 2012). The strict segregation of land uses is known as Euclidean zoning after a court case that authorised the town of Euclid to create zones for single houses only and exclude multi-family houses (apartments) and home-based work activities (Hirt 2016). Besides creating extreme shortages of affordable housing, this form of zoning purposefully excluded lower-income residents and created swathes of unsustainable low-density housing in the USA (Silver 1997; Talen 2013; Alterman 2014; Glaeser 2017) and, to some extent, in South Africa (Musakwa and Van Niekerk 2014; Magidi and Ahmed 2019). Although zoning in South Africa is similar to the Euclidian version, residential areas in South Africa have more diverse uses and densities. Mixed-use developments are now encouraged to promote sustainability and social justice.

3.2.2 Territorial Governance in South Africa Before 1994

Prior to the colonisation of South Africa, most indigenous people were herders with cattle featuring strongly in the beliefs and design of settlements. Most settlements were dispersed (to ensure adequate grazing land), although large settlements did occur, particularly in the north western areas of the country occupied by the Tshwana. The 'Central Cattle Pattern' was a defining feature of most traditional settlements, even those that included numerous households (Dreyer 1992), where the circular huts (the equivalent of rooms in Western houses) were arranged around the byre (kraal). The exact arrangement of huts varied between clans. Stone was used in central South Africa (Dreyer 1992), but grass, timber supports and clay were used in other areas. Where more than one household occupied and homestead or settlement, there were communal grazing areas and individual plots for cultivation of maize, other grains and vegetables. According to Frescura the Western missionaries believed that round huts were uncivilised and encouraged rectangular shapes, particularly in the Eastern Cape occupied by the Xhosa.

Before diamonds and gold were discovered in South Africa, most settler towns were developed as administrative centres for the church or government, while some were established for defensive purposes (Floyd 1960). From the mid-1900s towns or their extensions were usually planned according to elements of neighbourhood planning (Ben-Joseph 2005).

Even before the infamous Land Acts of 1913 and 1936, the colonial government began limiting the land available to indigenous people through 'betterment policies' focussed in the Eastern Cape. Restricting land meant restricting the heads of cattle each household could support and hence their wealth. Additional taxes forced people to enter the cash economy and find work in the cities. However, the Land Acts further restricted most Black Africans to reserves or *homelands*, while the Group Areas Act of 1950 designated areas for Asian/Indian and Coloured people, usually on the periphery of the urban area. These laws produced South Africa's distorted spatial form. Changing this form has been a key focus of the African National Congress (ANC) government since 1994.

As the apartheid government began loosening control over the movement and citizenship of Black people in the late 1980s (Cilliers 1986), the territorial governance legislation regarding spatial planning and land-use management saw little change from the segregationist policies of the previous seven decades, although the government did accept the reality of Black urbanisation in 1986 (Tomlinson 1988).

The first formal spatial planning action undertaken was the report by the Social and Economic Planning Council in 1944. It suggested regional planning and regional planning institutions similar to the Tennessee River Valley Authority. Although these ideas were not implemented immediately, they did inform the Physical Planning Act of 1967 (Oranje and Merrifield 2010; Van Wyk 2012). The National Physical Development Plan (NPDP) was published in 1975. This plan excluded the 'homelands' based on the notion that they were no longer part of South Africa. It also sought to re-distribute the population away from the metropolitan regions. The NPDP was followed by guide plans that were intended to integrate transport and infrastructure planning with land uses. Van Wyk (2012: 42) notes the many problems that accompanied the implementation of the guide plans. The NPDP was never approved by the Cabinet for "fear of a negative response from voters located in areas which were not to be boosted/favoured" (Oranje and Merrifield 2010: 35).

The Physical Planning Act of 1967 was amended by another act of the same name in 1991 and the revised act remained on the statute books until repealed by the SPLUMA in 2013. There were no obligations on provinces or municipalities to prepare spatial plans to inform development applications or to integrate the fragmented landscape. Many municipalities took it upon themselves to prepare spatial plans at an appropriate level of detail to assist in planning future development along with the necessary social and engineering infrastructure (Fig. 3.1).

Provinces were responsible for 'development control' (land use management) in white urban areas (Van Wyk 2012). Provincial Townships[2] Boards decided on all applications requiring an amendment of the town planning scheme or removing restrictive title conditions. Many municipalities in the previous Transvaal province were authorised to decide on applications, yet the provincial townships boards decided appeals against municipalities' decisions. Provinces also had the sole authority to approve town planning schemes applicable to white areas (de Visser 2016). National government departments administered the Black Communities Development Act 1986, which applied to urban townships.[3] Separate legislation applied to the former homelands administered by their 'governments' or a national department (Abrahams and Berrisford 2012). Although the homelands were incorporated into the country in 1994, their planning laws remained applicable until repealed, which created a complicated and confusing situation (Van Wyk 2012).

---

[2]Townships in this context refers to new layouts for new suburbs or the extension of existing suburbs.

[3]Townships in this context refers to the housing areas development as dormitory suburbs with few amenities for Black Africans

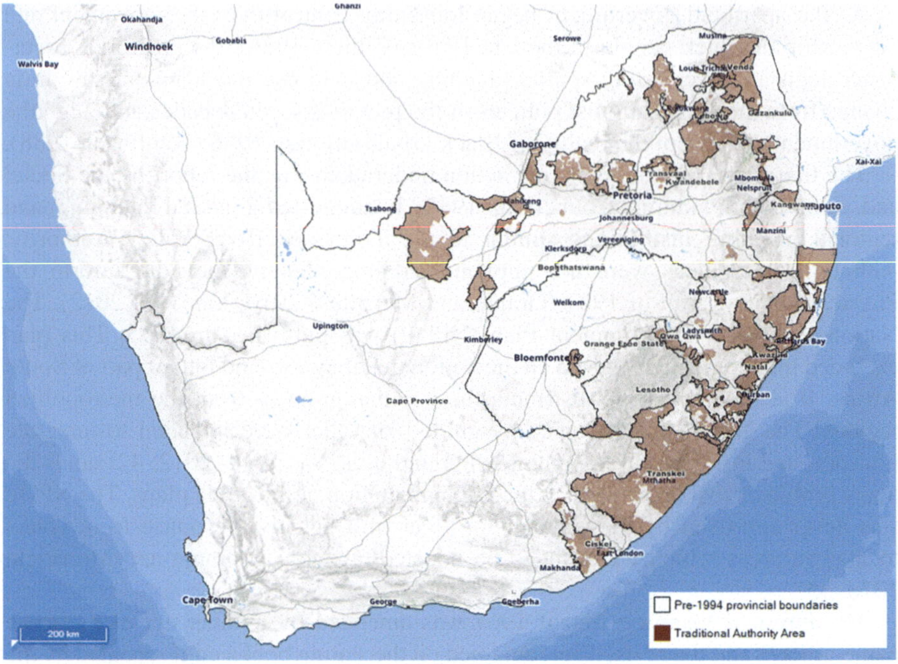

Fig. 3.1 Pre-1994 provincial boundaries (Municipal Demarcation Board) and traditional authorities (National Geo-Spatial Information)

3.3 Territorial Governance After 1994

This section describes the changes in territorial governance after 1994, starting with the spatial planning component. Thereafter, we describe the changes in the land-use management component. However, it must be noted that the two aspects are complementary in territorial governance. Spatial plans inform land development decisions while pressure in an area with many land development applications may require a new or amended spatial plan.

3.3.1 Changes to the Spatial Planning Policy Framework After 1994

The spatial planning policy framework established after the 1994 election in South Africa was heavily influenced by the work of Dewar and Uytenbogaardt, who published their ideas in 'South African Cities: A Manifesto for Change' (1991). They identified four key elements for the transition of South Africa's cities with a more compact city form as the first precondition for change. The second important element was the multifunctional use of space and facilities, which were deemed economically essential. The third element of their work focused on increased densities, while the fourth was integrating land uses and urban activities (Dewar and Uytenbogaardt 1991). These concepts became central to South Africa's spatial planning policy post-1994 and have found expression in two key processes. The first pertains to critical national spatial planning policy and legislative frameworks, and the second to applying these spatial planning goals and concepts in municipal and sub-municipal SDFs.

The White Paper on the Reconstruction and Development Programme (RDP) was the first policy framework in 1994. Although a 'watered down' version of the 1993 ANC election manifesto of the same name, the RDP's purpose was to establish a framework for implementing the government's transformation strategy and socio-economic aims. Its goal was to coordinate the various state organs to achieve the RDP's objectives (RSA 1994). Although the RDP did not explicitly prioritise spatial planning, it emphasised the importance of locating land for development to ensure adequate access to social amenities and economic opportunities.

Following the introduction of the RDP, the Development Facilitation Act (DFA) (1995) was enacted to facilitate the efficient implementation of development projects and programmes related to land (Berrisford 1997). It was also the first step towards a new planning system. It required Land Development Objectives—a spatial representation of the municipality's proposed investment in land development. Additionally, the DFA included a set of principles that emphasised the integration of the administrative, environmental, social, and economic dimensions of land development. It advocated for improved spatial integration of employment and residential areas, optimal utilisation of land and its physical attributes, services infrastructure, transport networks and social amenities, encouraged mixed land uses, promoted compact city structures, the rectification of historical spatial patterns, and adoption of environmentally sustainable land development practices (RSA 1995). Before the SPLUMA was enacted in 2013, the DFA principles served as the de facto standard for urban spatial planning in South Africa.

In 1997, the Urban Development Framework was published to provide national guidelines for the planning and development of human settlements in South Africa (RSA 1997). Later versions of urban development directives were introduced, including the Breaking New Ground housing strategy (DoH 2004), and later the Integrated Urban Development Framework (IUDF) in 2016. The IUDF outlined a comprehensive vision for the sustainable development of human settlements, including guidelines and programs for achieving the defined objectives, efficiency of urban transport, and integrating environmental management with development processes.

A Draft Green Paper on planning and land-use management was produced by the Development and Planning Commission (1999) established by the DFA. It was followed by a White Paper on Spatial Planning and Land Use Management in 2001 (Department of Land Affairs 2001). These documents had a similar approach to spatial planning but differed regarding land-use management (Nel and Denoon-Stevens 2024). The principles in the DFA were consolidated into five development principles in the White Paper: sustainability, equality, efficiency, integration, and fair and good governance.

The Municipal Systems Act (RSA 2000) and the subsequent Local Government: Municipal Planning and Performance Management Regulations (RSA 2001) brought significant reforms to local governance following the 1994 election. It began a new era of so-called 'joined-up' governance. It introduced the concept of municipal IDPs. These reforms sought to ensure comprehensive spatial and temporal planning frameworks for delivering various initiatives and programs across different sectors of local government (Harrison 2006). Section 26(e) of the municipal Systems Act explicitly requires a spatial development framework as a component of municipalities' IDPs. The Regulations further specified the content of the SDF.

In 2010, the National Planning Commission (NPC) was established to provide recommendations on the country's long-term development. It was also tasked with developing and implementing the National Development Plan (NDP) Vision 2030 in 2012. The NDP recognises that while many of the elements of its spatial vision were in place in 1994 the challenge is "to translate the vision into implementation and meaningful spatial outcomes". The NDP confirmed the critical elements of the spatial vision for urban areas: limiting urban sprawl, supporting affordable high-density housing, creating sustainable settlements, focusing investment along public transport corridors and economic nodes, integrating poor urban communities into the economy by, for example, developing diverse economic nodes in townships, in-situ upgrading of informal settlements, promoting mixed-income housing, and enhancing urban public spaces.

The 2016 Integrated Urban Development Framework (Department of Co-operative Governance and Traditional Affairs 2016) elaborates on chapter Eight of the National Development Plan, focusing on sustainable and integrated human settlements, transforming the space economy, and developing resilient and liveable settlements. The IUDF has four strategic goals for achieving spatial transformation: spatial integration, social integration, inclusion and access to services and opportunities, sustainable economic growth, and improved governance. The IUDF recognises that integrated urban communities require an extensive spatial planning framework that enables an integrated transport system and efficient land governance. Other critical components are financial reform and the empowerment of communities. To achieve its aims, the IUDF proposes the following: core public transport nodes and corridors, renewing and redeveloping inner cities, supporting densification, redeveloping townships, tenure upgrading and the provision of services in informal settlements, and enhancing the design and quality of urban public spaces.

The enactment of the SPLUMA in 2013 was the most significant development from a territorial governance perspective since 1994. SPLUMA sets out five principles for spatial planning and land use (similar to those in the 2001 White Paper) that require all spatial planning and land use management actions to promote efficiency, spatial justice, resilience, sustainability and good administration. SPLUMA also requires the formulation of SDFs with implementation plans, targets, and monitoring indicators, specifically by local governments. It also calls for land-use schemes to implement the SDFs.

From the above, it is clear that the post-1994 urban planning doctrine aligns with common urban planning objectives, including environmentally sustainable development, access to amenities and economic opportunities in new urban development, greater integration and mix of land uses, optimal utilisation of existing infrastructure, promotion of compact urban form, the rectification of historical spatial patterns, support for local economic growth, and introducing efficient public transport systems.

3.3.2 Current Issues with Municipal Spatial Planning

Although South Africa has a clear framework for municipal spatial planning, the country's cities continue to be plagued by inefficiencies, making them among the least effective urban environments globally (Du Plessis and Landman 2002). Recognising the urgency for new strategies and methods (Serfontein and Oranje 2008), practitioners and academics have called for a fresh approach to municipal spatial planning in South Africa (Todes et al. 2010). Nevertheless, several key challenges have emerged, significantly impacting the planning and outcomes of municipal areas since the country's political transformation.

One of the primary and long-standing challenges facing South African municipalities is the need for improved coordination and collaboration. The lack of coordination and alignment between various functions and spheres of government was already identified as a serious hindrance in the first rounds of their development (Harrison 2001). Municipalities still face difficulties implementing their spatial plans due to insufficient coordination with other planning activities (DRDLR 2014). Moreover, the lack of guidance and detail provided by municipal SDFs—due to the large areas they cover—has resulted in a hierarchical approach to planning, with more detailed plans at lower spatial levels. The overlapping responsibilities between provincial governments and local municipalities, particularly concerning housing (human settlements) and environmental and land-use management, further complicate vertical alignment. The implementation of various national departments' policies with spatial implications also makes it challenging to align investment programs and achieve a cohesive approach towards national priorities (Turok and Parnell 2009; Nel and Minnie 2022).

Consequently, it is crucial to align and coordinate the institutional entities involved in planning and implementing urban spatial development frameworks. Horizontal coordination of planning and development activities among different local government units, along with the development of collaborative action and expansion of urban spatial planning, can contribute to overcoming these challenges (Du Plessis 2019). To this end, the Department of Co-operative Development has proposed a 'One Plan' or district development model (DDM) to coordinate plans and budgets. How the alignment of national and provincial sector departments, district and local municipalities will work in practice remains to be seen (Nel and Minnie 2022).

The second challenge concerns spatial integration in South Africa's planning system which has been a fundamental goal of spatial planning and development in the country since the 1990s. Various policy documents, such as the RDP (RSA 1994) and the National Development Plan (NPC 2013), emphasise the importance of unifying and densifying urban areas to address historical segregation. Despite this emphasis, the concept of integration lacks a clear definition and criteria to measure change.

A document containing guidelines for Spatial Development Frameworks was issued to distinguish between socio-economic and functional integration (DRDLR 2014). However, limited progress has been made in socio-economic integration spatially, primarily due to resistance from financial institutions and communities. As a result, socio-economic segregation levels have increased in various neighbourhoods despite changes in ethnic composition in higher-income areas (Badenhorst et al. 2005; Turok 2018). To better address integration, both qualitative and quantitative methods to evaluate the complex integration processes are recommended (Horn 2005). Additionally, spatial plans must include more quantitative analyses dealing with socio-economic and functional integration proposals (Du Plessis 2019).

The third challenge is the need for a comprehensive understanding of the space economy in South Africa and specifically at a municipal level. Balancing the need for growth and competitiveness while addressing the inequitable distribution of wealth in prosperous cities presents a significant challenge (Jenkins and Wilkinson 2002; Todes and Turok 2018). The lack of comprehensive understanding of the space economy hinders addressing community tensions. Although progress has been made in spatial economic analysis, applying such analyses in urban areas remains challenging (Todes 2008; van Huyssteen et al. 2009). This limited understanding and application of knowledge of the spatial economy within the planning system further hampers development outcomes (Harris 1990; Serfontein and Oranje 2008; National Planning Commission 2013). Thus, there is a need for a thorough analysis of factors influencing urban development and investment trends in South Africa's urban areas (Du Plessis 2019), which has implications for the training of planners and urban geographers. It also speaks to the availability of reliable economic data at local levels.

Strengthening capital investment and infrastructure planning is the fourth challenge that must be tackled. The availability of reliable infrastructure is essential for urban development, and integrating capital investment and infrastructure development into spatial planning processes is crucial (Doxiadis 1970; Healey 2004). However, challenges include the lack of coordination between public and private sectors and the dominance of private-sector projects (Todes 2008). A comprehensive strategy focusing on specific areas and policies to address infrastructure needs is proposed to overcome these challenges (Turok and Parnell 2009).

The fifth vital challenge to consider is pursuing sustainability through spatial planning. Although the importance of sustainability was recognised during the 1992 Earth Summit, persistent challenges in South Africa include a lack of understanding of sustainability principles, limitations in using environmental data, the emphasis on housing and infrastructure over the environment and the exclusion of sustainability priorities in national planning departments (Harrison et al. 2007). Local governments prioritise housing and essential engineering services over environmental concerns due to performance management systems focusing on project aspects while lacking tools and resources to monitor sustainability. Spatial development frameworks in South Africa consider sustainable development but lack implementation techniques or clear targets with sufficient evidence for evaluating the sustainability of their proposals (Du Plessis 2019). SPLUMA, while including 'spatial sustainability' as a development principle, merely requires SDFs to "take cognisance of environmental instruments" (section 12(1)(m)).

The sixth challenge to address is informality in South Africa's urban landscape. The concept of urban informality encompasses various aspects of the urban landscape, including land, housing, infrastructure, and decision-making processes (Roy 2005). The informal economy is growing, and the number of households living in backyard structures and informal settlements is expanding rapidly (Ligthelm and Masuku 2003; Fourie 2018; Scheba and Turok 2020). Policymakers and planners recognise the informal sector's significance and focus on formalising and developing it, but reconciling informal activities with planned cities remains a challenge (Roy 2005). While the National Development Plan supports upgrading informal settlements to provide affordable housing and land for the poor, a better understanding of the mechanisms and strategies of informality is necessary to work with it (National Planning Commission 2013).

The last challenge that requires attention is the need for a comprehensive set of spatial statistical indicators to inform decision-making processes (Du Plessis 2019). Currently there is insufficient economic and other quantitative data to develop sensible targets and evaluate the changes in a city's structure due to new planning policies (DRDLR 2014). The ability of a spatial plan to position itself within political spheres and effectively communicate progress indicators and strategies is crucial (Healey 2006), yet the support capabilities for planning and statistical analysis remain underdeveloped in South Africa.

3.3.3 Land Use Management After 1994

As explained in section 2.2, the land use management system in South Africa was highly fragmented by 1994, and it largely remained unchanged until 2015, when SPLUMA came into effect. The DFA was the first statute on planning matters under the new regime and was enacted as 'emergency legislation' to expedite the construction of housing and services under the RDP. Besides the development principles and the land development objectives (LDOs), the DFA introduced a new system for land development applications (RSA 1995). It enabled applicants to submit one application for the subdivision, rezoning and removal of title deed restrictions of a property instead of two or three separate applications. It also enabled the creation of provincial development tribunals to decide on applications and appeals against municipal decisions. While applying the development principles and preparing LDOs was mandatory, provinces could choose whether or not to use the DFA application process (Berrisford 1997). Provinces with many areas under traditional authority (homelands) were among the first to use the DFA to manage applications. Most provinces began preparing new provincial planning and land use management acts to create one land use management law for the province. KwaZulu Natal and the Northern Cape both enacted Planning and Development Acts. However, the DFA did not address routine land use management or rationalise the pre-1994 planning laws and regulations (Berrisford 1997).

The DFA established a Development and Planning Commission to advise the government on planning laws relating to various facets of development. The commission published a draft Green Paper on Planning and Development in 1991. The green paper was followed by a White Paper on Spatial Planning and Land Use Management in 2001 (Department of Land Affairs 2001). As the Municipal Systems Act and its regulations had already provided for spatial development frameworks, the Department prepared a draft Land-Use Management Bill In 2001. However, this Bill and subsequent versions were referred back to the Department.

Meanwhile, the DFA application process in Gauteng was (mis)used to obtain approvals from the DFA Development Tribunals, where applicants believed the municipality would not approve their applications. Thus, applications contradictory to the SDF and IDP were approved (de Visser 2016). The City of Johannesburg (CoJ) challenged the Gauteng Development Tribunal on this issue. When the Gauteng High Court ruled in favour of the Tribunal, the CoJ appealed to the Supreme Court of Appeal, which ruled in favour of the CoJ, stating that land-use management was an integral part of 'municipal planning' as set out in Part B of Schedules 4 and 5 of the Constitution (South Africa 1996) and thus Chapters V and VI of the DFA were unconstitutional. The Constitutional Court concurred with the Supreme Court of Appeal, declaring the chapters unconstitutional. It suspended the declaration of invalidity for 24 months to enable the government to prepare new legislation. This decision was the impetus for the preparation of SPLUMA (Van Wyk 2020).

The Constitution Court's decision also meant that municipalities, and not provinces, are now responsible for all decision-making regarding land development applications. Provinces that had previously administered land development applications now had to devolve the function to local governments. They also had to assist local municipalities to develop the systems, procedures and capacity to process and decide on the applications. The onus was on the Department of Agriculture, Land Reform and Rural Development[4] (DALRRD) and provincial departments to provide training and materials to municipal staff and develop systems to assist the municipalities in this function (such as the e-Lodgement system in the Free State) (Nel et al. 2017). The emphasis was on enabling municipalities to process applications rather than empowering them with the knowledge and skills to evaluate applications.

[4]Then the Department of Rural Development and Land Reform.

In the absence of provincial legislation dealing with the processing and assessment of land development applications, standard municipal by-laws were prepared so that each municipality could adapt to its needs. The unanticipated outcome is that municipalities' requirements regarding documents and procedures differ, partially defeating the ideal of one consistent set of regulations throughout South Africa.

While the first four chapters of SPLUMA are transformative, those dealing with land-use management are less so. Section 24 is the only section that contains requirements in line with the development principles (section 7), namely that land-use schemes must include the entire municipal area, enable affordable housing, and implement the IDP and SDF as well as provincial and national policies. The remainder of chapters five and six are similar to the 'old order' provincial ordinances; however, with far less detail regarding the particulars of a land-use scheme. It is no wonder that some municipalities consider a land-use scheme that consolidates previous schemes and covers the entire municipality, with large areas zoned as agriculture or traditional areas, as 'SPLUMA-compliant'.

Unfortunately, most land-use schemes and municipal by-laws have not transformed. They still meet the expectations of a small elite while ignoring the needs of the majority. Primarily designed for formal towns and suburbs, they are not flexible enough for the needs of the diverse population and landscapes of South Africa. Essentially, the nature and content of land-use schemes need to be revised if they are to promote the development principles, while the processing of development applications must be simplified to reduce the costs of development.

The post-1994 era in South Africa introduced a new planning paradigm, yet the efficacy of spatial planning in addressing the complexities of spatial development has been notably limited. The prevailing approaches, which are normative and driven by conceptual frameworks, have not significantly influenced the issues prevalent in these environments, nor have they facilitated the required changes and spatial transformations. Despite increasing advocacy for a shift towards an evidence-based approach to spatial planning that incorporates complex analyses utilizing quantitative, spatial, and statistical techniques, there remains a persistent tendency among planners to depend excessively on anecdotal evidence and assumptions, rather than on rigorous empirical assessments.

The predominant methodologies in spatial planning have consequently led to a significant challenge in monitoring and evaluating the impact and success of spatial plans in achieving their objectives and intended spatial outcomes. This issue stems largely from the spatial frameworks' failure to establish practical and measurable targets for evaluating the effectiveness of a plan in meeting its defined goals at a local level.

In light of these shortcomings, the following section posits that a potential solution may lie in an integrated approach that combines configurational methods and urban morphology assessments with the transect concept. Such a synthesis could bridge the gap in current knowledge and practice, offering a more robust framework for assessing and guiding urban spatial development.

3.4 Moving Forward

While a solid framework for territorial governance has been created through SPLUMA, much more work is required to achieve spatial transformation. Further guidance and detail are required. The current emphasis on normative planning needs to be underpinned by reliable and current data along with the necessary analyses of the natural and built environment, social, economic and institutional aspects of the area. Due to the lack of adequate data on the existing situation, many SDFs use demographic data that is over 10 years old or propose excessive areas for industrial and commercial purposes (Denoon-Stevens and Nel 2017). Others may neglect to consider the impact of the proposed development on the area's infrastructure.

In response, new approaches to territorial governance are required (Du Plessis 2014; Nel and Denoon-Stevens 2024). Todes et al. (2010) emphasised the need for innovation and flexibility in the spatial planning process, with others (Todes 2008; Du Plessis 2019) calling for a more evidence-based spatial planning approach with a greater focus on more complex analysis using quantitative, spatial, and statistical techniques. The rapid development of modern computer technology in geographic information systems (GIS), spatial data collection through remote sensing applications and artificial intelligence (AI) may provide the opportunity to overcome the current limitations and challenges faced by territorial governance in South Africa (Du Plessis 2014).

3.4.1 Configurational Approaches to Spatial Planning

Urban morphology and spatial configurational approaches to spatial analysis may be of value in analysing space. According to Sevtsuk (2010), Configurational Analysis is a methodology used in urban geography, transportation research, architecture, and urban planning. It focuses on understanding the relationships between multiple spatial elements within a complex environment and is commonly applied in configurational analysis (Hillier et al. 1984, 1987). The analysis emphasises the interconnections between various spatial elements and their influence on people's movement within a space (Kropf 2018).

A critical concept is that of natural movement which proposes that peoples' movement influences the distribution of economic and social activities in a community. Configurational analysis helps understand the unique characteristics of urban areas and how the movement of people and economic opportunities are related to the overall structure and configuration of a city (Hillier et al. 1993; Sevtsuk 2010; Porta et al. 2012; van Nes 2021). The analysis is rooted in network or graph theory (Marshall 2004; Kropf 2018) and examines street networks on a large scale in urban environments (Hillier 2007; Cooper and Chiaradia 2015). Various measures like network centrality, betweenness centrality, and connectivity are used to understand the significance of nodes in the network and potential flows within it.

One well-known approach in configurational analysis is the Space Syntax method. However, this method has faced many criticisms. In response to these criticisms, new spatial configurational tools have emerged. One of these is the Place Syntax tool that incorporates accessibility (Ståhle et al. 2005) and point or polygon features as origins and destinations (Berghauser Pont et al. 2019), and can calculate integration and choice centrality, the distance between an origin and a destination, attraction reach, and attraction betweenness. These tools show great promise for spatial planning (Figs. 3.2 and 3.3).

Other tools have been developed. The Spatial Design Network Analysis tool (sDNA) adopts a non-linear approach to examine urban networks, retaining their physical and geographical characteristics (Cooper and Chiaradia 2015). MIT's Urban Network Analysis (UNA) tool focuses on analysing intersections within a network and allows for multimodal transportation analysis (Sevtsuk and Mekonnen 2012). Higgins (2019) has developed an accessibility tool for R and ArcGIS that expands on the reach and gravity features of the UNA framework. This tool enables the creation of a 3D network with points or polygons serving as origins or destinations. It also provides additional gravity-based access methods not included in the UNA framework (Fig. 3.4).

Despite recent advancements in these tools, challenges remain in adapting to long-term disruptions and adaptations. Each tool has strengths and limitations, and their suitability depends on the specific analysis requirements. However, ongoing debate continues about their effectiveness in long-term planning.

3.4.2 The Transect: A Place-Type Approach for Territorial Governance

Another promising approach that can enhance municipal spatial planning is the concept of the transect promoted by urban planners associated with New Urbanism (Duany and Talen 2002). By allocating rural attributes to appropriate rural locations and urban attributes to urban areas, this approach emphasises the significance of intensity and character in each (Talen 2015). In doing so, it reflects the fundamental principle that urban systems are interconnected within a broader natural continuum, advocating for their design and development accordingly (Duany and Talen 2002; Talen 2002).

The concept of a transect, as a cross-sectional analysis of changing habitats or environments, is not new. Early planning theorists like Ebenezer Howard and Patrick Geddes recognised the potential of the transect as an analytical tool for regional planning (Hall 2002; Talen 2002). Howard envisioned a gradient of densities in his garden cities, ranging from urban core to agricultural fringes. Geddes promoted a 'valley section' that encompasses natural and urban habitats (Hall 2002). These visionary ideas laid the groundwork for the transects application in urban planning.

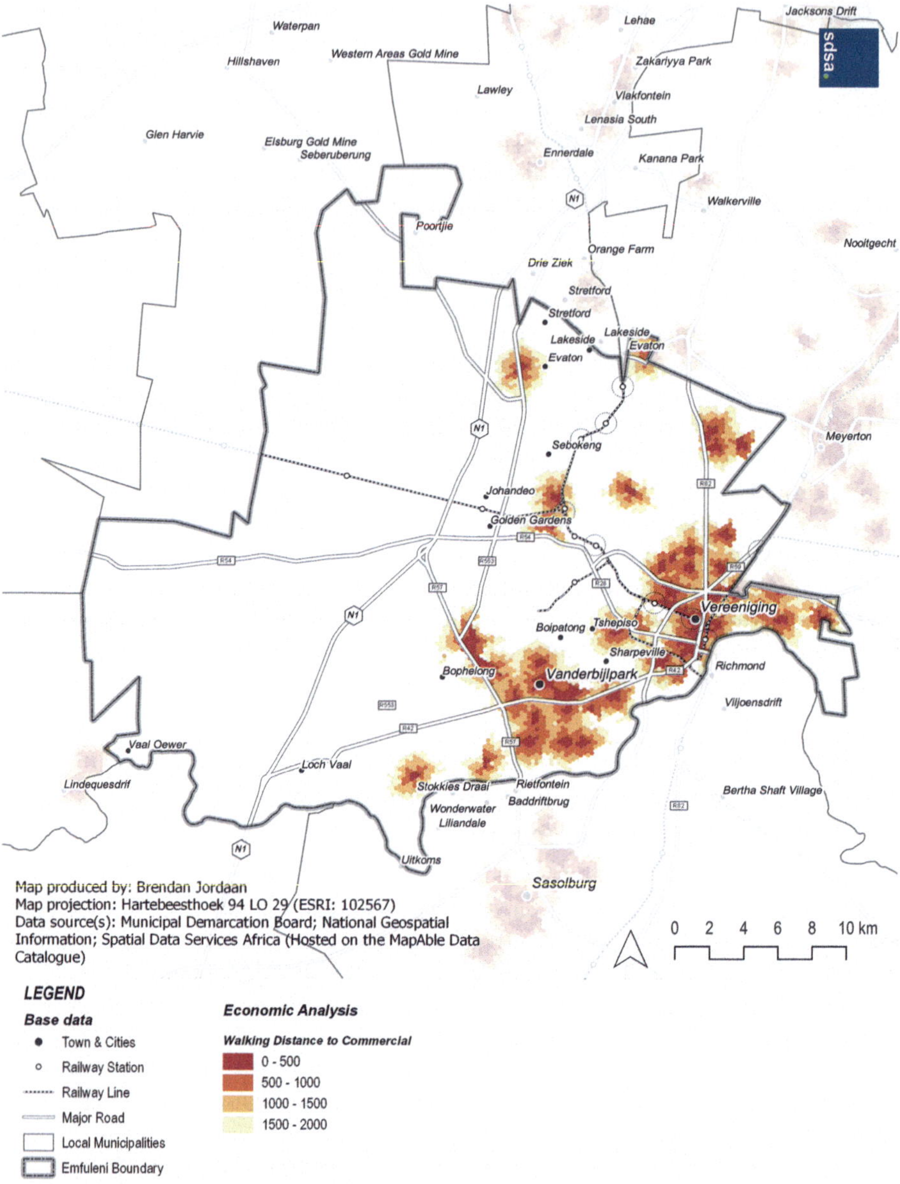

Fig. 3.2 Attraction distance analysis to commercial points using the PST (City of Emfuleni 2023)

A hallmark of the urban-rural transect concept is the Smart Code (Duany et al. 2000), a well-crafted framework that categorises the transect into six distinct zones that range from the serene rural to the bustling urban core. These zones offer a strategic framework for allocating different urban elements harmoniously (Talen 2002).

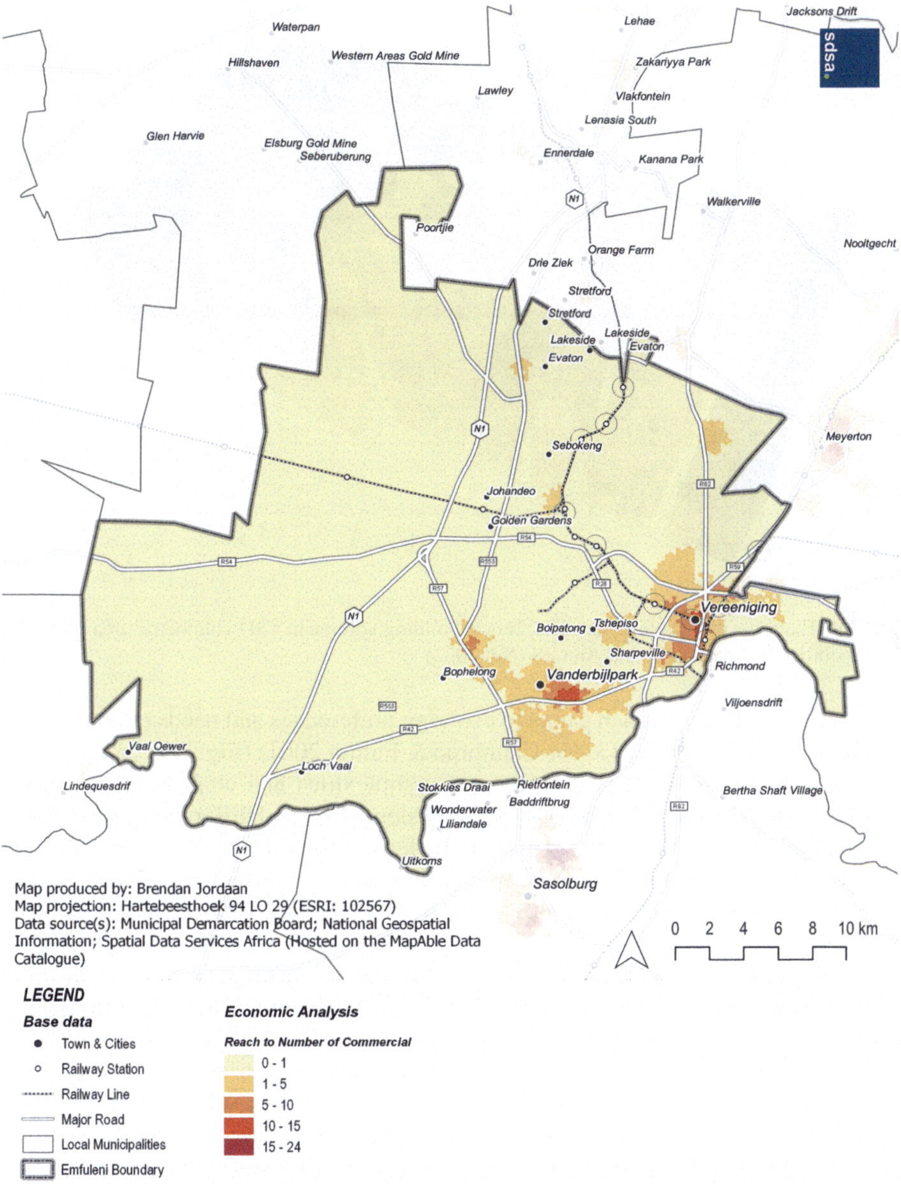

Fig. 3.3 Attraction reach analysis to commercial points using the PST (City of Emfuleni 2023)

The transect promotes mixed land uses, prioritises public spaces, and emphasises pedestrian access—cornerstones of not only contemporary urban planning but also traditional neighbourhood development and transit-oriented development (Carlton 2009; Calthorpe 1993). By creating vibrant cities that integrate diverse

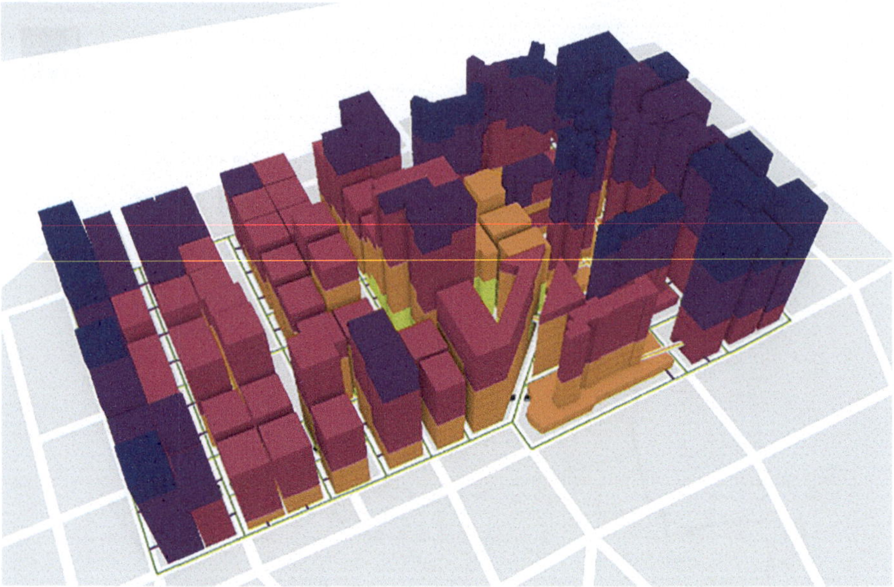

Fig. 3.4 Example of an assessment of access to building volume in a 3D volumetric area using the Accessibility Toolbox for R and ArcGIS (Nel 2022)

environments, this approach caters to evolving preferences and needs (Congress for the New Urbanism 2000: 13–69; Calthorpe & Fulton 2001) (Fig. 3.5).

Although the urban-rural transect has complexities and challenges, it can be adapted to local circumstances (Duany and Talen 2002) and different aspects of territorial governance (Nel 2016). Through the lens of the transect, planners can craft communities that bridge the gap between urban and rural, which may weave in and out of the urban fabric in South African cities (Beall 2004; Drummond and Nel 2021).

The transect approach can also be used as a land-use management tool. The transect zones can form the land-use zones that SPLUMA requires. In the South African context at least 13 zones would be required: natural and protected areas; agriculture (which could be sub-divided into extensive and intensive); traditional areas; small towns, peri-urban areas; informal settlements; 'townships'; suburbs; city centres; light industry; heavy industry; and possibly a transport zone for airports, stations and harbours. This form of zoning will be far more flexible than the current land-use zones and can be crafted to promote sustainability, regenerative development and support livelihoods. The transect can be a better tool to promote the SPLUMA development principles of spatial justice, spatial sustainability, spatial resilience, efficiency and good administration than the existing land-use schemes. A spatially just land use scheme will include provisions that facilitate the survival strategies of low income households, including informal housing and enterprises, unlike some schemes which make it difficult to operate a business from home (Charman et al.

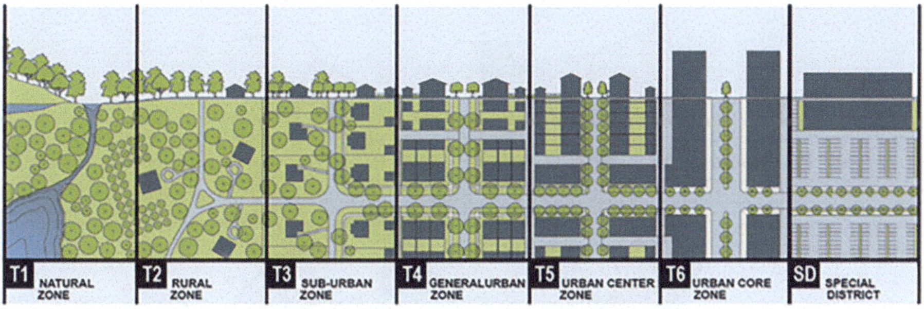

Fig. 3.5 Transect zones (Centre for Applied Transect Studies n.d.)

2017; Turok et al. 2017). Providing secure tenure (not necessarily ownership) is possible in many places using modern GPS (Williams-Wynn 2021) and will stimulate investment in the settlement, creating social and economic assets improving resilience. Unlike the existing South African land-use management system—a colonial legacy—a just and sustainable LUS should align with the values and beliefs of the majority of residents (Fig. 3.6).

3.5 Conclusion

This chapter has explored the intricate landscape of territorial governance and planning in South Africa, shedding light on the challenges and potential solutions that have recently emerged. The journey toward achieving spatial transformation has been marked by significant hurdles, leading to calls for new approaches. This conclusion reflects on the multifaceted issues discussed and the innovative strategies that hold promise for the future of territorial governance in the country.

South Africa has made significant strides in reshaping its urban landscape, transitioning from apartheid-era policies to more inclusive and sustainable approaches. Key legislative milestones, such as the DFA, the Municipal Systems Act of 2000 and SPLUMA, have laid the foundation for spatial transformation. However, despite these regulatory frameworks, challenges persist. Coordination and collaboration among government entities remain a hurdle, hindering the efficient implementation of spatial plans as does the lack of coordination between public and private sectors in infrastructure planning. Socio-economic and functional integration has not lessened but increased. Tackling this challenge demands a better understanding the space economy in order to balance growth with equitable wealth distribution. Sustainability principles are appreciated but require more explicit implementation techniques and targets.

The persistent inefficiencies in the urban environment have compelled practitioners and academics to seek fresh approaches to territorial governance. This chapter

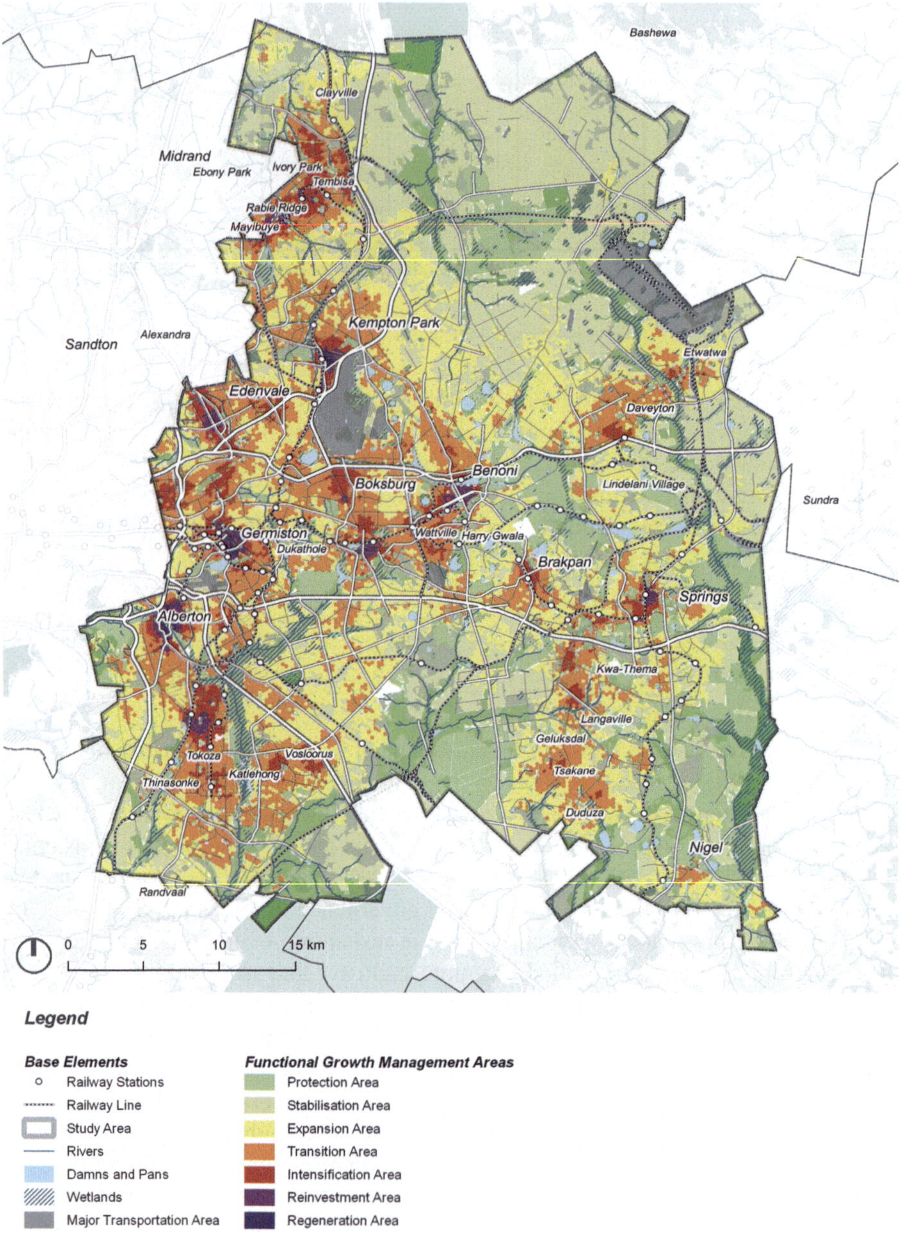

Fig. 3.6 The adaptation of the transect concept for municipal spatial planning through the application of the Place Syntax Tool (City of Ekurhuleni 2022)

has evaluated some of these challenges and explored the potential solutions that could usher in a new era of territorial governance in South Africa.

The historical perspective offered in this chapter sheds light on the roots of South Africa's spatial disparities, tracing them back to discriminatory legislation like the Land Acts of 1913 and 1936 and the Group Areas Act of 1950. These policies contributed to the socio-spatial disparities. However, territorial governance in South Africa has also evolved, with significant shifts in responsibilities and legislative frameworks. The DFA introduced sound principles, but its implementation faced challenges. The subsequent shift of responsibilities from provinces to municipalities has created a need for capacity building.

South Africa's spatial planning system faces significant challenges that demand fresh approaches to improve coordination and collaboration, increase integration, develop a comprehensive understanding of the space economy, strengthen capital investment and infrastructure planning, pursue sustainability, address informality, and bolster empirical evidence and support. Tackling these challenges to develop effective territorial governance strategies for South Africa will require collaborative efforts from practitioners, academics, policymakers, and communities.

A critical aspect is the need for territorial governance to promote justice and inclusivity. Land-use schemes (LUS) must align with the development principles and cater to the needs of the marginalised, including those in informal settlements and enterprises. Providing secure tenure, even without ownership, is a viable option that can stimulate investment and create social and economic assets. Moreover, territorial governance should respect local religions and cultural practices, ensuring that these are accommodated within planning and regulation of space. Recognising and embracing the diverse values and beliefs of South Africa's population is essential for creating inclusive and just urban environments.

In response to these challenges, this chapter has explored potential solutions. Configurational approaches in spatial planning integrate network science and complexity theory, providing a scientific, objective framework. This methodology is grounded in three principles of urban morphology: the city's physical components, the resolution of analysis, and the temporal element of change. Emphasis is placed on measuring urban form through connectivity, diversity, and particularly density, crucial for sustainable development and improved access. These measures are vital as they dovetail with existing spatial planning policies. Additionally, the transect concept, aims to combat unsustainable urban growth by promoting compact, pedestrian-friendly, and sustainable urban forms that harmoniously integrate with natural ecosystems. Collectively, these strategies imply a transformative potential for spatial planning in South Africa, steering it towards sustainability, inclusivity, and harmonious urban-nature integration. Despite the new planning paradigm since 1994, South Africa's spatial planning has struggled to address challenges. Current approaches lack the needed impact due to a reliance on anecdotal evidence instead of empirical assessment, resulting in irrelevant plans and ineffective monitoring and evaluation. Current research suggests that merging the configurational approaches with the transect concept might bridge this knowledge gap and promote successful spatial transformation in South Africa as both a spatial planning and land use management tool, enabling improved analysis, the development of better spatial form, enhanced decision making and providing a framework for monitoring and evaluation.

In this ever-evolving landscape of territorial governance, collaboration among practitioners, academics, policymakers, and communities is paramount. Together, we can chart a course toward a more equitable, sustainable, and inclusive urban future for South Africa. By embracing innovation, respecting diversity, and addressing the complex challenges head-on, South Africa can forge a path toward spatial transformation that truly serves all its residents.

References

Abrahams G, Berrisford S (2012) Addressing the crisis of planning law reform in South Africa. South African Cities Network. http://sacitiesnetwork.co.za/wp-content/uploads/2014/06/ADDRESSING-THE-CRISIS-OF-PLANNING-LAW-REFORM-IN-SOUTH-AFRICA.pdf

Alterman R (2014) Planning laws, development controls, and social equity: lessons for developing countries. World Bank Legal Rev 5:329

Badenhorst MS, Van Helden P, Schoonraad MD (2005) Post-apartheid Pretoria: Verskuiwings in die sosio-ruimtelike landskap, 1996–2001. Stads-En Streeksbeplanning Town Reg Plan 2005(49):1–16

Beall J (2004) Navigating tradition: traditional authorities and governance in eThekwini Metropolitan Municipality, South Africa. Int Dev Plan Rev 26(4):457

Ben-Joseph E (2005) The code of the city: standards and the hidden language of place making. MIT press

Berghauser Pont M, Ståhle A, Marcus L, Koch D, Fitger M, Legeby A, Stavroulaki I, Karlström A, Miranda Carranza P, Nordström T (2019) PST. DiVA. https://doi.org/10.13140/RG.2.2.25718.55364

Berrisford S (1997) Implementation of the Development Facilitation Act. Stads-En Streeksbeplanning Town Reg Plan 1997(42):57–59

Calthorpe P (1993) The next American metropolis (Vol. 23). New York: Princeton Architectural Press

Calthorpe P, Fulton WB (2001). The regional city: Planning for the end of sprawl

Carlton I (2009) Histories of Transit-Oriented Development: Perspectives on the Development of the TOD Concept, Working Paper, No. 2009,02, University of California, Institute of Urban and Regional Development (IURD), Berkeley, CA

Centre for Applied Transect Studies (n.d) https://transect.org/

Charman A, Tonkin C, Denoon-Stevens S, Demeestere R (2017) Post-Apartheid spatial inequality: obstacles of land use management on township micro-enterprise formalisation. Sustainable Livelihoods Foundation

Cilliers S (1986) Demise of the Dompas: from influx control to orderly urbanisation. Indic S Afr 4(1):95–98

City of Emfuleni (2023) Spatial Development Plan https://emfuleni.gov.za/wp-content/uploads/2025/02/EMFULENI_SDF_2023-2024.pdf Accessed 1/8/2024

City of Ekurhuleni (2022) https://www.ekurhuleni.gov.za/wp-content/uploads/2022/11/City-of-Ekurhuleni-Growth-Management-Strategy-Document.pdf

Congress for the New Urbanism (2000) Charter of the new urbanism. Bulletin of Science, Technology & Society, 20(4), pp.339–341

Cooper C, Chiaradia AJ (2015) sDNA: how and why we reinvented Spatial Network Analysis for health, economics and active modes of transport. Presented at: GISRUK, Leeds, 15–17 April 2015. Published in: Malleson, Nick ed. GIS Research UK (GISRUK) 2015 Proceedings de Visser J (2016) Devolution by court injunction: the case of land use planning and management in South Africa. Ugandan J Manag Public Policy Stud 10(1):84–99

Denoon-Stevens S, Nel V (2017) Spatial planning for Postmasburg. In: Burger P, Marais L, van Rooyen D (eds) Mining and community in South Africa. Routledge, New York

Denoon-Stevens S, Nel V, Mphambukeli T (2018) Spatial planning for Postmasburg. In Lochner Marais, L, Burger, P and van Rooyen , D. (eds). Mining and Community in South Africa (pp. 95–114). Routledge

Department of Co-operative Governance and Traditional Affairs (2016) Integrated Urban Development Framework (IUDF)

Department of Housing (2004) Breaking new ground: a comprehensive strategy for the development of integrated sustainable human settlements. Department of Housing. https://www.dhs.gov.za/sites/default/files/documents/26082014_BNG2004.pdf

Department of Land Affairs (2001) Wise land use: spatial planning and land use management white paper. Government Printer. https://www.gov.za/documents/spatial-planning-and-land-use-management-white-paper

Development and Planning Commission (1999) Draft green paper on development and planning. National Development Planning Commission. https://www.gov.za/sites/default/files/gcis_document/201409/devplan1.pdf

Dewar D, Uytenbogaardt RS (1991) South African cities: a manifesto for change. Urban Problems Research Unit, University of Cape Town

Doxiadis CA (1970) Ekistics, the science of human settlements: ekistics starts with the premise that human settlements are susceptible of systematic investigation. Science 170(3956):393–404. https://doi.org/10.1126/science.170.3956.393

DRDLR, D. R. D. and L. R (2014) Guidelines for the development of provincial, regional and municipal Spatial Development Frameworks and precinct plans. Final draft. DRDLR, Pretoria

Dreyer J (1992) The Iron Age archaeology of Doornpoort, Winburg, Orange Free State. Navorsinge van Die Nasionale Museum: Researches of the National Museum 8(7):263–269

Drummond J, Nel V (2021) Mahikeng: where traditional leadership and development frameworks collide. In: South African urban change three decades after apartheid. Springer, pp 197–214

Duany et al, 2000. Suburban nation: The rise of sprawl and the decline of the American dream. Macmillan

Duany A, Talen E (2002) Transect Planning. Journal of the American Planning Association, 68(3), 245–266. https://doi.org/10.1080/01944360208976271

Du Plessis DJ (2014) A critical reflection on urban spatial planning practices and outcomes in post-apartheid South Africa. Urban Forum 25(1):69–88. https://doi.org/10.1007/s12132-013-9201-5

Du Plessis DJ (2019) The impact of spatial planning on the structure of South African cities since 1994. Stellenbosch University, Stellenbosch

Du Plessis D, Boonzaaier I (2015) The evolving spatial structure of South African cities: a reflection on the influence of spatial planning policies. Int Plan Stud 20(1–2):87–111

Du Plessis C, Landman K (2002) Sustainability analysis of human settlements in South Africa. Prepared by CSIR Building and Construction Technology. Programme for Sustainable Human Settlement

Floyd TB (1960) Town planning in South Africa, 1st edn. Shuter and Shooter

Fourie F (ed) (2018) The South African informal sector: creating jobs, reducing poverty. HSRC Press South Africa

Glaeser E (2017) Reforming land use regulations. Brooking Institute. https://www.brookings.edu/research/reforming-land-use-regulations/

Hakim BS (2001) Julian of Ascalon's treatise of construction and design rules from sixth-century Palestine. J Soc Archit Hist 60(1):4–25

Hakim BS (2008a) Generative processes for revitalizing historic towns or heritage districts. Urban Des Int 13(3):210

Hakim BS (2008b) Mediterranean urban and building codes: origins, content, impact, and lessons. Urban Des Int 13(1):21–40

Harris N (1990) Urbanisation, economic development and policy in developing countries. Habitat Int 14(4):3–42. https://doi.org/10.1016/0197-3975(90)90002-I

Harrison P (2001) The geneology of South Africa's integrated development. Third World Plann Rev 23(2):175
Harrison P (2006) Integrated development plans and third way politics. In: Democracy and delivery: urban policy in South Africa. HSRC Press, pp 186–207
Harrison P, Todes A, Watson V (2007) Planning and transformation: learning from the post-apartheid experience. Routledge
Healey P (2004) The treatment of space and place in the new strategic spatial planning in Europe. Int J Urban Reg Res 28(1):45–67. https://doi.org/10.1111/j.0309-1317.2004.00502.x
Healey P (2006) Urban complexity and spatial strategies: towards a relational planning for our times. Routledge
Higgins CD (2019) Accessibility toolbox for R and ArcGIS. Findings. https://findingspress.org/article/8416-accessibility-toolbox-for-r-and-arcgis Accessed November 2023
Hillier B (2007) Space is the machine: a configurational theory of architecture. Space Syntax
Hillier B, Hanson J, Peponis J (1984) What do we mean by building function?. E & FN Spon Ltd
Hillier B, Hanson J, Graham H (1987) Ideas are in things: an application of the space syntax method to discovering house genotypes. Environ Plann B Plann Des 14(4):363–385
Hillier B, Penn A, Hanson J, Grajewski T, Xu J (1993) Natural movement: or, configuration and attraction in urban pedestrian movement. Environ Plann B Plann Des 20(1):29–66
Hirt SA (2016) Rooting out mixed use: revisiting the original rationales. Land Use Policy 50:134–147
Horn A (2005) Measuring multi-ethnic spatial segregation in South African cities. S Afr Geogr J 87(1):58–72. https://doi.org/10.1080/03736245.2005.9713827
Jenkins P, Wilkinson P (2002) Assessing the growing impact of the global economy on urban development in Southern African cities. Cities 19(1):33–47. https://doi.org/10.1016/S0264-2751(01)00044-0
Kotkin J (2005) The city: a global history, vol 21. Modern Library Chronicles
Kropf K (2018) The handbook of urban morphology. Wiley
Ligthelm AA, Masuku T (2003) Size, structure and profile of the informal retail sector in South Africa. Bureau of Market Research, University of South Africa
Magidi J, Ahmed F (2019) Assessing urban sprawl using remote sensing and landscape metrics: a case study of City of Tshwane, South Africa (1984–2015). Egypt J Remote Sens Space Sci 22(3):335–346. https://doi.org/10.1016/j.ejrs.2018.07.003
Marais L, Nel V, Rani K, Van Rooyen D, Sesele K, Van der Watt P, Du Plessis L (2021) Economic transitions in South Africa's secondary cities: governing mine closures. Politics Gov 9(2):381–392
Marshall S (2004) Streets and patterns. Routledge
Matebesi S, Marais L, Nel V (2024) Local responses to mine closure in South Africa: dependencies and social disruption. Taylor & Francis
Musakwa W, Van Niekerk A (2014) Monitoring urban sprawl and sustainable urban development using the Moran Index: a case study of Stellenbosch, South Africa. International Journal of Applied Geospatial Research (IJAGR) 5(3):1–20
National Planning Commission (2013) National development plan vision 2030
Nel D (2022) Morphologies of resilience: Towards an approach to the study of urban spatial resilience. PhD thesis, Hong Kong Polytechnic University. https://theses.lib.polyu.edu.hk/handle/200/11665
Nel V (2016) A better zoning system for South Africa? Land Use Policy 55:257–264. https://doi.org/10.1016/j.landusepol.2016.04.007
Nel V, Denoon-Stevens SP (2024) Land-use management to support sustainable settlements in South Africa. Routledge
Nel V, Minnie S (2022) Can the district development model deliver development in South African local governments? Rozwój Regionalny i Polityka Regionalna 60s:39–53. https://doi.org/10.14746/rrpr.2022.60s.05

Nel V, Nel D, Minnie F (2017) Making small cities smart in South Africa. In: Smart communities, vol 13. ISOCARP. https://isocarp.org/product/review-13-smart-communities-e-book/
Oranje M, Merrifield A (2010) National spatial development planning in South Africa 1930–2010: an introductory comparative analysis. Town and Reg Planning 56:29–45
Porta S, Latora V, Wang F, Rueda S, Strano E, Scellato S, Cardillo A, Belli E, Cardenas F, Cormenzana B (2012) Street centrality and the location of economic activities in Barcelona. Urban Stud 49(7):1471–1488
Reader J (2005) Cities. Vintage
Rodriguez RB (2009) The law of Indies as a foundation of urban design in the Americas. Leadersh Archit Res 395
Roy A (2005) Urban informality: toward an epistemology of planning. J Am Plan Assoc 71(2):147–158. https://doi.org/10.1080/01944360508976689
RSA, R. of S. A. (1994) Reconstruction and development programme. In: Government Gazette, vol 353, issue 16085
RSA, R. of S. A. (1995) Development Facilitation Act 67 of 1995. The Presidency, Pretoria
RSA, R. of S. A (1997) Urban Development Framework. The Department of Housing
RSA, R. of S. A. (2000) Local Government: Municipal Systems Act 32 of 2000. The Presidency, Pretoria
RSA, R. of S. A. (2001) Local government: municipal planning and performance management regulations. Government Gazette no. 7146, 24 August
RSA, R. of S.A. (2013) Spatial Planning and Land Use Management Act 16 of 2013. The Presidency, Cape Town
Scheba A, Turok I (2020) Informal rental housing in the South: dynamic but neglected. Environ Urban 0956247819895958. https://doi.org/10.1177/0956247819895958
Serfontein K, Oranje M (2008) Spatial connections and disconnections: a critique on the conceptualisation of Tshwane (2000 to 2004). Stads-En Streeksbeplanning, Town Reg Plan 52:20–31
Sevtsuk A (2010) Path and place: a study of urban geometry and retail activity in Cambridge and Somerville, MA. Massachusetts Institute of Technology
Sevtsuk A, Mekonnen M (2012) Urban network analysis. Revue Internationale de Géomatique–n 287:305
Silver C (1997) The racial origins of zoning in American cities. In: Manning T, Ritzdorf J, Ritzdorf M (eds) Urban planning and the African American community: in the shadows, vol 38. Sage Publications. http://racialequitytools.org/resourcefiles/silverracialoriginsofzoning.pdf
South Africa (1996) Constitution of the Republic of South Africa, 1996. Government Printer
Ståhle A, Marcus L, Karlström A (2005) Place Syntax: geographic accessibility with axial lines in GIS. In: Fifth international space syntax symposium, pp 131–144
Talen E (2002) Help for urban planning: the transect strategy. Journal of Urban Design. 7(3), 293–312. https://doi.org/10.1080/1357480022000039349
Talen E (2012) Zoning and diversity in historical perspective. J Plan Hist 11(4):330–347. https://doi.org/10.1177/1538513212444566
Talen E (2013) Zoning for and against sprawl: the case for form-based codes. J Urban Des 18(2):175–200. https://doi.org/10.1080/13574809.2013.772883
Talen E (2015) Do-it-yourself urbanism: A history. Journal of Planning History, 14(2), pp.135–148
Todes A (2008) Rethinking spatial planning. Stads-En Streeksbeplanning, Town Reg Plan 2008(53):9–13
Todes A, Turok I (2018) Spatial inequalities and policies in South Africa: place-based or people-centred? Prog Plan 123:1–31
Todes A, Karam A, Klug N, Malaza N (2010) Beyond master planning? New approaches to spatial planning in Ekurhuleni, South Africa. Habitat Int 34(4):414–420
Tomlinson R (1988) South Africa's urban policy: a new form of influx control. Urban Aff Quart 23(4):487–510. https://doi.org/10.1177/004208168802300402
Toynbee A (1967) Town-planning in the ancient Greek world. Ekistics 24(145):445–448

Turok I (2018) Worlds apart: spatial inequalities in South Africa. In: Smith MN (ed) Confronting inequalities in South Africa. Jacana Media, pp 159–182
Turok I, Parnell S (2009) Reshaping cities, rebuilding nations: the role of national urban policies. Urban Forum 20:157–174
Turok I, Visagie J, Scheba A (2017) Reducing spatial inequalities through better regulation. Report to the High-Level Panel on the assessment of key legislation and the acceleration of fundamental change. HSRC
van Huyssteen E, Biermann S, Naudé A, le Roux A (2009) Advances in spatial analysis to support a more nuanced Reading of the South African space economy. Urban Forum 20(2):195–214. https://doi.org/10.1007/s12132-009-9061-1
van Nes A (2021) The impact of the ring roads on the location pattern of shops in town and city centres. A space syntax approach. Sustain For 13(7):3927
Van Wyk J (2012) Planning law, 2nd edn. Juta
Van Wyk J (2020) Planning law, 3rd edn. Juta
Whitehand JWR, Gu K (2006) Research on Chinese urban form: retrospect and prospect. Prog Hum Geogr 30(3):337–355. https://doi.org/10.1191/0309132506ph611oa
Williams-Wynn C (2021) Applying the fit-for-purpose land administration concept to South Africa. Land 10(6):602

Chapter 4
Citizen Participation, Network Building and the Space In-Between: The Importance of NGOs in Community and Spatial Transformation

Anele Horn

Abstract One of the most profound changes to the territorial governance processes in South Africa after 1994, was the recognition of the importance of public participation in the preparation of various planning instruments. Since 1994, there has been deliberate efforts towards stakeholder participation emphasised in post-apartheid legislation and guidelines for Integrated Development Plans and Spatial Development Frameworks. These frameworks prescribe in detail how such engagements can and should take place, however most stakeholder engagement processes still seem unable to produce plans that appreciate and reflect community dynamics that can lead to community social and spatial transformation. Civil society organisations such as NGOs, have become critical role players in spatial and social transformation. Their proximity to communities often affords them a better understanding of local opportunities and challenges than municipal authorities. By taking advantage of this proximity, civil society organisations have discovered other meaningful forms of stakeholder participation, such as network and partnership-building approaches. These approaches build social capital and lead to increased resources and a capacity for transformation. This chapter reviews the lessons learnt from traditional consultative approaches applied in territorial governance processes in South Africa since 1994, considers the impact of these processes on spatial transformation in general, and reflects on the potential value of novel approaches such as community-based networks that have emerged.

Keywords Network governance · Participatory planning · Community involvement · Spatial transformation

---

A. Horn (✉)
Department of Geography and Environmental Studies, Stellenbosch University, Stellenbosch, South Africa
e-mail: anelehorn@sun.ac.za

D. du Plessis (ed.), *Territorial Governance and Spatial Transformation in Post-apartheid South Africa*, Local and Urban Governance,
https://doi.org/10.1007/978-3-031-85489-7_4

4.1 Introduction

In many countries with stable democracies, a growing disillusionment with democratic institutions is becoming evident. This is especially noticeable in the decline of electoral turnout, lower levels of support and trust in politicians and political institutions, a rise in authoritarianism and populism, as well as a growing gap between citizens and decision-makers. As a result of these observations, there is mounting interest in increasing the legitimacy of democratic decision-making and strengthening deliberative and participatory models of democracy (Dalton 2004).

The involvement of citizens in urban planning processes has been a cornerstone of modern urban planning frameworks and processes for the last three decades. This reformation, pioneered by theorists like John Forester, Judith Innes, and Patsy Healey, is referred to as communicative planning and dominated theorising around fair and equal policymaking between the 1980s and 1990s. The foundation of communicative planning for policy-making stems from the work of social scientist Jurgen Habermas (Taylor 1998), who pioneered the theory of communicative rationality for critical, interpretive, and empirical communication processes (Habermas 1976). Healey focused on understanding how participation towards rearranging government and institutional systems could change how planning was executed (Healey 1992, 1996). The theorising around involving citizens in urban planning resulted in a large body of research dedicated to creating successful stakeholder participation by using planning processes for transformative change. The emphasis in communicative planning is on the establishment of trust and dialogue whereupon social relationships between interdependent stakeholders can then be created. Stakeholders and citizens who participate are subsequently capacitated to provide input into policy decisions and conceptualise shared visions towards addressing community development. In this way the formation of collective action stemming from participation in communicative planning is expected to lead to increased social capital (Booher and Innes 2002; Healey 2003; Olsson 2009).

South Africa has an active and vocal citizenry as witnessed by citizen participation in service delivery protests and political debate. Yet, apart from the input and commentary on policies and frameworks, the active participation of citizens in addressing community-level challenges has largely been neglected. The National Treasury (RSA 2012) thus recognised the increased support and incentivising of citizen engagement as priorities for future development. This recognition is not new, as the intention to involve citizens in local urban planning matters has, since 1994, been entrenched in post-apartheid policies and legislative frameworks. There are, however, doubts about the value added by current citizen participation approaches and standards practised at local government level. At the present participation processes fail to encourage community networks and increase their ability to undertake bottom-up action towards solving local challenges and strengthening social capital. The purpose of this chapter is to compare the current approach to citizen participation as prescribed by South African legislation and policy frameworks to more action-oriented network-building participation mechanisms used by many organisations outside of local government. This research demonstrates new or active forms of citizen participation that are taking place outside the mandate of local authorities and that are proving to be useful alternatives for the capacitation of local communities. This demonstration is achieved by considering the manifestos and approaches of South African non-governmental organisations (NGOs) involved in spatial planning and community transformation projects.

4.2 Community Development and Spatial Transformation

In 2015 the United Nations announced the global Sustainable Development Goals (SDGs) with the explicit purpose of striving for a world that is just, equitable and inclusive while putting people who are committed to working together, at the centre of the process to promote sustained and inclusive economic growth, social development and environmental protection to the benefit of all. This agenda, through goals 11 and 17 (Sustainable Cities and Communities; Partnerships for Goals) suggest an increase in the participation of citizens who want to transform their livelihoods from a social, economic, and spatial point of view. In theory, participation can be analysed in terms of the level of decentralisation, execution of democracy, and the extent of collective action (Howard and Wheeler 2015). Decentralisation contributes to greater empowerment and agency in society through the participation of citizens in a decision-making process. In addition, a firmly established democracy enables citizens to openly engage and contribute to decisions that affect their lives (Chikerema 2013), while collective action assists citizens in pursuing collective goals within a community (Meinzen-Dick et al. 2004).

As a result of intimate knowledge of the functioning of neighbourhoods as well as the ability to actively support the transformation of planning practices, the participation and input of citizens in local spatial planning issues are key to creating cities that are lively, sustainable, safe, and healthy. Planning processes have demonstrated that when citizens were consulted in planning decisions, it had a positive effect on them supporting the planning proposals and changes (Blanchet-Cohen 2015), therefore yielding a greater probability of policy frameworks being implemented. Actively involved citizens who are informed and committed often lead to improved community development as those citizens become contributors and implementers of planning processes, rather than reactive receivers (Ife and Tesoriero 2006).

In 1969, Arnstein published the well-known typology of participation indicating eight levels of participation in terms of differential empowerment, from the lowest form of participation, namely informing citizens, to the highest form of participation being community control. This typology has since been widely used as a tool with which to benchmark practices of public participation and for evaluating levels of cooperation between authorities and citizens. The suggested ideal level of participation is at level six, described as partnerships (Blanchet-Cohen 2015).

Community development can be defined as:

> "a way of working which essentially starts with the needs and aspirations of groups of disadvantaged people … and which struggles, first of all, to articulate and organise politically around those needs and aspirations, placing them at the front rather than the end of political debate." (Craig 1998: 15)

Early social scientists recognised community-driven development, or a bottom-up development process grounded in community development principles as a social action process consisting of two components operating at community level: firstly, consensus-building, and secondly resource mobilisation (Fraenkel 1977). Such a social action process could be mandated by a local authority or an external donor where NGOs would play a brokering role (Howard and Wheeler 2015).

Mobilizing citizens to participate in community development processes usually requires individuals to come to an agreement regarding the desirability of community improvement as well as controlling the resources (financial, human, networks) to undertake improvements (Richards and Dalbey 2006). Thus, for transformation to take place in community development, there must firstly be mutual understanding and input from local community members into future plans and policies and, secondly, engagement of the local community members that go beyond input and dialogue. This important second step in community transformation will therefore lie in the action of local resource identification and mobilisation through network building.

4.3 Citizen Participation, Network Building and the Role of NGOs

A general definition of citizen participation refers to the involvement of communities in matters relating to institutional planning and decisions that affect their daily lives. This involvement can range from active dialogue to written or verbal input or collaboration with specific individuals or groups within a community. The stated objective of citizen participation is to empower individuals and communities to influence the policies and plans that have a bearing on their daily lives (Devas and Grant 2003). There are many advantages to participatory processes, including increased transparency and accountability on the part of local authorities, as well as the promotion of social justice and democratic principles because of the representation of the needs and preferences of an entire community (Devas and Grant 2003). Whereas citizen participation emphasises verbal dialogue and input, the concept of network building and partnerships moves beyond conversation into a greater level of action. Network building generally refers to the establishment of connections, stakeholder collaborations, alliances between individuals, and partnerships within and across communities. These connections are established to foster relationships and gain access to resources that can potentially be shared. As connections develop into networks, community members are enabled to gain access to a broader range of resources, such as financial capital, human capital, technical expertise, skills, experiences, and social connections. By pooling these resources and leveraging collective strengths, communities can undertake larger-scale projects, implement innovative initiatives, and address complex challenges more effectively.

In contrast to citizen participation which involves direct verbal or written engagement with community members, network building is a more indirect approach to connecting stakeholders and building trust among communities. Network building processes therefore involve a different approach to citizen participation, and usually take place during networking events, business forum meetings, and social media and other non-formal participation events (Devas and Grant 2003). Citizen participation and network building are often complementary, as citizen participation can be improved by strong networks and partnerships, and networks can be built during effective and inclusive participation processes. The establishment of networks and partnerships within communities furthermore strengthens social cohesion, trust, and solidarity among residents. By promoting inclusive participation and network building, stakeholders can build social capital and gear local action towards community improvement and transformation.

The dissatisfaction with and perceived inadequacy of government-led civic engagement mechanisms opened new opportunities for the growth and expansion of NGOs during the last two decades. The ability of NGOs to connect with marginalised regions as well as their proximity to local communities position them favourably as innovators in the development process (Banks et al. 2015). NGOs are furthermore considered to function as non-political links between the government and citizens as they could provide effective channels of communication between government and local communities in instances where communities don't have technical or institutional knowledge on a particular aspect (Shettigar and Mayya 2015). The functions of NGOs in community development are thus to help communities develop social, capital, and human resources, improve knowledge and skills, and encourage people to participate in activities, while also acting as a link between communities and systems.

The advantages of NGOs in community development processes relate to their ability to quickly adapt to changing community needs, to experiment with innovative approaches, and to tailor interventions to specific contexts without being constrained by bureaucratic red tape or political considerations. It is however noted that the ability to aptly respond to challenges in some instances are unfortunately made possible by an absence of appropriate checks and balances, ensuring that processes are executed ethically and responsibly. In addition to NGOs' responsiveness to local challenges, unlike local authorities that are mostly perceived as distant or serving political ideologies, NGOs often have well-established connections and are trusted by the communities where they function. Within these communities, NGOs encourage people to identify their challenges and ultimately take ownership of their development processes. Grassroots linkages and proximity to beneficiaries are seen to give NGOs a comparative advantage in providing effective targeted aid and ensuring that programmes are designed in a bottom-up manner "reflecting local contexts, needs, and realities and are not subject to commercial or political capture" (Koch et al. 2009: 906).

In 2014, Statistics South Africa (Stats SA) (2014) reported that 200,000 NGOs and 127,000 non-profit organisations (NPOs) were registered in South Africa (NGOs refer to voluntary groups or institutions with a social mission, which operates independently from the government, whereas NPOs are required to register with the Department of Social Development and are established for a public purpose (NGO Law 2023)). These figures represent an increase of over 200% in NGO and NPO registrations since 2001 that have been exacerbated by the vulnerability of the South African economy which led to a decrease in government spending and resulted in limited or no service delivery. This situation created an important gap that allowed NGOs to start delivering the often defunct (government) services. This strengthened the critique of the growing number of NGOs as it demonstrated the failure of the state with NGOs that had become agents for basic services by extension. This additional burden of having to deliver basic services detracts from the critical role that NGOs should be playing, namely that of advancing social change and transformation (Radebe and Nkonyeni 2020).

4.4 Stakeholder Engagement in South African Plans and Policies Post-1994

South Africa has a well-established legislative framework that defines the institutional context of community participation and inclusion at national, provincial, and local government levels. The Constitution of South Africa (Act 108 of 1996) which forms the basis of government responsibility to involve communities in local planning states in section 151(1) that "Municipalities are obliged to encourage the involvement of communities and community organisations in local government" (RSA 1996). The White Paper on Local Government (RSA 1998a) delves into more detail in terms of activating said involvement by mandating municipalities to undertake citizen participation at four levels, namely (a) as voters—to ensure maximum democratic accountability of the elected political leadership for the policies they are empowered to promote; (b) as citizens—who express via different stakeholder associations, their views before, during and after the policy development process to ensure that policies reflect community preference as far as possible; (c) as consumers and end users—who expect value for money, affordable services, courtesy and responsive services; and (d) as organised partners—who are involved in the mobilisation of resources for the development of businesses, non-governmental organisations and community-based institutions. Strengthening these prescriptions are the Batho Pele principles formulated in 1998, to encourage citizens to participate in policymaking and ultimately render accountable and transparent development within the public administration.

In terms of Chapter 5 and Section 27 of the Municipal Systems Act (MSA) (2000), municipalities are required to prepare and adopt framework plans indicating how municipalities will align their integrated development plans (IDPs) with the plans of other spheres of government. The objectives of the local municipality framework must, amongst others, include that the municipality would (a) involve and integrate all relevant stakeholders in planning; (b) ensure that all local authorities fulfil the responsibilities entrusted to them by legislation in the form of powers and functions; and (c) ensure that the needs of communities and interest groups are identified, acknowledged and addressed. The objects of the Municipal Finance Management Act 56 of 2003 (MFMA) are aligned with those of the MSA and together they give effect to constitutional imperatives. Sections 21–23, read together with Chapter 4 of the MSA, make provision for public participation processes regarding the municipal budget, integrated development plan, performance management processes, and policy development. Such public participation processes are to be concluded within set time frames following the specifications set by the MFMA.

In 2007 the National Policy Framework for Public Participation was announced. In terms of the Policy Framework, public participation is regarded as people who participate as individuals, interest groups or people in general. The Policy Framework further considers existing civil society groupings and district forums as a starting point to strengthen community-based involvement in municipal decision-making. These interest groups or sectors are thus encouraged to come together during policy formulation or any decision-making process in a municipality.

In 2013 the Spatial Planning and Land Use Management Act (SPLUMA) was promulgated. Section 7(e) of SPLUMA prescribes the principles of good administration and particularly emphasises that "the preparation and amendment of spatial plans, policies, land use schemes as well as procedures for development applications include transparent processes of public participation that allow all parties to provide inputs on matters affecting them" (RSA 2013).

More recently the Integrated Urban Development Framework (IUDF) put forward initiatives to create a space where government and civil society can engage for participation in local-level planning (RSA 2016). These local-level initiatives include the following:

- Completing the National Framework on Participatory Governance, which encourages properly funded, citizen-led, neighbourhood planning processes.
- Developing models for civic education, exploring training options and equipping community members with the necessary skills.
- Establishing and maintaining public participation forums at various levels.
- Building institutional capacity within municipalities (and the government in general) to engage with civil society.
- Exploring co-production mechanisms for finding solutions to better deliver local government services.
- Developing a social media strategy for broad-based information and engagement by leveraging new and accessible technologies.
- Exploring mechanisms that allow civil society to play a bigger role in social services delivery.

To facilitate the practical implementation of public participation at local government level as is required in various pieces of legislation, the government has, since

1994, implemented various initiatives. These initiatives include *Izimbizos* (Political leaders' engagements with communities), ward committees, community development workers, satisfaction surveys and citizen forums. Despite the positive intentions of these initiatives, studies researching their effectiveness revealed many challenges to the operational realities and functioning of these structures. Gray and Mubangizi (2010) identified mistrust between government officials and community development workers as one of the biggest obstacles to achieving meaningful dialogue. The proximity of community development workers and ward councillors to local communities would in many instances be threatening to government officials, whereas community workers and ward councillors in turn would feel that officials used them to function as a buffer between local governments and communities who were dissatisfied with service levels. Literature notes that where politicians and government officials were directly involved in participatory structures, the nature of their engagements would be politically charged and would often lead to further conflict and alienation rather than constructive dialogue (Marais and Botes 2006; Mnguni 2018).

Ward committees are the appropriate legislated forums where community or ward-level challenges can be heard and discussed by the attending ward councillors as elected in local government elections. Ward committees have been created in terms of the Municipal Structures Act (117 of 1998) and are statutory bodies. The purpose of ward committees is to assist the democratically elected representative of a ward (the councillor) to carry out their mandate. Ward committee members are members of the community who should represent the needs of the people in the areas where they live. These ward committees must advise the municipal council of the needs and concerns of the community and must keep the community informed of developments in the municipal council. Ward committees can make recommendations to the municipal council but do not have decision-making powers (Seitlholo and Diedericks 2018). However, in studies evaluating the success of ward committees as participatory mechanisms, various challenges have been identified, such as training, education, the transparency of ward councillors (Seitlholo and Diedericks 2018; Naidoo and Ramphal 2018) and the (often) arbitrary establishment of ward boundaries which renders the election of one ward councillor as a true representative almost impossible.

A citizen forum is another platform used for stakeholder engagement. These forums are used during the preparation of IDPs and Spatial Development Frameworks (SDFs)—processes that must be performed every 5 years. Citizen forums serve as mechanisms to convey information to citizens while also gaining their input. In most instances, citizens are invited to submit written comments on the proposals made by the municipality in respect of either the IDP or SDF. During the consultation process for the municipal SDF, a municipality may decide to engage citizens per region or district to convey and gather more detailed information on or about a specific area. Outside of the 5-year consultation processes for IDPs and SDFs, the expectation is that communities can relay their challenges and expectations regarding community development through their elected ward committee councillors who represent communities forming part of their ward at regular council meetings with the local municipality.

Despite numerous laws that encourage community participation in the IDP, Mamokhere and Meyer asserted that:

> "service delivery challenges persist as a result of poor community participation in the IDP process, primarily in local municipalities, and are deeply rooted in the disaster of implementing decentralisation, mismanagement, corruption, inability to deliver to communities, and lack of community participation in major municipal decision-making". (Mamokhere and Meyer 2022: 290)

In the same study, the findings indicated that most respondents did not view an IDP as an instrument for improved community participation and a mechanism to resolve service delivery backlogs confronting their community, and as a result, they refrained from actively participating (Mamokhere and Meyer 2022). The legislative requirements and mechanisms for citizen participation in South Africa, are depicted in Fig. 4.1 that demonstrates the relationship between legislative prescriptions and the corresponding level of citizen participation as per Arnstein's ladder (Arnstein 1969). From this figure it is evident that citizen participation as is currently prescribed, rarely moves beyond tokenism (levels 3–5), and only the provisions of the IUDF introduced opportunities and mechanisms for network building and citizen power represented by levels 6–8. The IUDF demonstrates a departure from traditional thinking about what public participation should be and recognises the importance of civic society organisations and the role they should play in community transformation.

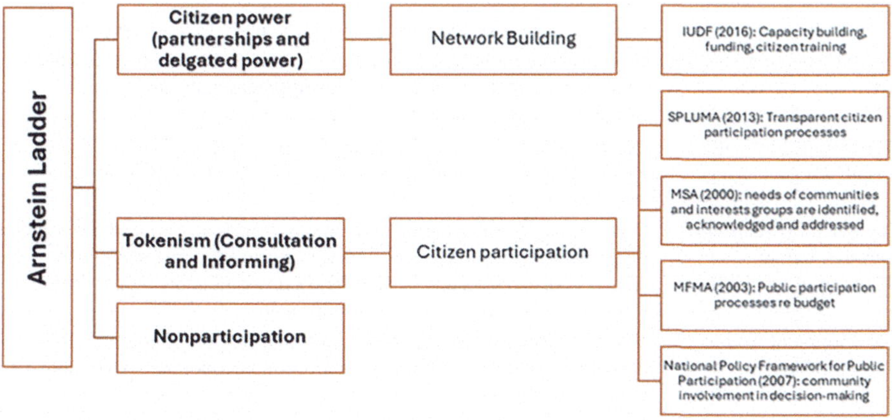

Fig. 4.1 Citizen participation vs network-building approaches in South African policies and legislation (original to the author)

4.5 Emerging Participatory Approaches and NGOs in South Africa

In South Africa, several NGOs participate in urban planning and place-making as part of spatial transformation practices. Placemaking as a process aims to add value and meaning to public spaces and neighbourhoods through collaborative and participatory projects that draw inspiration from local values, histories, natural environments, and cultures (Lepofsky and Fraser 2003; Akbar and Edelenbos 2021). Placemaking stands in contrast to traditional comprehensive planning that tends to follow a top-down, institutionally led approach, as placemaking rather reflects on place-focused development that includes participation and collaboration with targeted beneficiaries (Beza 2016; Eckenwiler 2016; Patriquin and Halpenny 2017). Since a placemaking process creates a platform for transformative dialogues, the process results in the social construction of the place which encourages communities to increase their attachment to the space. Place attachment presents insights into people's feelings about their communities and actions in different spaces.

As alluded to earlier in this chapter, more than 200,000 NGOs are currently registered in South Africa. There are many distinct categories of NGOs, and each category of NGO also functions within a specific sector of development. This chapter focuses specifically on NGOs undertaking community projects relating to urban planning and spatial transformation. Based on these criteria, six South African NGOs have been identified and listed in Table 4.1. These NGOs are all located in Cape Town and were identified and studied because of ease of proximity for the author. Table 4.1 provides a broad overview of the individual approach to urban

Table 4.1 Urban planning and spatial transformation NGOs

Name	Approach
Isandla Institute	Building networks and partnership approaches to development
Social change assistance trust	Empower communities, strengthen civil society, and promote social change in partnership with local development agencies which function as community-driven responses to social justice
Habitat for humanity South Africa	Election and formation of a body of community leaders Partner with people and organisations who can help provide the necessary resources
Violence Prevention through Urban Upgrading (VPUU)	Form strategic partnerships with NGOs, CBOs and other development organisations that want to make a difference by transforming the spatial and social legacy of apartheid into sustainable neighbourhoods through negotiated and joint development
Development Action Group (DAG)	Build community capacity and coalitions, support, and advocate for community-led development by addressing economic, social, and spatial imbalances
Ranyaka Community Transformation	Focus on place-based collaboration and geared towards restoring the eco-system within which socio-economic challenges affect community well-being

planning and spatial transformation as obtained from each of the NGOs' official web content.

An investigation into the projects, mission, vision and approach of each of these NGOs revealed that the following concepts reflected across their organisations (with the number of NGOs indicated in parenthesis), namely building partnerships (6), leading community-led projects (4), building capacity (6), following an eco-system approach (3), building networks (5) and building social capital (4). These concepts reflect deeper values of social and community theory as well as network building as opposed to the principles of traditional citizen participation as prescribed by South African legislation. It is evident from the manifestos of the NGOs that the emphasis is on capacitating a community so that they can help themselves in a sustainable and planned manner. Reliance on donor funding or institutional support is in most instances discouraged, and the focus is on assisting communities and community leaders in identifying internal strengths and resources. An example from Table 4.1 where the manifesto demonstrates all six of these values of network building is Ranyaka Community Transformation (RCT).

RCT was established in 2013 to develop transformative strategies for towns and neighbourhoods. RCT engages with residents, civil society entities, faith-based institutions, businesses, municipal and provincial authorities, academic institutions, and local change agents to undertake community development and transformation. This NGO currently works in 28 communities in South Africa, including eThekwini Metro, the City of Tshwane, Paarl, Stellenbosch, and Nelspruit. Within these communities, the focus is on rolling out their main initiative called "Proud of my Town" which initiative is funded by Nedbank. In addition to and supplementing the "Proud of my Town" initiative, RCT actively engages in programmes related to community safety, the development of public spaces, supporting early childhood development centres, promoting land development for small-scale farming and agri-entrepreneurship, building small businesses, and establishing sports clubs and informal sports leagues.

The network-building approach that RCT implements is based on challenges that RCT identifies, namely (1) a fragmented project and investment approach to socio-economic issues; (2) the absence of a common agenda for a specific neighbourhood; (3) limited local ownership of interventions; (4) a lack of cross-sectoral collaboration; (5) a lack of social cohesion within communities; and (6) the need for funders to increasingly support a systemic approach to achieve sustainable change. This approach, called a Collaborative Community Development Model, consists of the steps detailed in Table 4.2.

The collaborative community development model detailed in Table 4.2 demonstrates the extent of civic engagement necessary for community transformation. In each of the steps in the development model, the emphasis is on the identification of resources and the strengthening of relationships to enable the optimal use of these resources. It is a clear developmental model that prioritises the building of networks and the formation of partnerships.

Table 4.2 Collaborative community development model (Ranyaka Community Transformation 2023)

Community profiling	RCT collaborates with local stakeholders across multiple sectors through data collection and analysis, dialogue and workshops to identify role players, challenges, resources, and opportunities in an area. The 'well-being' of the community is measured through a set of indicators, asset-mapping and geo-referencing of issues. This provides the first baseline that is measured on an ongoing basis throughout implementation.
Co-design plans	Community profiling information is used to co-design a sustainable neighbourhood plan based on sound spatial planning principles, followed by a development strategy designed to address the area's challenges, unlock opportunities, activate local resources, and facilitate smart investment. Local stakeholders are key players in this process.
Implement programmes and projects	Collaborating with local partners, programmes and projects are implemented that address priority needs, identify available resources, and align with the broader development strategy. Network building, forging of strong partnerships, and community activation are ongoing. Funders and donors are connected to smart investment opportunities.
Manage	Ongoing process management of the broader strategy, programmes, projects, stakeholder networks, and local coordination teams are undertaken by RCT coordinators. Regular stakeholder engagement enables the continuation of strong relationships, anticipation of potential risks, and identification of new opportunities for growth and development. RCT typically works in a community for a minimum period of 3 years.
Measure	Measurement and evaluation are ongoing and include regular stakeholder feedback, collecting data for heat maps, measuring Social Return on Investment, and creating a platform for sharing impact stories. The over-arching plan is constantly revised based on measurement outcomes.
Unlock investment opportunities	Investment opportunities for funders and donors who pursue sustainable, measurable change are identified. RCT's approach to community development allows for the multiplier effect as the initial investment achieves more than the funding of a single intervention as it unlocks doors for sustainable transformation.
Build partnerships	The golden thread throughout this process is the building of strong partnerships with stakeholders across multiple sectors. These include residents, community forums, faith-based organisations, schools and training institutions, NGOs, NPOs, businesses, ward councillors, municipalities and other levels of government, potential investors, and community champions.

4.6 The Role of South African Municipalities vs NGOs

The previous section demonstrated the efforts that NGOs put into building networks and partnerships, which go beyond the mandated levels of citizen participation practised by local authorities in South Africa. An understanding of the role and capacity of local authorities could prove useful at this point. In this section it is argued that, instead of NGOs and local authorities functioning in competition or conflict with each other, these institutions can play contributing roles to spatial transformation in communities.

South African municipalities are suffering from institutional decline, a situation that is illustrated by municipalities' poor service delivery and limited capacity for the financial and spatial upgrading of towns and cities (SACN 2021). As indicated earlier, South African municipalities have been mandated to undertake IDP and SDF planning on a 5-year basis. The point of departure for both these planning frameworks is the establishment of baseline information, however, in most instances municipalities do not possess the financial or human capacity to undertake detailed data collection among their citizens or spatially within their boundaries (Todes et al. 2010). Therefore, municipalities mostly rely on census data gathered and distributed by Statistics South Africa for the detailed socio-demographic and socio-economic data of their inhabitants. These national censuses are however only undertaken at 10-year intervals, and the detailed spatially disaggregated results (useful for planning purposes at municipal level) are generally only released 2–3 years after completion of the census. This implies that municipalities in their planning processes often must rely on information that is more than a decade old. Given the exceptionally dynamic nature of immigration and population growth in many cities and towns (Nkabinde et al. 2018), this reliance on outdated lower-level data is extremely problematic. The NGOs mentioned that are making use of local inhabitants' capacity, strengthened by NGO leadership, and who possess the time and resources to undertake much more detailed analyses and surveys of their communities can thus play a key role in informing planning frameworks and policies.

The co-design of spaces at neighbourhood level is not a mandated function of municipalities; it is usually an endeavour practised in selected spaces and because of the availability of external funding or community self-funding (Karuri-Sebina 2014). While municipal authorities and planning departments can take ownership of neighbourhood-level plans and initiatives, they often do not have the capacity or expertise to devote to co-producing plans with selected communities. In addition, the municipal IDP represents an integrated approach to the development of a municipal area, but it does not necessarily consider the integration of different sectoral challenges that make up the 'eco-system' of a neighbourhood at a lower level of spatial aggregation. When neighbourhood or community planning is undertaken for a specific area, such plans are usually championed by sector departments (for example Health and Safety, Community Services, Spatial Planning). The complexity of interrelated challenges that lead to community development challenges is therefore not addressed in a single, detailed approach.

The establishment of partnerships with non-governmental agencies and service providers is a well-established practice in South African local authorities. These partnerships are, however, awarded based on a service provider's ability to deliver a specific need, such as refuse removal, security, or childcare. There is limited scope and capacity to extend these partnerships to actions in favour of community and individual enablement that use internally mobilised resources towards a community benefit. Thus, the need for youth training programmes, aftercare facilities for children and recreational sports facilities are some of the community development functions that must often be identified and initiated by community organisations.

4.7 Reflection

This chapter has suggested that NGOs can potentially play a key role in furthering citizen involvement in urban planning and transformation beyond traditional legislatively prescribed participation mechanisms. Current forms of citizen participation as prescribed by the MSA, MFMA and National Policy Framework on Public Participation predominantly focus on consultation processes between local authority planners and communities to gather communities' input into government plans and policies. This tradition of citizen participation is inadequate and has to date failed to result in meaningful transformation at a local community level.

The Integrated Urban Development Framework introduces objectives to further strengthen the role of citizens in shaping their livelihoods and environments. These concepts of building partnerships and networks and involving community-based organisations and community-led developments as reflected in the IUDF, present the opportunity for expanding practices of citizen participation in South Africa beyond mere consultation.

To this end, many NGOs have been registered in South Africa during the past two decades. While many of the NGOs focus on the delivery of services in specific sectors, such as Early Childhood Development, Health Services or Community Facilities, various operational NGOs follow urban planning approaches for community and spatial transformation. The manifestos, visions, missions, and approaches of NGOs were analysed in this chapter and identified a common language of participation beyond mere consultation. In most cases they seek to build partnerships, networks, social capital, and capacity while following an eco-systems approach to community-led development. These approaches have moved beyond consultation to deeper levels of understanding of the communities within which the NGOs operate, and in doing so are assisting communities to help themselves in terms of identifying challenges and resources to address such challenges. Not only are NGOs uniquely and closely positioned in communities to build social capital and trust, but they also often possess the capacity to undertake deeper levels of participation in the form of network and partnership building. These forms of citizen participation are often challenging to local authorities bound by institutional and legislative frameworks and limited resources.

A critical success factor is that NGOs and local authorities should not be seen as working against each other or as infringing on institutional territories. The activities undertaken by ethical and responsible NGOs can in fact contribute to achieve spatial transformation among local communities more swiftly than that allowed by some of the tedious processes offered by local municipalities. Citizen consultation processes are helpful, especially for participatory and comprehensive plan-making processes in local municipalities. These processes could further be strengthened by developing networks to sustain social capital and communicative planning processes which can go beyond plan creation (Olsson 2009; Proli 2020) and could prompt a communicative turn to shift from 'planning spaces' to promoting community institutional capacities.

References

Akbar PNG, Edelenbos J (2021) Positioning place-making as a social process: a systematic literature review. Cogent Soc Sci 7(1):1–29
Arnstein SR (1969) A ladder of citizen participation. J Am Inst Plann 35(4):216–224
Banks N, Hulme D, Edwards M (2015) NGOs, states, and donors revisited: still too close for comfort? World Dev 66(C):707–718
Beza BB (2016) The role of deliberative planning in translating best practice into good practice: from placeless-ness to placemaking. Plan Theory Pract 17(2):244–263
Blanchet-Cohen N (2015) Igniting citizen participation in creating healthy built environments: the role of community organizations. Commun Develop J 50(2):264–279
Booher DE, Innes JE (2002) Network power in collaborative planning. J Plan Educ Res 21(3):221–236
Chikerema AF (2013) Citizen participation and local democracy in Zimbabwean Local Government System. IOSR J Human Soc Sci 13(2):87–90
Craig G (1998) Community development in a global context. Commun Develop J 33(1):2–17
Dalton R (2004) Democratic challenges, democratic choices: the erosion of political support in advanced industrial democracies. Oxford University Press, Oxford
Devas N, Grant U (2003) Local government decision-making—citizen participation and local accountability: some evidence from Kenya and Uganda. Public Adm Dev 23:307–316
Eckenwiler LA (2016) Defining ethical placemaking for place-based interventions. Am J Public Health 106(11):1944–1946
Fraenkel RM (1977) Community development goals and citizen participation. Commun Develop J 12(3):177–185
Gray M, Mubangizi B (2010) Caught in the vortex: can local government community development workers succeed in South Africa? Commun Develop J 45(2):186–197
Habermas J (1976) Communication and the evolution of society. MIT Press, Cambridge
Healey P (1992) A planner's day: knowledge and action in communicative practice. J Am Plan Assoc 58(1):9–20
Healey P (1996) The communicative turn in planning theory and its implications for spatial strategy formations. Environ Plan B Plan Design 23(2):217–234
Healey P (2003) Collaborative planning in perspective. Plan Theory 2(2):101–123
Howard J, Wheeler J (2015) What community development and citizen participation should contribute to the new global framework for sustainable development. Commun Develop J 50(4):552–570
Ife J, Tesoriero F (2006) Community development: community-based alternatives in the age of globalisation, 3rd edn. Pearson Education Australia
Karuri-Sebina JW (2014) Applying the systems of innovation approach to neighbourhood planning: assessing local development analysis through an appreciative study of two South African townships. Doctoral dissertation, University of the Witwatersrand
Koch DJ, Dreher A, Nunnenkamp P, Thiele R (2009) Keeping a low profile: what determines the allocation of aid by non-governmental organizations? World Dev 37(5):902–918
Lepofsky J, Fraser JC (2003) Building community citizens: claiming the right to placemaking in the city. Urban Stud 40(1):127–142
Mamokhere J, Meyer DF (2022) Including the excluded in the integrated development planning process for improved community participation. Int J Res Busi Soc Sci 11(4):286–299
Marais L, Botes L (2006) Income generation, local economic development and community development: paying the proc for lacking business skills? Commun Develop J 42(3):379–395
Meinzen-Dick R, DiGregorio M, McCarthy N (2004) Methods for studying collective action in rural development. Agric Syst 82:197–214
Mnguni S (2018) The role of public participation in facilitating integrated development planning for improvement of services in uMhlathuze Local Municipality. Doctoral dissertation, University of Zululand

Naidoo C, Ramphal RR (2018) The factors that affect public participation for effective municipal service delivery: a case of ward committees. S Afr J Ind Eng 29(4):82–93

NGO Law (2023) Understanding the differences between NPO, NGO and NPC and the governance policies. Retrieved 22 May 2024 from www.ngolaw.co.za

Nkabinde B, Lekhanya LM, Dorasamy N (2018) The rural immigration effects on urban service delivery in South Africa (SA). J Econ Behav Stud 10(6(J)):11–24

Olsson AR (2009) Relational rewards and communicative planning: understanding actor motivation. Plan Theory 8(3):263–281

Patriquin DL, Halpenny EA (2017) Building consensus through place: place-making as a driver for place-based collaboration. Cogent Soc Sci 3(1):1300864

Proli S (2020) Communicative turn in spatial planning and strategy. In: Filho WL, Azul AM, Brandli L, Özuyar PG, Wall T (eds) Sustainable cities and communities. Springer, London, pp 91–100

Radebe K, Nkonyeni N (2020) NGOs today: competing for resources, power and agency. Retrieved 10 Dec 2023 from https://mg.co.za/analysis/2020-03-05-ngos-today-competing-forresources-power-and-agency

Ranyaka Community Transformation (2023) Retrieved 8 Jan 2024 from www.ranyaka.co.za

Republic of South Africa (RSA) (1996) Constitution of the Republic of South Africa. Retrieved 31 Jan 2024 from https://www.gov.za/documents/constitution/constitution-republic-south-africa-1996-04-feb-1997

Republic of South Africa (RSA) (1998a) White Paper on Local Government. Department of Cooperative Affairs and Local Government. Retrieved 7 Jan 2024 from https://www.cogta.gov.za/cgta_2016/wp-content/uploads/2016/06/whitepaper_on_Local-Gov_1998.pdf

Republic of South Africa (RSA) (1998b) Local Government: Municipal Structures Act 117 of 1998. Retrieved 31 Jan 2024 from https://www.gov.za/documents/local-government-municipal-structures-act

Republic of South Africa (RSA) (2012) National development plan 2030. National Planning Commission. Retrieved 10 Dec 2023 from https://www.gov.za/documents/other/national-development-plan-2030-our-future-make-it-work-15-aug-2012

Republic of South Africa (RSA) (2013) Spatial Planning and Land Use Management Act 16 of 2013. Retrieved 10 Dec 2023 from https://www.gov.za/documents/spatial-planning-and-land-use-management-act

Republic of South Africa (RSA) (2016) Integrated Urban Development Framework. Department of Cooperative Governance and Traditional Affairs. Retrieved 10 Dec 2023 from https://iudf.co.za/

Richards L, Dalbey M (2006) Creating great places: the role of citizen participation. Community Dev 37(4):18–32

Seitlholo S, Diedericks M (2018) Challenges facing the effective functioning of ward committees as a mechanism to promote local democracy in the former Tlokwe Local Municipality (TLM). Politeia 37(1):1–17

Shettigar C, Mayya S (2015) Role of non-governmental organisation in rural development: a study in Dakshina Kannada District. SUMEDHA J Manag 4(4):27–38

South African Cities Network (SACN) (2021) State of South African cities report. www.sacities.net. Accessed 22 May 2024

Statistics South Africa (Stats SA) (2014) Registration of NGOs and NPOs. Government Press, Pretoria

Taylor N (1998) Urban planning theory since 1945. Sage, London

Todes A, Karam A, Klug N, Malaza N (2010) Beyond master planning? New approaches to spatial planning in Ekurhuleni, South Africa. Habitat Int 34(4):414–420

Chapter 5
Institutional Change and Transformation of the Planning Profession

Martin Lewis and Verna Nel

Abstract Planners should drive spatial transformation. However, this requires a transformed planning profession with the knowledge, skills, and desire to improve the existing situation. This chapter addresses three areas of change; the planning institutions, the content of planning programmes, and the knowledge and abilities planners need. It first describes the reform of planning institutions and their role in transitioning from a Whites-only to a multiracial profession. Thereafter, the standards and competencies project are explained. As it identified the core knowledge and skills required by planners, it has influenced planning qualifications at universities and ensures that graduates can contribute to transformation. Furthermore, the 'internship' period ensures that planners can apply the skills in practice, while the continuing professional development requirements emphasise the need to remain up to date with theory and practice. Although acknowledging these achievements, there are still areas that require attention to encourage and possibly compel the planning profession to transform practice, thinking and space.

Keywords Spatial transformation · Planning profession · Institutional change · Planning institutions

5.1 Introduction

In February 1990, following a momentous speech by President FW de Klerk, South Africa started its journey towards democracy. Later that year, Professor Peter Robinson challenged the planning profession to respond progressively to the changing environment of planning, including the new political climate and rapid urbanisation. He emphasised the need for increased participation of communities in planning,

M. Lewis (✉) · V. Nel
Department of Urban and Regional Planning, University of the Free State, Bloemfontein, South Africa

D. du Plessis (ed.), *Territorial Governance and Spatial Transformation in Post-apartheid South Africa*, Local and Urban Governance,
https://doi.org/10.1007/978-3-031-85489-7_5

greater accountability to a wider public, and the need to proactively promote integrated settlements. Furthermore, he called for widening the scope of planning, placing less focus on control while giving more attention to development, and creating opportunities for communities to move towards their goals (Robinson 1990).

Seven years later, Professor John Muller again challenged the profession to transform by providing leadership while maintaining clear ethical standards, seeking community participation in the development process, and adopting a normative approach—as opposed to a narrow technical stance—regarding the nature and purpose of urban and regional planning. He also noted that while planning operates in the "bumpy field of politics" it can make a unique contribution based on the "knowledge, competencies and skills that characterise and are peculiar to the planning community" (Muller 1997: 55). Both professors' challenges included the ability to adapt to the rapidly changing social and political environment, as well as accountable and ethical practice. They also emphasised the empowerment of disadvantaged communities through collaborative planning.

In 2016, Sihlongonyane and Lewis (2016) put forward a 'Conceptual Framework for Transformation' that makes an appeal to individual planners for a change in attitudes, behaviour and practices. It calls for the transformation of the mind, of space, of practice, and of society at large. This chapter explores the transformation of the urban/town and regional planning profession and its institutions over the past century and will measure it against the following criteria: the adaptability of the profession, its ethical practice, the empowerment of the disadvantaged, and collaborative planning with communities. The chapter first considers the role and nature of professional bodies—whether voluntary such as guilds, associations and institutes—or professional councils created through legislation. It then traces the history of the profession from its initial foundations to its status in 2023. The final section of the chapter debates the extent of the transformation of the institutions and the planning profession.

5.2 Role of Professional Institutions

A 'Professional body' is defined in the South African National Qualifications Framework Act, 2002, and by the South African Qualifications Authority (SAQA) as "any body of expert practitioners in an occupational field, and includes an occupational body" (RSA 2008: 4; SAQA 2022: 5).

Professional bodies set out to develop, support, regulate, and promote professional standards for technical and ethical competence to enhance the knowledge and practice of their respective professions. These bodies are concerned with the public benefit as well as the reputation of professionals. They aim to maintain and develop professionalism, thereby securing high quality professional services for society (Professional Association Research Network 2023). Friedman and Afitska (2023), and Ramphal and Rajcoomar (2019) highlight similar functions. Mokele and Chetty (2022) indicated that "professional bodies play a valuable role in the education and training systems of the countries under review. ... professional bodies are amongst the key stakeholders that are included in engagements, consultation, collaboration, and joint work in relation to qualification systems..." (Mokele and Chetty 2022).

In a survey conducted by SAQA in 2021 on the roles of professional bodies,[1] the actions listed in Table 5.1 were identified as key to their role as professional bodies by those who participated in the survey.

The role of professional bodies to encourage and facilitate continuing professional development (CPD) had the highest proportion of responses (96.4%) as depicted in Table 5.1. This was followed by the requirement to communicate with stakeholders in the profession(s) (95.2%), and the promotion and protection of the profession (94.0%).

The following roles attracted more than 80% of responses: encourage professional practice in the context of the South African democracy, equity and equality (88.0%); investigate, in accordance with the relevant legislation, complaints relating to professionals (86.7%); encourage transparency in the profession(s) (85.5%); control the professional registration of persons (81.9%); and to set standards for the training of intended practitioners (81.9%).

Table 5.1 Roles of professional bodies as identified in the SAQA survey (SAQA 2021)

Item	% of responses
Encourage and facilitate continuing professional development (CPD)	96.4%
Communicate with stakeholders in the profession(s)	95.2%
Assist in the promotion and protection of the profession(s)	94.0%
Encourage professional practice in the context of the South African democracy, equity and equality	88.0%
Investigate, in accordance with the relevant legislation, complaints relating to professionals	86.7%
Encourage transparency in the profession(s)	85.5%
Control the professional registration of persons	81.9%
Set standards for the training of intended practitioners	81.9%
Develop, administer and govern policy relating to the profession(s)	72.3%
Encourage community involvement	68.7%
Set standards for the examinations required	65.1%
Control the practice of the profession(s)	61.4%
Advise the Minister on the amendments or adaptation of the legislation to strengthen professional practice; democracy; transparency; equity; accessibility; and community involvement	51.8%
Advise the Minister concerning any aspect falling in the scope of the legislation	43.4%
Investigate, in accordance with the relevant legislation, complaints relating to students	41.0%

[1] A total of 78 of 97 professional bodies that were provided with the survey link participated in the survey. This represents an overall response rate of 80.4%. Of the professional bodies that participated in the survey, 17 were statutory bodies and 61 were non-statutory bodies.

It is evident that the requirement to continually enhance the professionals as well as the profession's standing are central to the role of professional bodies. These institutions also seek to ensure that a professional's competencies and skills are continually improving and transforming, usually through the training of the professionals, CPD, communication with professionals and well as the general public. Ethical practice is a hallmark of how a professional person engages with clients and communities; thus professional organisations seek to ensure that their professionals and the public are protected.

This was evident in a list of functions of professional bodies provided by the Open University (n.d). These include setting and assessing professional examinations, providing support for CPD to augment the capacity of a professional and boost the standing of the profession. Instituting a code of conduct to guide professional behaviour, procedures to deal with complaints against professionals and implementing disciplinary measures against persons acting unprofessionally speak to the promotion of an ethical and responsible profession. Professional bodies represent their members in lobbying government in relation to relevant legislation and promoting fairer access to the profession to people from all backgrounds, supporting a just, transformative and equitable profession.

In addition to the above, the Open University also listed publishing professional journals or magazines (which could fall within the ambit of communication with the relevant sector); providing networks for professionals to meet and discuss their field of expertise; and providing career support and opportunities for students, graduates and people already working in the profession (The Open University, nd).

There are eight statutory professional bodies established through Acts of Parliament responsible for the different professions within the broader built environment in South Africa. The legislation that governs them provides, amongst others, for the establishment of a Council as a juristic person, the composition of the Council, powers with regards to registration and the categories of registration, powers with regards to education (e.g. accreditation of programmes, development of competencies and standards, examinations, continuous education/professional development etc.), recognition of voluntary associations, identification of work to be done only by a registered person, and professional conduct (including the investigation of improper conduct) (RSA 2000a, b, c, d, e, f, 2002, 2013). Two of these professional bodies' Acts, SACPLAN and the SAGC, explicitly require them to address and achieve transformation and specifically transform the profession (RSA 2002, 2013).

The councils are: the South African Council for the Architectural Profession (SACAP), the South African Council for the Landscape Architectural Profession (SACLAP), the Engineering Council of South Africa (ECSA), the South African Council for the Property Valuers Profession (SACPVP), the South African Council for Project and Construction Management Professions (SACPCMP), the South African Council for the Quantity Surveying Profession (SACQSP), the South African Council for Planners (SACPLAN), and the South African Geomatics Council (SAGC). Six of these are directly related to the physical construction processes within the built environment and have an overarching body namely, the Council for the Built Environment.

In the process of understanding the functions, roles and responsibilities of the different professional bodies, their vision and mission and their main focus areas for programmes and projects as captured on their websites were evaluated. Table 5.2 sets out the vision and mission of the differing bodies.

Table 5.2 reveals commonalities among the professions' visions and missions such as being an inclusive and transparent profession that promotes diversity and transformation. Establishing standards regarding the education and registration of professionals is also shared along with regulating compliance with the profession's ethical code. Actions to ensure upholding and achieving standards, excellence and integrity through high educational standards for training members and CPD also feature in the professions' vision and missions. Some of the professional bodies stress the need to serve the public and promoting the interest of the public. SACPLAN's mission is achieving transformation through innovative planning and providing social and environmental justice. It is the mission of most of the professional bodies is to be recognised globally. These visions and missions, although grand statements, are not always achieved or are difficult to achieve due to internal issues as well factors beyond the control of the relevant professional body.

In an article blog on *Intuto*[2] entitled "The Biggest Challenges Facing Professional Associations in 2023." the writer described the following as the biggest challenges: (1) adapting to changing technology, (2) diversity, equity, and inclusion, (3) advocacy and public relations, (4) financial sustainability, (5) relevance and differentiation, and (6) member engagement (AJ 2023).

Adapting to changing technology relates to how professional bodies incorporate the new and rapidly changing technologies into their operation to remain effective. The technologies range from the use of social media to online learning platforms. Diversity, equity, and inclusion addresses the requirement for professional bodies to be inclusive and transformative for all people.

Advocacy and public relations pertain to how professional bodies represent their respective professions to the public and the enhancement and advocacy of members' interests. The blog states that "Professional associations must also be proactive in shaping public opinion about their industries." (AJ 2023: np). Financial sustainability concerns finding the balance between budgets and 'value for money' for its members. Professional bodies must also stay relevant. The last challenge relates to the engagement with members which is crucial for the success and continued existence of professional bodies.

---

[2] Intuto is a learning management platform that combines a learning management system and content production support service to help associations create engaging online education. It develops certified professional development (CPD) online learning. Transform current in-person training sessions and webinars into organised, interactive online courses and provide simple-to-administer industry-wide benchmark assessments. Intuto has more than 500 customers and in excess of 680,000 enrolments. (https://www.intuto.com/)

Table 5.2 Vision and mission of the different professional institutions within the built environment in South Africa

SACAP[a]	SACLAP[b]	ECSA[c]	SACPVP[d]	SACPCMP[e]	SACQSP[f]	SACPLAN[g]	SAGC[h]
Vision: "Transformed Architectural leaders serving society in a sustainable built environment." (SACAP 2023).	*Vision*: "To establish, sustain, advance and govern the Landscape Professions in the interest of the public and to the benefit of the environment." (SACLAP 2023).	*Vision*: "An effective regulator assuring engineering excellence." (ECSA 2023).	*Vision*: None listed.	*Vision*: "To be a progressive regulatory body for the Built Environment Management Professions." (SACPCMP 2023).	*Vision*: "To be an appropriate regulatory body for a dynamic, successful and ever-evolving Quantity Surveying Profession who are leaders in the development of the built environment." (SACQSP 2023).	*Vision*: "A Regulatory and Transformative Council that advances Spatial Transformation, Sustainability, Redress and Societal Impact." (SACPLAN 2023).	*Vision*: None listed.
SACAP's mission is summarised as a professional body promoting an inclusive and transparent approach, with the Architectural Profession being recognized as a global leader in the built environment (SACAP 2023).	SACLAP's mission is summarised as a body to serve the public, establish a stable Council, transform the landscape professions through competency recognition, and establish influence within the Built Environment (SACLAP 2023).	ECSA's mission summarised is to be a professional body aiming to establish engineering standards, develop a globally recognized profession, enforce compliance, maintain a competent workforce, educate the public on quality standards, regulate environmental protection, and engage with government (ECSA 2023).	Mission not listed.	SACPCMP's mission is summarised as a professional body promoting the advancement and transformation of Built Environment Management Professions through advocacy, research, and best practices (SACPCMP 2023).	SACQSP's mission is summarised to be a professional body aiming to uphold standards, achieve excellence and integrity in the Quantity Surveying profession, and protect the public in an engaging environment (SACQSP 2023).	SACPLAN's mission is summarised to be a professional body aiming to elevate the planning profession to global standards, promoting diversity, ensuring high education standards, promoting social and environmental justice, and protecting registered professionals' interests. It sets to achieve transformation through innovative spatial planning and education standards (SACPLAN 2023).	Mission not listed.

[a] South African Council for the Architectural Profession
[b] South African Council for the Landscape Architectural Profession
[c] Engineering Council of South Africa
[d] South African Council for the Property Valuers Profession
[e] South African Council for Project and Construction Management Professions
[f] South African Council for the Quantity Surveying Profession
[g] South African Council for Planners
[h] South African Geomatics Council

During an event held in July 2018 by the Independent Professional Body Forum, formed in 2015 by CEOs from the profession body space within South Africa, the risks facing professional bodies were presented by the Institute of Risk Management South Africa. A consideration of the top ten risks facing professional bodies during 2017 and 2018 identified during both years are: (1) membership decrease due to economic pressures (ranked first in 2018 and 2017); (2) event/training/sponsorship registration slow down due to budget constraints (ranked second in 2018 and forth in 2017); (3) dependence on key external resources—strategic service providers (ranked fourth in 2018 and second in 2017); (4) innovation—keeping ahead of competitors and meeting needs of new generation (ranked fifth in 2018 and seventh in 2017); (5) relevance of services and value to stakeholders—organisation's services/objectives not meeting stakeholders needs (ranked sixth and eight in 2018 and 2017); (6) lack of buy-in (ranked seventh and sixth respectively in 2018 and 2017); (7) quality of service (ranked eighth in 2018 and ninth in 2017); (8) lack of value proposition (ranked ninth in 2018 and third in 2017); and (9) attraction and retention of key employees (ranked tenth and fifth respectively in 2018 and 2017) (RMSA 2018).

A large part of the challenges identified by AJ (2023) are echoed in the risks identified by the RMSA (2018) confirming the impact of factors influencing professional bodies in achieving their vision and mission. Aspects such as a decrease in membership impact on the financial sustainability of a professional body. In addition, where professional bodies are reliant on government grants, the increasing pressure on government funding further exacerbated this aspect.

Given the roles, expectations and challenges facing professions, how has the planning profession fared in achieving its goals? In the next section the history and transformation of the planning profession is explored.

5.3 The Planning Profession

5.3.1 Beginnings

The planning profession in South Africa had its beginnings in 1919 with the establishment of the Transvaal Town Planning Association. The Association sought to promote planning by disseminating knowledge about planning, internally within the emergent profession and externally through conferences and seminars. Improving the city and dealing with slums was high on their agenda. The Association also lobbied for town planning legislation to prevent the chaotic development of suburbs (Muller 1993).

At much the same time, the Public Health Act empowered the Minister to enable municipalities to undertake planning and development control. Planning was added to Provincial powers in 1924. The first planning and development control regulations were provincial ordinances modelled on British planning legislation (Oranje

1997). During the 1930s, planning was defined as the planning of towns in an orderly manner and emphasising health and hygiene, traffic management and amenities (Bowling 1936; Muller 1993). The ordinances were focused on control—probably due to the rate of urbanisation that followed the development of industries in the country. They were very technical; political policy, community participation, power, and moral concerns did not feature (Mabin and Smit 1997).

Concomitantly, the profession was debating new ideas such as modernism and radicalism. Modernism was a range of ideas on reshaping the way we view and represent the world: a change from preindustrial art, architecture, and planning. It sought a bright new world that did away with the old and its problems (Fishman 1977). Modernism influenced art and design, seeking sleek, efficient and minimalist products and furniture instead of the ornate designs of the past. The machine became a motif. New building materials such as concrete and plate glass created opportunities for different built forms in architecture and urban planning. Howard's garden cities (translated into suburbs) and le Corbusier's vertical cities set in parks (tower blocks of apartments) transformed many cities. Radicalism refers to neo- Marxist concerns with the socially blind city that ignored injustices and inequalities, and voices of communities (Taylor 1998). Planning was not about buildings and modern forms of transport, but the people who lived in the buildings and depended on the transport, about social justice, inclusion of the 'other' and planning with affected people (Harrison et al. 2008).

Although the enthusiastic planners had lofty goals, the reality on the ground was different. According to Col Bowing—who had a great influence on planning—the purpose of town planning was to "make provision for the future expansion of towns in order that the needs of the new population, as supplied by the engineer, may be met in the most economical and efficient manner" (Bowling 1936: 109).

The Southern African Branch of the Town Planning Institute of Great Britain was established on 19 January 1944 and included planners in British colonies in Southern Africa. This institute laid the foundation of the South African Institute of Town and Regional Planning.

New ideas of neighbourhood planning that arose after the Second World War were translated into segregationist models by the Smuts and later the Apartheid governments. The Smuts government consolidated the power to formulate city-wide plans in the Natural Resources Development Council. Municipalities had to apply for permission to undertake spatial planning (Mabin and Smit 1997).

It was only in the 1950s that planning became an identifiable profession. The South African Institute of Town and Regional Planners (SAITRP) was established in 1954. However, from the outset, it was viewed as subservient to land surveying and engineering professions, whose services were in high demand for the development of infrastructure (Oranje 2014). With their low status, the role of town planners was limited to 'development control' of the town planning schemes introduced by the provincial ordinances, predominantly in the wealthier white areas or planning further extensions of cities (Mabin and Smit 1997; Oranje 2014).

5.3.2 Apartheid Era

Given its focus on controlling development, it is unsurprising that planning became very technocratic in its approach, ignoring the broader social issues of apartheid policies, discrimination, fragmentation of spaces and poverty (Muller 2000; Sandercock 1998). Thus, the profession retained its technical focus and amoral and apolitical stance. While the apartheid government proceeded with racial segregation and the fragmentation of cities, only a few young planners raised objections to these policies. The remainder of the planners were complacent, "*turning a blind eye to the social consequences attaching to apartheid legislation, the indifference to the plight of the black majority*" (Muller 1991: 22). In time, the planning profession became associated with advancing apartheid planning, which further reduced its standing.

The seething discontent and anger of the subjugated black majority boiled over in the 1976 riots that were the catalyst for change leading up to the democratic elections in 1994. For the first time, there were non-white speakers at the SAITRP planning conference in 1977 and a shift towards poor and marginalised communities. Muller (1991) noted a growing tension between community development and private sector property developments. Opposing viewpoints emerged between those who supported public participation and the right of communities to be informed of proposed developments and those who believed it slowed the already wieldy application approval process. Additionally, the more liberal planning schools increasingly supported non-governmental and civic organisations with technical advice, giving rise to Planning non-governmental organisations (NGOs) in the 1980s (Oranje 1997). During the 1985 SAITRP conference, there were heated debates about planners' contributions to developing and uplifting the most disadvantaged communities. The same conference revealed deep dissatisfaction with the profession (Mabin and Smit 1997).

Several planning institutions existed in 1986. The first was the South African Institute of Town and Regional Planners (SAITRP), formed in 1954 to advance the profession and the science and art of planning (Mabin and Smit 1997). The Town and Regional Planners Act 1984, established a second planning institution, the South African Council for Town and Regional Planners (SACTRP). This body was charged with regulating the profession by registering qualified individuals and accrediting planning schools. Later, Professor John Muller aided the formation of a third institution, the Development Planners Association, comprising black planners and progressive white planners who rejected the SAITRP (Harrison et al. 2008).

By 1994, the profession had grown in stature. Yet its blinkered vision of the role of planners prevented it from leading South Africa into the democratic era. It held onto its technical approach to planning and only a few ventured into the development and upliftment of marginalised (Harrison et al. 2008). "*The tainted, morally bankrupt, internally fractured state of the profession meant that ... the profession as an entity was not leading or giving strategic guidance*" (Oranje 2014: 5). The planning profession had reached its nadir.

5.3.3 Club Mykonos Accord

The road to transformation of the profession began with prodding by the Department of Land Affairs—then responsible for the planning profession—which arranged a workshop at Club Mykonos in the Western Cape Province in 1995 (Lewis and Nel 2020). It was a turning point for the profession. After some soul-searching and hand wringing agreements were reached on transforming the profession, as captured in the Club Mykonos Accord. This Accord committed planners to adopt the principles of the Reconstruction and Development Plan, review the ethical foundations of the profession, create a new institute, review the planning education system and reconstitute the SACTRP to be more representative of the profession and communities (Harrison et al. 2008). Among the outcomes of the accord were (1) the merger of the Development Planners Association and the SAITRP to form a new, non-racist institute, the South African Planning Institute (SAPI), (2) and the amendment of the Town and Regional Planning Act, 1984 to create a more inclusive and transformative Council. This led to the Planning Profession Act, 2002 (PPA) which created the South African Council for Planners (SACPLAN).

5.3.4 South African Planning Institute (SAPI)

SAPI was formally created on 1 July 1996. Its purpose is to "*enhance the art and science of sustainable local, regional and national human and physical development planning, and the theory and practice relating thereto*" (SAPI 2023). It aims to represent planners, promote the profession, ethical conduct and the standard of education, and encourage practices that support sustainable use of resources. It is now recognised by and associated with SACPLAN as a Voluntary Organisation and its members are often asked to partake in the accreditation of planning schools.

Between 1990 and 2001 planners were uncertain of their roles and future in the country and possibly feeling overwhelmed by the legislative and organisational changes (SAPI 2016). However, among the plans required by the Local Government: Municipal Systems Act, 2000 (MSA), are municipal integrated development plans and spatial development frameworks that created a demand for planners, which has contributed to increased dignity of the profession.

The Planning Africa Conferences held by SAPI also contributed to stabilising and strengthening planning. The conferences have enabled planners from municipal and other government sectors, the private sector and academia to interact and share ideas or discuss mutual problems. The conferences have been held every second year—excluding 2004, 2020 and 2022—usually in the international conference centres in Durban (eThekwini), Johannesburg or Cape Town.

Regional branches are the foundation of SAPI. Membership is not limited to professional planners, but anyone who has an interest in planning. The branches are supported by a national executive and a small administration. Prior to 2006, administrative activities were carried out by volunteers. Later, SAPI shared an office and membership administration with the South African Institution of Civil Engineering, and in 2010 it established its own infrastructure. As with so many organisations, the COVID-19 pandemic detrimentally affected the work of SAPI. In 2023 it began re-establishing its branches and activities. The KZN Branch has initiated the arrangements for another Planning Africa conference for 2024.

5.3.5 *Planning Schools*

The first urban and regional planning qualification was a post-graduate diploma offered by the University of the Witwatersrand in 1944. Many land surveyors registered for these diplomas so that they could undertake town planning work to supplement their incomes. Undergraduate Town and Regional Planning courses were introduced in the 1960s at the Universities of the Witwatersrand and Pretoria (Oranje 1997). These were followed by other universities over time. Currently, the 11 planning schools in South Africa offer 13 active planning qualifications accredited by SACPLAN for purposes of registration[3] (see Table 5.3). There are seven undergraduate programmes and six post-graduate programmes. Some of the programmes are coupled to enable a graduate to achieve all the required competencies through the relevant qualifications. This has also been referred to as a 3 + 1[4] or a 1 + 1[5] combination. A list of the accredited qualifications as in December 2023 is provided in Annexure A. This list includes qualifications that have been phased out over time.

5.3.6 *Changing Constitution of Planners*

As indicated in Table 5.4 and Fig. 5.1, the number of registered planners has grown from just over 400 carried over from the SAPI database (Before the formation of the South African Council for Town and Regional Planners, the SAITRP undertook the registration of planners.) in 1984/5 to approaching 6000 in 2023 (SACPLAN registration data). The number of registered planners has roughly doubled every decade since 1985 as indicated in Fig. 5.2.

---

[3] Graduates with a three-year diploma qualification can register as a Technical Planner after the required practical training period, while graduates with an accredited 4-year Bachelor's, Bachelor Honours or a Master's in planning can register as a Professional Planner after their practical training.

[4] The '3 + 1' combination refers to a three-year Bachelor's degree (on an NQF 7) tied to a one-year Bachelor Honours degree (on an NQF 8).

[5] The '1 + 1' combination refers to a Bachelor Honours degree or a Post Graduate Diploma (on an NQF 8) tied to a one-year Master's degree (on an NQF 9).

Table 5.3 Accredited programmes at planning schools per province and level (SACPLAN 2023)

Institution with a Planning School/Planning programme	**Province**	**Undergraduate**	**Post-graduate**
University of Cape Town	Western Cape		■
University of Stellenbosch	Western Cape		■
Cape Peninsula University of Technology	Western Cape	■	
Durban University of Technology	KwaZulu-Natal	■	
University of KwaZulu Natal	KwaZulu-Natal		■
University of the Free State	Free State		■
North-West University	North West	■	
University of Johannesburg	Gauteng	■	
University of the Witwatersrand	Gauteng	■	■
University of Pretoria	Gauteng	■	■
University of Venda	Limpopo	■	

Table 5.4 Cumulative number of registrations per decade (SACPLAN registration data)

Year	Male				Female			
	Black	Asian	Coloured	White	Black	Asian	Coloured	White
1984	1	3	0	375	0	0	0	25
1985–1994	13	7	0	802	4	0	0	142
1995–2004	70	24	1	1136	32	9	3	308
2005–2014	567	65	11	1461	385	37	12	534
2015–2023	1629	155	11	1461	1495	100	33	775

The profile of the profession has also changed as indicated in Fig. 5.1. From about 2005 (shortly after the establishment of SACPLAN) the number of Black African planners began to exceed that of white planners and is now greater than the number of white planners in respect of both males and females. There is also a marked increase in the number of Black female planners. This change of profile is a clear indication of the transformation of the profession from white male-dominated to one that is more representative of the South African population. The active planners as of November 2023 are indicated in Tables 5.5 and 5.6. Table 5.5 indicates the number of registered planners in 2023 and Table 5.6 expresses the number of registered planners as a percentage of the total number of planners.

However, the data in Tables 5.5 and 5.6 refer only to registered planners; it does not include qualified planners who are not registered. These planners may be in local

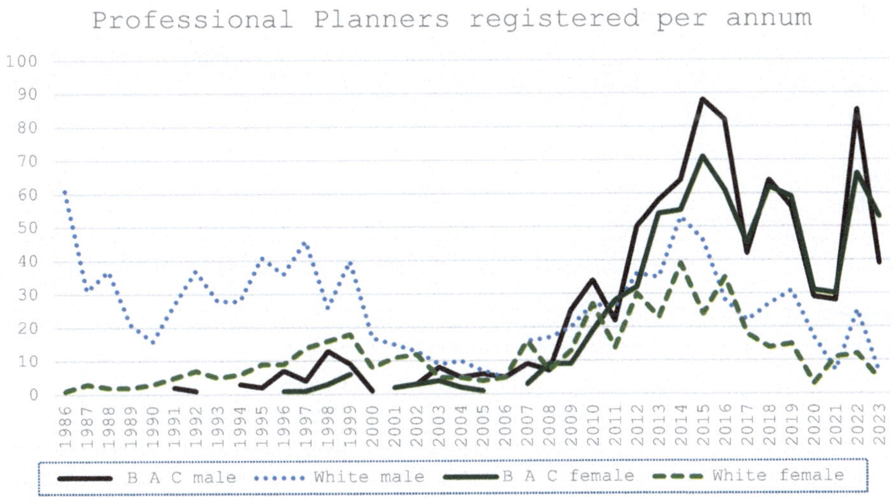

Fig. 5.1 Number of Professional Planners registered per annum 1986–2023 (SACPLAN registration data; Todes 2009; original to the author). *BAC* refers to Black, Asian, and Coloured planners

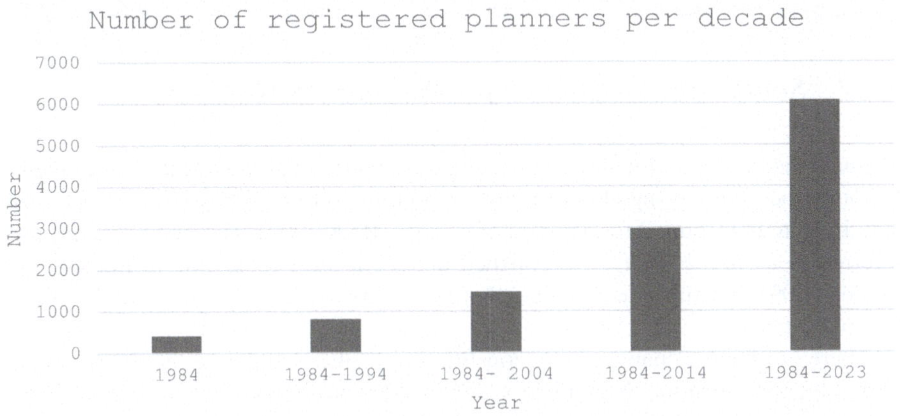

Fig. 5.2 Number of registered planners per decade (SACPLAN registration data) (original to the author)

government where the occupation-specific dispensation[6] does not apply and there is no incentive to register as a planner, or they may be qualified planners who are employed in other industries. Nonetheless, it is most likely that the general trend indicated in Fig. 5.1 will not change significantly when non-registered planners are taken into account.

[6]The Occupation Specific Dispensation (OSD) enables qualified and registered professionals in national or provincial government to qualify for salary increments. This has been an incentive to register with SACPLAN.

Table 5.5 Registered planners 2023 (SACPLAN registration data)

	Black			Indian/Asian			Coloured			White			
Category	M	F	T	M	F	T	M	F	T	M	F	T	Total
Candidate student	83	121	204	3	3	6	1	0	1	11	11	22	233
Candidate graduate	569	630	1199	32	25	57	19	11	30	171	118	289	1575
Technical	128	32	160	24	0	24	3	2	5	65	3	68	257
Professional	722	613	1335	84	63	147	23	18	41	929	386	1315	2838
Total	**1502**	**1396**	**2898**	**143**	**91**	**234**	**46**	**31**	**77**	**1176**	**518**	**1694**	**4903**

Table 5.6 Registered planners 2023 (SACPLAN registration data)

	Black		Indian/Asian		Coloured		White	
Category	Male	Female	Male	Female	Male	Female	Male	Female
Candidate student	1.69%	2.47%	0.06%	0.06%	0.02%	0.00%	0.22%	0.22%
Candidate graduate	11.61%	12.85%	0.65%	0.51%	0.39%	0.22%	3.49%	2.41%
Technical	2.61%	0.65%	0.49%	0.00%	0.06%	0.04%	1.33%	0.06%
Professional	14.73%	12.50%	1.71%	1.28%	0.47%	0.37%	18.95%	7.87%
Total	**30.63%**	**28.47%**	**2.92%**	**1.85%**	**0.94%**	**0.63%**	**23.99%**	**10.56%**
% Per Race	**59.11%**		**4.77%**		**1.57%**		**34.55%**	

5.3.7 South African Council for Planners (SACPLAN)

Through the PPA, the Minister of Agriculture, Land Reform and Rural Development[7] appoints the Council members through a nomination process. Members are then appointed to the Council for a period of 4 years (RSA 2002). The first Council (also referred to as SACPLAN 1) established under the PPA took time to find its feet (Todes 2009), but subsequent Councils have improved the system for the registration of planners.

There has been a change in the constitution of planners due to the democratisation of higher education. Nonetheless SACPLAN has also been active in transforming planning. This ongoing process includes holding workshops and participating in conferences. SACPLAN co-hosted the 2016 ISOCARP conference in eThekwini, participated in formulating the New Urban Agenda at the United Nations Conference on Housing and Sustainable Urban Development (Habitat III) in Quito, Ecuador, 2016, and collaborated with the Town and Regional Planning Journal on Special Issues with a topical thematic focus.[8] SACPLAN also embarked on a Competency and Standards generation project, which is discussed later in this chapter.

[7] Previously the Department of Land Affairs and later the Department of Rural Development and Land Reform and now the Department of Agriculture, Land Reform and Rural Development (DALRRD)

[8] The special issues covered the land question in Southern Africa: Planning and built environment perspectives (Volume 75, December 2019); on building resilient cities and communities in the face

SACPLAN instituted a process for dealing with complaints about planners and their work, and in December 2013 produced Rules and a Code of Conduct for registered planners (DRDLR 2013). By December 2023 a total of 62 complaints had either been resolved or dismissed. The biggest hurdle in this process is finding people to undertake the work of investigating officers into complaints. Another difficulty relates to complaints against persons that are not registered with SACPLAN. SACPLAN has worked closely with the Department of Rural Development and Land Reform on amendments to the PPA in this regard.

5.3.8 Standards and Competencies

The Standards and Competencies project has probably been one of the most transformative actions undertaken by SACPLAN. The Standards and Competency project was initiated to not only align with the processes within the Council for Higher Education (CHE), the South African Qualifications Authority (SAQA), the Higher Education Qualification Council (HEQC), and the Higher Education Qualifications Framework (HEQF) and Sub-Framework (HEQSF), but also to transform and "re-invent" the planning profession (SACPLAN 2014). It further informed the Occupation Specific Dispensation (OSD) of the Department of Public Services and Administration (DPSA) and the job specification processes of the Department of Cooperative Governance and the National Municipal Capacity Coordination and Monitoring Committee (NMCCMC). As the Standards and Competencies project has been well documented by Lewis and Nel (2020) and the process and stakeholders by Sihlongonyane (2018) these are not repeated here. The Standards and Competencies developed by SACPLAN impacts planning education, registration, and practice.

The first set of planning competencies was set in 2000 when the Heads of Planning schools came together in Bloemfontein to formulate a set of competencies to guide planning education in the 'new' South Africa (Harrison et al. 2008). These competencies guided the curricula of planning schools until 2014. After many failed attempts at setting standards and competencies, a new project was launched in 2012 (Lewis and Nel 2020; Sihlongonyane 2018). The project was undertaken in two phases. The first phase entailed a comprehensive background report, the proposed standards and competencies for curriculum reform, registration guidelines, and proposals for continuing professional development (CPD) and recognition of prior learning (RPL).[9] The second phase was focused on implementing the CPD and RPL proposals.

---

of climate change (Volume 77, December 2020); on Cities for all / The right to the city (Volume 79, December 2021); and on climate change and resilient cities (Volume 80, December 2022), and are available from https://journals.ufs.ac.za/index.php/trp/issue/archive

[9] The documents are available on the SACPLAN website: https://sacplan.org.za/competency-and-standards-project/org.za/planning-education

The standards and competencies for curricula were discussed with the planning schools and other stakeholders in a series of workshops and amendments were made to the initial proposals following inputs from stakeholders (Lewis and Nel 2020; Sihlongonyane 2018). The competencies are grouped into three sets: generic (such as leadership and teamwork), core knowledge and skills, and functional competencies that entail the application of the generic abilities and core knowledge in preparing plans of various types and scales. Besides the standards and competencies, a set of accreditation criteria was prepared to assist planning schools in preparing for the accreditation visits required by the PPA. While every planning school must include the planning competencies in their curricula, there is freedom for the schools to place their unique stamp on their programmes. Accreditation of programmes depends on matters such as the curricula and the staff, finances and infrastructure available to present the qualifications. It has been the experience of the authors in dealing with the accreditation of planning schools and planning programmes that the quality of the self-evaluation reports have improved over time.

5.3.9 Other Changes from the Standards and Competencies Project

As with the academic standards and competencies, the registration criteria were debated in a series of workshops across the provinces (Sihlongonyane 2018). These criteria determine the knowledge and practical experience required by graduates to enable them to register as professional or technical planners. Since the approval of the guidelines, and based on concerns from graduate programmes during workshops, further policies on the supervision and mentoring of candidates have been implemented to assist graduates in obtaining the experience and knowledge they require to register. Additionally, an examination system has been proposed that contains theory, practical and ethical elements.

In the second phase of the project, the details of the continuing professional development (CPD) process and requirements were deliberated with planners in a series of 'roadshows' across the country. Based on the comments received from planners, the initial CPD proposal was amended, and it has been implemented since 1 April 2018.

The RPL process enables people who may not have the full set of academic competencies to apply for recognition of prior learning. This policy has been implemented on 15 January 2020. The process may involve an interview and an examination. Policies concerning people with foreign qualifications are also in place.

Following the standards and competencies project, several amendments to the PPA have been proposed. These have been discussed with planners across the provinces. However, the most contentious issue is the identification of planning work which will be discussed in the following section.

5.4 Transformation of Planning Practice

5.4.1 Local Government

The transformation of the country is an ongoing process. The period from 1990 to about 2000 was tumultuous as there were numerous changes to the structure of local and provincial government along with a multitude of new laws and the repeal of discriminatory legislation. The first planning-related legislation was the Development Facilitation Act, 1995, an 'emergency law' to drive post-apartheid integration and development. Firstly, it created a new set of principles to guide planning practice and decision-making on applications. Secondly, it required municipalities to prepare and implement land development objectives to focus attention on linking plans and budgets to ensure the desired development occurred. Thirdly it created an alternative process for the submission, processing and deciding on land development applications. The latter was intended to be a 'fast track' process to accelerate the development of sites and housing for disadvantaged communities (Berrisford 1997).

Municipalities are probably the biggest employers of planners (Todes 2009) and the changes experienced during the transformation of local government since 1993 have affected the profession. During the first of three phases (pre-interim) of the transition from racially defined local governments to democratic municipalities, Local Negotiating Forums decided on municipal councils until the first elections could take place.[10] The negotiations included the extension of municipal boundaries to include the functional areas and racially based settlements of the majority of the workforce of the municipality (De Visser 2009). The first non-racial municipal elections in 1994/5 constituted the interim phase. The White Paper on Local Government (1998) was the foundation for a series of laws to prepare the way for the final phase. The Local Government: Municipal Structures Act and the Local Government: Municipal Demarcation Act were enacted in 1998, and the Local Government: Municipal Systems Act followed in 2000, in preparation for the final phase towards democratic local government that began in December 2000. The Municipal Finance Management Act followed in 2003 and the Municipal Property Rates Act in 2004. However, it was probably the Municipal Systems Act, that had the greatest impact on municipal planning practice with its requirement for an integrated development plan which includes a spatial development framework. This created work for municipal planners and planners in private practice contracted to prepare these plans if the local government does not have the capacity to do so.

Land use management was devolved from provinces to local government following a Constitutional Court judgement in 2010 that confirmed that municipalities are responsible for the full ambit of municipal planning (De Visser 2016). The Spatial Planning and Land Use Management Act, 2013 (SPLUMA) provided for this change. It includes a revised set of principles (as many of the DFA principles have been included in other legislation) and expands on the requirements for spatial

[10] This phase was governed by the Local Government Transformation Act, 1993.

development frameworks. Many municipalities have difficulty managing this responsibility. They have adopted municipal spatial planning and land use management bylaws to guide the process but struggle to implement them and link the municipal planning functions.

Despite SPLUMA's requirements, spatial plans at a municipal level are often too broad to guide decision-making and budgeting as the links between the Integrated Development Plan, the spatial plan and capital projects are tenuous. Planners, in their role of integrators between disciplines and silos can play a critical part in coupling the plans for more effective and transformative development.

Informality is still disparaged by many, with Western and colonial ideals of order conflicting with the realities of poverty and the need to create work and housing where the formal system has failed. The quality of most public spaces in 'townships' and informal settlements has changed little since 1994. Transformation in this arena requires changing mind-sets of politicians, government officials and private sector developers. Such a change extends far beyond the profession.

While SPLUMA has addressed the process and content of spatial development frameworks, the land use management component has not received as much attention. The nature and content of many land use schemes have changed little from those prior to SPLUMA or democracy; many are merely an extension of their scheme boundaries from urban spaces to cover the entire municipal area. Land use schemes have to a great extent remained the same as was prevalent during the apartheid area. It does not make provision for the existence of informality, nor does it cater for land under traditional leadership (Nel and Denoon-Stevens 2024). Planners should take responsibility for the transformation of this area of practice. Land-use schemes must change to meet the needs of communities, becoming more flexible and work towards regulations for health and safety without depriving households of the opportunity of generating livelihoods and improving the quality of their lives (Nel and Denoon-Stevens 2024).

Nonetheless, as planning deals with space, places and land, it is a highly contested and politicised arena. Power—political or financial—often has more influence on development than planners. Planners may have some influence, but little power (Denoon-Stevens et al. 2022).

5.4.2 Professionalisation of Planning

There are several ongoing debates relating to planning practice. The loudest is the issue concerning people who are not registered planners doing 'planning work'. The complaints relate to geographers, social scientists, and other professions (e.g., land surveyors, architects and lawyers) who submit land development applications. SACPLAN cannot institute disciplinary action against people not registered with it. Although this also relates to the issue of the professionalisation of the profession, it also refers to the challenge of transformation within the profession as well as of the profession's ability to address the transformation of space. The only recourse at present is in those municipal land use management bylaws that require a registered planner to sign off on certain applications. One component of the proposed amendments to the PPA is the 'identification of planning work'. In the public participation process concerning the amendments, it was clear that the consulting town planners desired the widest range of work reserved for planners. However, the land surveyors who apply for subdivisions were insistent that they should also be permitted to continue with these applications.

The second debate pertains to the number of planners produced by universities that can be absorbed by the market. On the one hand, there are accounts of unemployed graduates who are unable to find employment to complete their practical training, while on the other hand, many municipalities face a shortage of qualified planners. Based on SACPLAN registration data over a period of 11 years (from 2013 to 2023), there were on average 211 Candidate Planner, 28 Technical Planner and 175 Professional Planner registrations per annum. Over the past 5 years, data supplied by the planning schools indicate that there were on average 133 National Diploma graduates and 350 Bachelor Honours, 4-year professional Bachelors or Masters graduates per year. This implies that less than 500 planning graduates enter the market each year. Many of those graduating from Masters programmes offered in a block week format are already employed, which reduces the number of potential job seekers as does the absorption of planners in other industries, retirements, emigration or death (Todes 2009). According to a report by UN Habitat (2013) there were approximately 3.3 planners per 100,000 population in 2011 within South Africa. Currently, the ratio of registered planners to population is about 0.12 per 100,000 people. This significant decline in the ratio of planners to population may suggest that there is a shortage of planners rather than over-provision.

5.5 Conclusion

This chapter has revealed how the planning profession began in an era of hope (Oranje 2014), but then stagnated under apartheid until reinvigorated nearly 80 years later (Nel and Lewis 2020). There is evidence that both SAPI and SACPLAN have transformed concerning the diversity of their members, and certainly in the case of SACPLAN, increased numbers of registrations There is evidence of adaptability, empowerment and action against unethical practice in the profession.

Planning has shown adaptability as a profession, despite the reluctance of some planners to embrace change (Nel and Lewis 2020). Legislative reforms such as the Municipal System Act and SPLUMA have contributed to the change, creating a need for planners while forcing them to expand their boundaries of knowledge and practice. The Standards and Competencies process has influenced planning education through curriculum reform and the developmental approach to the accreditation of programmes. The student body of most planning schools comprises predominantly previously disadvantaged students. The profile of the staff in planning schools is also changing, albeit more slowly.

SACPLAN's code of conduct sets out the expected conduct of planners and disciplinary processes are in place to address complaints about deviations from ethical conduct. Unfortunately, someone must register a complaint to activate the process; thus, the process is unable to deal with corruption, particularly in government departments, or shoddy applications by non-planners. The latter could be addressed by the proposed amendment of the PPA.

The introduction of CPD and the need to attain a minimum number of points to remain in good standing has had a positive impact. The quarterly spatial planning and land use management (SPLUM) forums presented by DALRRD and COGTA/SALGA are well attended. These forums are a platform for training and the dissemination of information. Since COVID-19 they have been presented online, substantially reducing the costs of holding, or attending the events.

The need for participation has been included in legislation such as the Municipal Systems Act and SPLUMA, but the extent and depth of participation led by local governments is often limited. Despite the many tools available for online participation (e.g., WhatsApp, Facebook, Instagram), there is little evidence that these are used. This is an area where the planning profession could improve, beginning with more emphasis on the skills required to conduct effective public engagements. There are, however, NGOs and community groups that represent specific communities' interests and have become their voice, such as *Abahlale baseMjondolo*, Afesis-corplan, asivikelane.org, Development Action Group (DAG) and Planact, to name a few.

Effective spatial transformation requires more than changes in the law or the transformation of planning institutions, but changes in practice, led by planners committed to social justice, inclusion, reducing poverty and improving lives. Planning schools strive to produce graduates with these values. However, it is in the ethos reflected in spatial plans, land use schemes and developments permitted that will ultimately lead to spatial transformation. The profession has not had the profound impact on spatial transformation desired. It is still too small, and its voice is not audible above that of politicians and wealthy developers. However, by improving its practice by working with communities, ensuring that SDFs do guide infrastructure investment, reimagining land-use schemes to be enablers of transformation and working towards integrating spatial plans, projects and budgets, it will become more transformative.

Annexure A

Accredited programmes at planning schools

Institution	Level	Accredited qualification	Active	Notes
University of Cape Town	Post-graduate	Master in City and Regional Planning (MCRP)—(SAQA Qualification ID 13855) (Minimum Credits—300)		Two-year Master's degree
	Post-graduate	Master of City and Regional Planning (MCRP)—(SAQA Qualification ID 94631)	Active	One-year—candidate must have completed the Bachelor of City Planning Honours (SAQA Qualification ID 94845) that feeds into the Master of City and Regional Planning (SAQA Qualification ID 94631)
	Post-graduate	Master in City Planning and Urban Design (MCPUD)		Was not accredited between September 2011 and September 2015
University of Stellenbosch	Post-graduate	Master in Town and Regional Planning (MTRP)		Discontinued—Phased out in 2004
	Post-graduate	Master in Urban and Regional Planning (MURP)—(Programme Code 11466-887)	Active	Two-year MURP programme (260 Credits) first offered in 2020—Offered by CRUISE
Cape Peninsula University of Technology	Undergraduate	National Diploma: Town and Regional Planning (ND TRP)		Phased out
	Undergraduate	Bachelor of Technology: Town and Regional Planning (BTech TRP)		Phased out
	Undergraduate	Diploma in Urban and Regional Planning (Dip URP)	Active	
Durban University of Technology	Undergraduate	National Diploma: Town and Regional Planning (ND TRP)		Phased out
	Undergraduate	Bachelor of Technology: Town and Regional Planning (BTech TRP)		Phased out

Institution	Level	Accredited qualification	Active	Notes
	Undergraduate	Bachelor of the Built Environment Honours in Urban and Regional Planning (BBEH URP)	Active	Candidate must have completed the BBE URP (SAQA Number 99018) that feeds into the BBEH URP (SAQA Number 112549)
University of KwaZulu Natal	Post-graduate	Master of Town and Regional Planning (MTRP)	Active	In the process of being phased out and replaced by an Honours and Masters degree combination
University of the Free State	Post-graduate	Master in Urban and Regional Planning (MURP)	Active	The Candidate must have completed the Bachelor of Spatial Planning: Honours (B Hons. SP) (BC460145 and 45345) that feeds into the MURP (BC470348 and 4762).
North-West University	Undergraduate	Baccalaureus Artium et Scientiae (Planning) (B Art et Scien)		Name changed
	Undergraduate	Bachelor of Science in Urban and Regional Planning (BSc URP)	Active	B Art et Scien qualification name changed from January 2018
University of Johannesburg	Undergraduate	National Diploma: Town and Regional Planning (ND TRP)		Phased out
	Undergraduate	Bachelor of Technology: Town and Regional Planning (BTech TRP)		Phased out
	Undergraduate	Bachelor of Urban and Regional Planning Honours (B URP Hons)	Active	Candidate must have completed the BURP (Prog. Code BUPOQ / SAQA ID 94666) that feeds into the BURP Hons (Programme Code H6URBQ / SAQA ID 108895)
University of the Witwatersrand	Undergraduate	Bachelor of Science Honours in Urban and Regional Planning (BSc Hons (URP))	Active	Candidate must have completed the BSc URP (SAQA 49918) that feeds into the BSc Hons URP(SAQA 49917).
	Undergraduate	Bachelor of Science in Town and Regional (BSc TRP)		Discontinued

Institution	Level	Accredited qualification	Active	Notes
	Post-graduate	Master of Science in Development Planning (MSc (DP))—(SAQA Qualification ID—9937). (Minimum 300 Credits)		Two year Master's degree
	Post-graduate	Master of Science in Development Planning (MSc DP)—(SAQA Qualification ID 9937) (180 SAQA credits)	Active	Duration One year. Candidate must have completed the Postgraduate Diploma in Planning (SAQA Qualification ID 96533–150 SAQA credits) that feeds into the MSc DP (SAQA Qualification ID 9937)
University of Pretoria	Undergraduate	Bachelor of Town and Regional Planning (BTRP)	Active	
	Post-graduate	Master of Town and Regional Planning (By Coursework) (MTRP)	Active	
University of Venda	Undergraduate	Bachelor of Urban and Regional Planning (BURP)	Active	

References

AJ (2023) The biggest challenges facing professional associations in 2023. March 7, 2023. https://www.intuto.com/blog/the-biggest-challenges-facing-professional-associations-in-2023. Accessed 12 Dec 2023

Berrisford S (1997) Implementation of the Development Facilitation Act. Stads-en Streeksbeplanning/Town Reg Plan (42):57–59

Bowling P (1936) Some aspects of town planning. Civil Eng = Siviele Ingenieurswese 1936(1):107–119

De Visser J (2009) Developmental local government in South Africa institutional fault lines. Commonw J Local Govern 2:7–25

De Visser J (2016) Devolution by court injunction: the case of land use planning and management in South Africa. Ugandan J Manag Public Policy Stud 10(1):84–99

Denoon-Stevens SP, Andres L, Nel V, Jones P (2022) Unpacking planners' views of the success and failure of planning in post-apartheid South Africa. Cities 130:103867

DRDLR (Department of Rural Development and Land Reform) (2013) Planning Professional Act (36/2002): South African Council for Planners: Rules and Code of Conduct. Government Gazette 37189, 23 Dec 2013

ECSA (Engineering Council or South Africa) (2023). https://www.ecsa.co.za/about/SitePages/Mission%20and%20Vision.aspx. Accessed 3 Dec 2023
Fishman R (1977) Urban Utopias: Ebenezer Howard and Le Corbusier. In: Campbell S, Fainstein S (2012 (3rd edn)) Readings in planning theory. Blackwell, Cambridge, MA, pp 27–53
Friedman AL, Afitska N (2023) Professional bodies. J Profess Organ 10(1):21–35
Harrison P, Todes A, Watson V (2008) Planning and transformation. Learning from the post-apartheid experience. Routledge, London
Lewis M, Nel V (2020) Setting standards and competencies for planners. In: Silva CN (ed) Routledge handbook of urban planning in Africa. Routledge, London, pp 162–176
Mabin A, Smit D (1997) Reconstructing South Africa's cities? The making of urban planning 1900–2000. Plan Perspect 12(2):193–223
Mokele T, Chetty Y (2022) An exploratory benchmarking review of professional bodies in national qualifications framework (NQF) systems. March 2022. SAQA. https://psetresearchrepository.dhet.gov.za/show-pdf/838. Accessed 12 Dec 2023
Muller J (1991) Ethics: theory and practice in South African planning. Stads-En Streeksbeplanning = Town Reg Plan 1991(31):17–25
Muller J (1993) Parallel paths: the origins of planning education and the planning profession. Plan Hist 15(2):5–11
Muller J (1997) The planning profession. Town Reg Plan 42:54–56
Muller J (2000) Deconstructing differences: the transformation of the planning profession in South Africa. In: Millennium conference of the South African history study group. Planning for reconstruction and development, 29–31 May 2000, University of Natal, Durban
Nel V, Denoon-Stevens SP (2024) Land-use management to support sustainable settlements in South Africa. Routledge, London
Nel V, Lewis M (2020) The resilience, adaptability and transformation of the South African planning profession. In: Silva CN (ed) Routledge handbook of urban planning in Africa, 1st edn. Routledge, London, pp 141–161
Oranje M (1997) The language game of South African urban and regional planning: a cognitive mapping from the past into the future. Doctoral thesis, University of the Pretoria
Oranje M (2014) Back to where it all began …? Reflections on injecting the (spiritual) ethos of the Early Town planning movement into planning, planners and plans in post-1994 South Africa. HTS Theolog Stud 70:01–10
Professional Association Research Network (2023). https://www.parnglobal.com/Public/Public/AboutUs/About_the_Professional_Body_Sector.aspx#:~:text=Professional%20bodies%20are%20dedicated%20to,as%20the%20reputation%20of%20professionals. Accessed 13 Dec 2023
Ramphal R, Rajcoomar A (2019) A framework for the management of professional bodies in South Africa. Soc Sci 14(1):29–38
Robinson P (1990) Facing the challenge of the new South Africa, annual report to the President. Off J SA Inst Town Region Plan 29:1–3
RMSA (The Institute of Risk Management South Africa) (2018) The risks facing professional bodies. Presentation to the independent professional body forum. 2 July 2018
RSA (Republic of South Africa) (2000a) Architectural Profession Act (Act 44 of 2000). Government Printer, Pretoria
RSA (Republic of South Africa) (2000b) Landscape Architectural Profession Act (Act 45 of 2000). Government Printer, Pretoria
RSA (Republic of South Africa) (2000c) Engineering Profession Act (Act 46 of 2000). Government Printer, Pretoria
RSA (Republic of South Africa) (2000d) Property Valuers Profession Act (Act 47 of 2000). Government Printer, Pretoria
RSA (Republic of South Africa) (2000e) Project and Construction Management Profession Act (Act 48 of 2000). Government Printer, Pretoria

RSA (Republic of South Africa) (2000f) Quantity Surveying Profession Act (Act 49 of 2000). Government Printer, Pretoria
RSA (Republic of South Africa) (2008) National Qualifications Framework Act, 2008 (Act 67 of 2008). Government Printer, Pretoria
RSA (Republic of South Africa) (2002) Planning Profession Act, 2002 (Act 36 of 2002). Government Printer, Pretoria
RSA (Republic of South Africa) (2013) Geomatics Profession Act, 2013 (Act 19 of 2013). Government Printer, Pretoria
SACAP (South African Council for the Architectural Profession) (2023). https://www.sacapsa.com/about. Accessed 3 Dec 2023
SACLAP (South African Council for the Landscape Architectural Profession) (2023). http://www.saclap.org.za/index.php?option=com_content&view=article&id=1&Itemid=135. Accessed 3 Dec 2023
SACPCMP (South African Council for Project and Construction Management Professions) (2023). https://sacpcmp.org.za/about-sacpcmp/. Accessed 3 Dec 2023
SACPLAN (South African Council for Planners) (2014) Consolidated report on competencies and standards. Document number 8/4/1/C&S/SQ/12-2014. https://sacplan.org.za/wp-content/uploads/Consolidated-Report-on-Competencies-and-Standards.pdf. Accessed 3 Dec 2023
SACPLAN (South African Council for Planners) (2023). https://sacplan.org.za/. Accessed 3 Dec 2023
SACQSP (South African Council for the Quantity Surveying Profession) (2023). https://www.sacqsp.org.za/page/vision_mission. Accessed 3 Dec 2023
SAPI (South African Planning Institute) (2023) About SAPI. https://sapi.org.za/about-sapi/. Accessed 3 Dec 2023
SAPI (South African Planning Institute) (2016) Reflections, 1996–2016. https://sapi.org.za/about-sapi/. Accessed 3 Dec 2023
SAQA (South African Qualifications Authority) (2021) A survey by The South African Qualifications Authority (SAQA) on the roles and responsibilities of professional bodies in South Africa—research report. 30 March 2021. https://www.saqa.org.za/wp-content/uploads/2023/02/The-roles-and-responsibilities-of-prof-bodies-in-SA-min.pdf. Accessed 2 Dec 2023
SAQA (South African Qualifications Authority) (2022) Policy and criteria for recognising a professional body and registering a professional designation for the purposes of the national qualifications framework Act, Act 67 of 2008. https://www.saqa.org.za/wp-content/uploads/2023/04/Professional-Body-Policy-and-Criteria-2023-final.pdf. Accessed 2 Dec 2023
Sandercock L (ed) (1998) Subversive histories: texts from South Africa. In making the invisible visible, a multicultural planning history. University of California Press
Sihlongonyane MF, Lewis M (2016) Urban transformation in South Africa. ISOCARP review 12. Envisioning future cities: ideas and examples. In: International Society of City and Regional Planners, pp 252–267
Sihlongonyane MF (2018) The generation of competencies and standards for planning in South Africa: differing views. Town Reg Plan 72:70–83
Taylor N (1998) Urban planning theory since 1945. London: Sage Publications
The Open University (n.d.) OpenLearn/ Functions of professional bodies. https://www.open.edu/openlearn/mod/oucontent/view.php?id=64020§ion=1.1 (Accessed 2 December 2024)
Todes A (2009) City planners. In: Erasmus J, Breier M (eds) Skills shortages in South Africa: case studies of key professions. HSRC Press, Pretoria
UN Habitat (2013) The state of planning in Africa: an overview. UN-Habitat, Nairobi

Part II
Influence of Territorial Governance on Urban Spatial Transformation

Chapter 6
Territorial Governance and Land Use Efficiency of the South African Urban System

Daniel du Plessis and Ryan Daniels

Abstract Population growth and land consumption are two key dimensions in evaluating Land Use Efficiency (LUE) from a spatial development perspective. This chapter analyses and compares the LUE of the different categories of towns and cities within the South African urban system over the period 1990–2020 by applying the United Nations SDG Indicator 11.3.1 as a global comparative measure to assess LUE. The primary indicators underpinning SDG Indicator 11.3.1 are Land Consumption Rate (LCR), Population Growth Rate (PGR), and ratio of Land Consumption Rate to Population Growth Rate (LCRPGR), as well as Land Consumption Per Capita (LCPC) as a secondary indicator. The results indicate that territorial governance instruments introduced since 1994 have contributed to an overall improvement in LUE, albeit highly uneven amongst different settlement categories and parts of the country. These results may be sobering for the future of spatial planning and policy direction in South Africa and may indicate the need to further refine the local spatial plans and policies taking into consideration the full range of perspectives of LUE within varying local contexts.

Keywords Land use efficiency · Land consumption rate · Population growth rate · Territorial governance · Spatial planning · South African cities

6.1 Introduction

The global urban population has increased rapidly over the past century with nearly 55% of the world's current population already residing in urban areas (United Nations 2018), and is expected to increase to 68% by 2050. Asia and Africa are

---

D. du Plessis (✉) · R. Daniels
Centre for Regional and Urban Innovation and Statistical Exploration, Department of Geography and Environmental Studies, Stellenbosch University, Stellenbosch, South Africa
e-mail: ddp@sun.ac.za

D. du Plessis (ed.), *Territorial Governance and Spatial Transformation in Post-apartheid South Africa*, Local and Urban Governance,
https://doi.org/10.1007/978-3-031-85489-7_6

expected to contribute more than 80% of global urbanisation (Seto and Pandey 2019). As urbanisation intensifies it becomes increasingly important to pursue efficient urban forms through urban planning and the use of relevant indicators to monitor LUE trends. The concept of LUE can generally be described as a measure of how effectively land is utilised to achieve desired outcomes, considering environmental, social, and economic factors (Ma et al. 2023). Population growth and land consumption are two key dimensions in the evaluation of LUE from a spatial development perspective. When any of these two components are increasing disproportionately this will result in unsustainable outcomes (Hunter 2000; UN-Habitat 2018). Goal 11 of the United Nations Sustainable Development Goals (UN SDGs) for 2030 recognises this challenge in the form of Indicator 11.3.1 which specifically deals with measuring how efficiently cities utilise land (UN-Habitat 2018). A global study of 200 cities (Schiavina et al. 2022) concluded that there has been an overall improvement in LUE over recent years. Moreover, Gambe et al. (2023) concluded that urban sprawl within Southern African Development Community (SADC) regions was generally low with larger cities experiencing more growth than small-medium cities. They also found that cities have been experiencing a decline in the urban expansion rate, which could potentially be indicative of land being used more efficiently.

The proportion of the South African population living in urban areas is expected to increase from 71% in 2030 to 80% in 2050 (DCOGTA 2016). The term 'urban areas' is however not a homogenous concept, and an accurate classification of different settlement types is essential for effectively measuring and comparing the LUE of different settlement categories within an urban system (Wineman et al. 2020). In the South African context, this official classification emanated from the development of a regional functional index by the Council for Scientific and Industrial Research (CSIR) as part of the National Spatial Trends Overview project initiated by the South African City Network (van Huyssteen et al. 2015). This index has become the cornerstone of settlement typologies within South Africa and has been integrated into the National Spatial Development Framework (NSDF) and applied in other urban development policies such as the Integrated Urban Development Framework (IUDF). The main purpose of this functional settlement topology is to provide evidence, analysis, and modelling outputs to assist government planning at regional, inter-regional, and intergovernmental levels (van Huyssteen et al. 2018).

The composition of the urban system is dominated by the city regions (Gauteng, Cape Town, eThekwini), cities, and large regional centres that together account for 72% of South Africa's formal national economic activity and 53% of the population. Other medium-sized towns, small towns, and rural service centres jointly account for a further 26% of the national population and 18% of economic activity (DALRRD and DPME 2022). The spatial structure of South African cities and towns has been strongly influenced by both the colonial era and the post-1948 apartheid spatial development policies. These characteristics include segregated and fragmented spatial form (Lemon 1991; Landman 2006), peripheral location of low-income settlements resulting in unequal access to economic and social opportunities

(Turok 2011); urban sprawl (Todes 2006; SACN 2016), comparatively low population densities compared with cities in other countries with similar incomes (SACN 2016) and inverted density profiles increasing with distance from the centre (Turok 2011). These spatial challenges are all to some extent related to the concept of LUE and the notion of 'spatial transformation' to address these challenges hence emerged as a key element of territorial governance after the 1994 transition to democracy in South Africa.

Although the NSDF provides a spatial classification of the national settlement system, it does not provide any comparative analysis of LUE measures across the entire urban system. The aim of this chapter is thus to analyse and compare the LUE of the different categories of settlements within the South African urban system over the period 1990–2020 by applying the UN SDG Indicator 11.3.1 for LUE. The specific objectives are to:

- Apply the UN SDG Indicator 11.3.1 to evaluate the LUE of the different categories of cities and settlements comprising the South African urban system.
- Compare the LUE patterns and trajectories of different categories of South African cities/settlements since 1990.
- Consider the potential influence of the post-1994 territorial governance policies on these trends and the implications for future territorial governance.

Section 6.2 starts by defining the concept of LUE and reviewing different approaches to measuring LUE, with a specific focus on the application of UN SDG Indicator 11.3.1 as a global comparative measure to assess LUE. Section 6.3 reviews the territorial governance framework and the emergence of a new post-1994 spatial planning policy framework in which the concept of spatial transformation plays a central role. Section 6.4 describes the research methodology whereafter the LUE patterns and trends in the South African urban system are analysed and discussed in Section 6.5. Section 6.6 offers some conclusions based on the findings of the research.

6.2 Defining and Measuring Land Use Efficiency (LUE)

The concept of LUE can be open to various interpretations and therefore needs to be clearly defined as a point of departure. Firstly, it is necessary to distinguish the terms 'land-use' and 'land-cover' which are both widely employed across various disciplines and sometimes used interchangeably. According to Nedd et al. (2021), 'land use' reflects human activities and interventions on the land, while 'land cover' refers to the physical structures occupying the land. 'Land use' is used to describe a particular land area based on its functionality which arises from human activities, whilst 'land cover' focuses on whether the land has been altered from its natural state (such as forestlands or grasslands). Land consumption as a key component of LUE involves both the physical structures occupying the land and the transformative impact of human intervention.

Secondly, 'efficiency' is normally understood from an economic point of view and therefore often linked and connected to monetary processes and concepts such as costs and benefits (Auziņš et al. 2013; Hubbell 2007). In simple terms 'efficiency' can be defined as occurring when operations achieve the necessary strategic outcomes, at an approved level of quality, for the lowest expenditure of a particular defined scarce resource (Hubbell 2007). In urban planning terms efficiency is generally associated with the use of land, and often expressed as economic output per defined area of land (Kuo and Tsou 2015; Meng et al. 2008; Wu et al. 2017; Zitti et al. 2015). Urban governance and planning are however not only concerned with the use of land but also the overall management of cities and settlements focusing on the spatial distribution of urban components and functions such as population and buildings. In this study, the term 'efficiency' is approached from an urban planning and sustainability perspective, focusing on optimising land development by using existing resources and infrastructure to minimise urban sprawl. LUE can thus be defined as a measure of how effectively land is utilised to achieve desired outcomes, considering environmental, social, and economic factors. LUE aims to optimise land use practices and maximise outputs or benefits while minimising inputs such as resources, infrastructure, and land.

A systematic review of LUE assessments by Ma et al. (2023) identified various alternative approaches to measuring LUE. The first perspective considers the ratio of land consumption rate to population growth rate over time, emphasising the need to balance land use with population dynamics. The second perspective evaluates economic efficiency by measuring the economic output generated per unit of land utilised. LUE can also address ecological efficiency by considering the material inputs, desirable and undesirable outputs, and production processes, intending to minimise environmental impacts (Ma et al. 2023). These different perspectives all share the common goal of not exceeding the ratio between input and output beyond a balanced degree.

The UN SDG Indicator framework provides a robust mechanism to review the implementation of the 2030 Agenda for Sustainable Development based on a framework of indicators and statistical data to monitor progress. SDG 11 aims to make cities and human settlements inclusive, safe, resilient, and sustainable, and it includes various distinct targets, each paired with one or more indicators to monitor progress in achieving these targets (Berisha et al. 2022; United Nations 2015). UN SDG Indicator 11.3.1 which aims to standardise the measurement of LUE is of specific relevance to this study. It provides a straightforward assessment of LUE by examining the land consumption rate, which is typically expressed as the annual increase in urban land area compared to population growth. While it may not provide the depth of analysis and nuanced insights that more complex models such as Data Envelopment Analysis (Ma et al. 2023), the Super-Efficiency Slack Based Model (Zhu et al. 2019), and the Stochastic Frontier Production Functional Model (Liu et al. 2020) can offer, it serves as a practical and globally comparable measure. The main advantage of UN SDG Indicator 11.3.1 is its standardised approach, which results in easily replicable and internationally comparable data. This standardisation makes it a valuable tool for monitoring LUE trends, especially in cases where comprehensive data or complex modelling may not be readily available or necessary, in turn making it accessible and applicable to a wide range of different contexts.

The primary indicators underpinning UN SDG Indicator 11.3.1 are Land Consumption Rate (LCR), Population Growth Rate (PGR), and the resulting ratio of Land Consumption Rate to Population Growth Rate (LCRPGR). LCR measures the rate at which land is used within urban areas, while PGR evaluates population growth, a significant driver of urbanisation. LCRPGR, often referred to as LUE, bridges these metrics, providing insights into the relationship between land consumption and population expansion.

The first two of the primary computations are LCR expressed as (UN-Habitat 2018):

$$LCR = \frac{LN\left(\frac{Urb_{t2}}{Urb_{t1}}\right)}{y}$$

where:

LN = Natural logarithm;
Urb_{t1} = Total urban built-up area in square km in the past/initial year;
Urb_{t2} = Total urban area in square km in the current/final year;
Y = Number of years between the initial/past and current/final year.

And for PGR expressed as:

$$PGR = \frac{LN\left(\frac{Pop_{t2}}{Pop_{t1}}\right)}{y}$$

where:

LN = Natural logarithm;
Pop_{t1} = Total population of an area in the past/initial year;
Pop_{t2} = Total population of an area in the current/final year;
Y = Number of years between the initial/past and current/final year.

LCR and PGR are then used to calculate the final primary computation, which is LCRPGR, expressed as (UN-Habitat 2018):

$$LCRPGR = \frac{Annual\,LCR}{Annual\,PGR}$$

LCR reflects the rate at which land is consumed by built-up surfaces over a defined period, while PGR calculates the rate at which the population is growing over the same defined time frame. The final primary computation is LCRPGR, which provides an indicator that determines whether a particular urban area of study is sustainable in terms of its built-up surface and population growth (UN-Habitat 2018). It measures the relationship between LCR and PGR and produces a ratio that indicates the annual rate at which these two variables or indicators are changing relative to one another. An LCRPGR < 1 (and larger than 0) implies that the LCR is lower than the PGR and suggests that a city is densifying or theoretically becoming more compact in nature, possibly indicative of increased LUE efforts. However, LCRPGR values of less than 1 (and larger than 0) below a certain threshold may also be indicative of some negative externalities where excessive densification might lead to increased congestion, reduced quality of life, and increased housing prices. Conversely, if LCRPGR > 1 it signifies that the built-up growth rate of a city or town is higher than the population growth rate, indicating potential urban sprawl tendencies, which is generally undesirable in a sustainable development context. Values closer to one are considered more desirable because they suggest that the rate of expansion is proportionate to population growth, indicating a balanced approach with neither excessive sprawl nor densification (UN-Habitat 2018). LCRPGR can also have negative values due to the dynamic nature of human settlements which can be influenced by factors such as high mortality rates due to disease or catastrophic natural disasters, or migration that can result in negative population growth rates or built-up surface areas over specific periods (Jalilov et al. 2021; UN-Habitat 2018).

In addition to the primary indicators, some secondary indicators also form part of UN SDG Indicator 11.3.1 to complement the primary metrics by offering additional dimensions and perspectives on urban development. One of these secondary indicators is Land Consumption Per Capita (LCPC), which quantifies the average land consumption or occupancy per person in a city during each year of analysis (UN-Habitat 2018). To compute LCPC, two inputs are required: the total built-up area and the urban population, both of which have already been generated and used in the primary indicators. The expression for LCPC is as follows:

$$LCPC_{t1} = \frac{UrBu_{t1}}{Pop_{t1}}$$

Where:

$UrBu_{t1}$ represents the total built-up area within the past/initial year's urban boundary.
Pop_{t1} represents the total population within the past/initial year's urban boundary.

With the LCPC for each year completed, the LCPC change between different years can be calculated in percentage terms. This resultant indicator provides additional insights into the rate at which the built area per person is changing over time. Although this indicator is informative, it only identifies whether a city or town is becoming less or more efficient in terms of its use of land and does not necessarily explain the reasons for these trends.

6.3 Territorial Governance and the Emergence of a New Post-1994 Spatial Planning Policy Framework

As indicated in Sect. 6.1 most of the spatial challenges of South African cities are either directly or indirectly related to LUE. The concept of spatial transformation is thus widely recognised in post-1994 territorial governance and policy frameworks. Despite this widespread use of the concept of spatial transformation, the interpretation and application of the term has however been described as "a concept with rather abstract and fluid meanings" (SACN 2016: 58). It is thus relevant to unpack the evolution of the components of LUE and its application in spatial transformation in the evolving policy framework after 1994 dealing with territorial governance structures and approaches to spatial planning. The White Paper on Reconstruction and Development (South Africa 1994) was the first post-1994 policy framework with the primary aim of achieving integrated and coherent socio-economic progress. It was not specifically focused on spatial transformation but on the transformation of the way in which the Government itself operates. The subsequent Development Facilitation Act of 1995 (South Africa 1995) represented the first practical steps towards a new planning governance system and introduced important principles that reflect the need for LUE such as discouraging urban sprawl, promoting densification and mixed-use development, and promoting the integration of residential and employment areas. This was followed by the Urban Development Framework of 1997 (South Africa 1997) concerned with both the transformation of governance structures influencing urban development, and the spatial transformation of the dysfunctional structure of South Africa's urban areas. Urban infill development and promoting urban densification were among its goals for overcoming the inefficiencies of urban settlements.

These early post-1994 policies highlighted the critical importance of ensuring temporal and spatial coordination of development and investment initiatives of all three spheres of governance as a key element of territorial governance. In response, the promulgation of the Municipal Planning and Performance Management Regulations (South Africa 2001) introduced a system of Integrated Development Planning (IDP) to give effect to the principle of 'joined-up' governance. It also introduced Spatial Development Frameworks (SDFs) as the primary instrument for giving spatial expression to the developmental vision and priorities of municipalities. Many of the principles and objectives forming part of the post-1994 territorial governance reforms (especially those related to spatial planning and land use management) were formally enacted by the promulgation of the Spatial Planning and Land Use Management Act (SPLUMA, Act 16 of 2013) in 2013. SPLUMA requires the preparation of SDFs at various levels of spatial aggregation ranging from a national level, down to a municipal level, and precinct level within an individual municipality. Of specific relevance to LUE are Sections 7(b) and 7(c) of SPLUMA which encourage the concentration of development in existing urban areas rather than promoting urban sprawl. By promoting higher-density developments and mixed land uses, SPLUMA aims to minimise the conversion of valuable agricultural and natural lands into urbanised areas.

Despite these various initiatives the country still lacked an overall coordinated long-term development vision within a national spatial development perspective and this informed the establishment of the National Planning Commission (NPC) in 2010 tasked with the preparation of an overall National Development Plan (NDP) for the country. The NDP (finalised in 2012) identified the transformation of human settlements as one of the key elements of the plan and called for the reconfiguration of towns and cities into more efficient and equitable urban forms informed by a robust set of indicators as part of a spatial governance evaluation framework (NPC 2012). The subsequent IUDF finalised in 2016 provides more detail on the intentions of Chapter 8 of the NDP dealing with the transformation of human settlements and the national space economy (DCOGTA 2016). The IUDF describes spatial transformation as changing existing city footprints to produce compact, coordinated and connected cities, using transit-oriented development (TOD) and other urban planning strategies (DCOGTA 2016). Its aim is to reduce urban sprawl and the encroachment of valuable agricultural and natural lands by encouraging higher-density development, mixed land use, and the revitalization of existing urban areas (DCOGTA 2016).

The NSDF of 2022 was the first formally adopted national spatial plan since 1994 and provides a strategic long-term spatial plan for South Africa towards 2050 (DALRRD and DPME 2022). Importantly it also provided the first officially adopted and unifying definition for the concept of spatial transformation (DALRRD and DPME 2022: 19–20): "The carefully and collaboratively planned and well-managed process of (1) locating infrastructure, (2) providing social services, and (3) allowing, placing and incentivising economic activities in settlements in such a way that the segregated spatial patterns inherited from colonial and Apartheid times are broken down, and the inefficiencies, injustices and inequalities in access to economic and other opportunities resulting from these past patterns are corrected.". One of the core principles of the NSDF is to promote compact and connected urban settlements and emphasises four essential requirements for settlement development across the country (DALRRD and DPME 2022):

1. Increasing development density,
2. Reducing urban sprawl,
3. Preventing the unsustainable use of productive land, and
4. Optimising investment in infrastructure networks.

These requirements are clearly of direct relevance to the concept of LUE.

This chronological summary of the evolution of territorial governance structures and resulting policies after 1994 indicates that several key principles inherent to LUE such as increased development densities, reducing urban sprawl, and preventing the unsustainable use of land were recognised and well established through national policy and legislation spanning a period of nearly 30 years. A multitude of SDFs and related policies and plans were prepared and revised at a municipal level since 1994 to pursue these principles. The influence of these various plans on spatial transformation in general and LUE specifically was however never adequately monitored and evaluated and has led to claims that SDFs and other policy instruments to guide urban spatial development were not sufficiently effective in achieving spatial planning goals such as LUE. The NPC recognised that densities have increased in some urban areas, but overall little progress has been made in reversing apartheid geography (NPC 2012). This view is also echoed by the IUDF which states that apartheid spatial patterns have largely not been reversed since 1994.

Despite the focus on LUE-related concepts such as compaction, densification and urban infill development there thus remained a surprising paucity of comprehensive temporal analysis of LUE of the South African urban system. As a result, the effectiveness and influence of territorial governance instruments in the post-1994 -era and its impact on the LUE of South African cities remained unclear.

6.4 Methodology

As discussed in Sect. 6.2, the settlement typology as used in the NSDF (Fig. 6.1) based on mesozone level classification was utilised as a point of departure for the comparative analysis and trends of LUE of different settlement categories within the South African urban system by applying the indicators forming part of UN SDG 11.3.1.

These indicators include LCR, PGR and the resulting ratio of LCRPGR, as well as LCPC. The calculation of these indicators requires both population and built-up surface information, and the study adopted an approach that relied on open-source data, known for its efficiency and effectiveness as confirmed in similar types of studies (Jalilov et al. 2021; Li et al. 2021; Melchiorri et al. 2019; Mudau et al. 2020). The analysis procedure used two primary datasets: the CSIR Mesozone dataset and the Global Human Settlements Layer (GHSL). The CSIR Mesozone dataset divides South Africa into a grid of 25,000 spatial units, each referred to as a mesozone, designed to be approximately uniform in size (around 50 km^2) (StepSA 2015). These mesozones were strategically established to align with existing municipal and significant geo-economic boundaries, accounting for natural features such as rivers and transitions between less populated regions (e.g., mountains) and areas with higher human presence (e.g., fertile valleys or urbanised zones) (StepSA 2015). The CSIR further enhanced this mesozone dataset by incorporating the functional settlement typology to better support spatio-temporal analyses (van Huyssteen et al. 2018). The resulting urban system classified into these eight functional settlement types was used as a point of departure in this analysis.

The second dataset utilised was the Global Human Settlement Layer (GHSL) dataset which provides a comprehensive, reliable, accessible, and openly available collection of global data concerning human habitation (Ehrlich et al. 2021; European Commission 2020; Melchiorri 2022; Schiavina et al. 2022). It offers standardised data on population and urban development spanning the years 1975–2030 (European Commission 2020). The GHSL represents a framework comprising data and tools

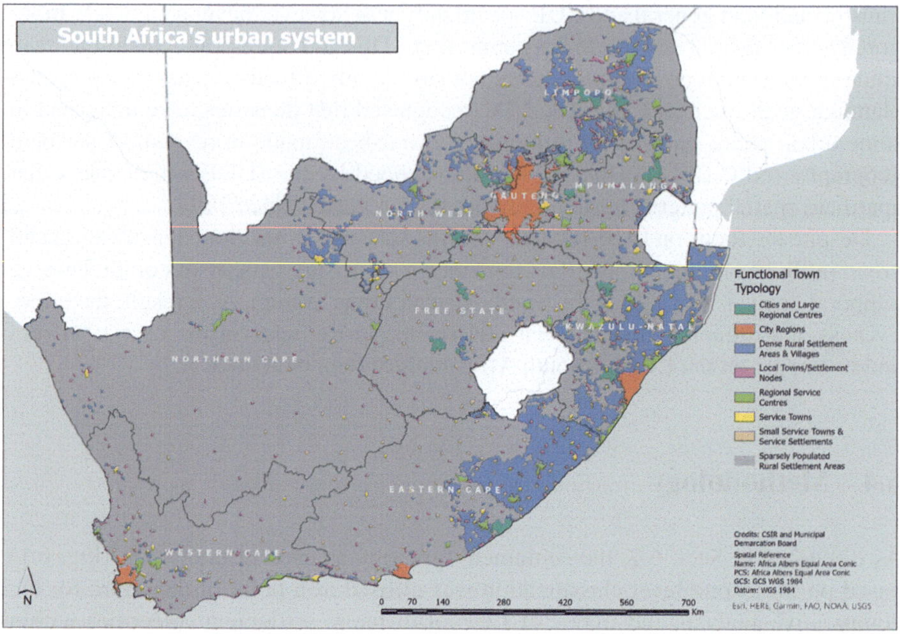

Fig. 6.1 South African settlement typology at mesozone level (original to the author)

developed through a combination of Earth Observation (EO), census data, and volunteered geographic information (Melchiorri et al. 2019). It encompasses various data components, including built-up surface (GHS-BUILT-S) and population (GHS-POP). The GHSL layers provide worldwide coverage and are openly accessible data resources (Melchiorri et al. 2019). Various studies dealing with urbanisation, disaster risk management, and analysis of human settlement growth and trends have used the GHSL dataset (Ehrlich et al. 2021; Mudau et al. 2020; Varis et al. 2022). The unavailability of recent census data (spatially disaggregated census 2022 results not available at the time of completion of this research) in South Africa has increased reliance on open-access datasets such as GHSL.

The GHS-BUILT-S and GHS-POP are spatial raster data at a resolution of 100 m (representing pixels of 10,000 m^2 in extent). Both datasets cover the years 1975–2030, with data available at 5-year intervals (European Commission 2023). From the perspective of LUE, it is important to recognise that urban form is resilient and does not change dramatically over short time periods and that longer intervals may be more appropriate to analyse LUE trends (Burton 2002). This study focuses on the post-1994 era in South Africa and the time periods used for the analysis were thus 10-year intervals from 1990 to 2020. The data analysis used the total built-up surface (encompassing all built-up areas).

After processing the data obtained from the mesozone and GHSL datasets within ArcGIS Pro, the outputs were subsequently transferred to Microsoft Excel to perform the necessary computations for the UN SDG 11.3.1 indicators. Following this, the results were analysed and then reintegrated into ArcGIS Pro for the spatial representation of the findings. The CSIR mesozones aggregated to functional town typologies (as illustrated in Fig. 6.1) were used as a base layer, aggregating GHSL data to obtain outputs measuring the total population and built-up surface within the dissolved mesozones based on settlement typology. To create an accurate reflection of 'urban extent' for each of the time periods used in the analysis, a combination of urban built up surface thresholds and population density thresholds were applied at an individual pixel level. A built-up threshold of GHS-BUILT-S ≥2500 per pixel was employed. From a population density perspective, the method to delineate cities, urban and rural areas for international statistical comparisons as suggested the European Commission and other international organisations was utilised (European Commission 2020). The lowest urban classification (semi-dense urban cluster) with a population density of ≥300 per km^2 was utilised. Pixels with values for GHS-BUILT-S ≥2500 and GHS-POP ≥3 was thus used as the 'urban extent' within the settlement typology.

6.5 Land Use Efficiency Patterns and Trends in the South African Urban System

6.5.1 Overall Population and Built-Up Development Trends

Figure 6.2 provides a comparative overview of the population and built-up[1] surface trends (based on individual pixel level thresholds for 'urban extent') in different settlement categories in South Africa over the period 1990–2020 and clearly depicts the dominance of the City Regions with a total built-up surface area of 1661.6 km^2 (representing 33.1% of the total built-up area in 2020) and 47.3% of the total population. At the other end of the settlement classification continuum, Dense Rural Settlement Areas and Villages also account for a significant proportion of the total built-up area, amounting to 1284.7 km^2 in 2020 representing 25.6% of the total built area, and 15.9% of total population. The other settlement categories exhibit a more even distribution of total built-up surface and population.

From a temporal perspective the proportion of total population in City Regions increased from 34.4% of total population in 1990 to 47.3% in 2020. Over the same period the total built-up surface increased from 1082 to 1662 km^2 which represents a proportional decrease of the total national built-up footprint from 38.4% in 1990 to 33.1% in 2020. This represents a population density of 15,644 people per km^2 in 2020. Over the same period Dense rural settlement areas and Villages increased its share of the total built-up footprint in both absolute (627 km^2 in 1990 to 1285 km^2 in 2020) and proportional (22.2% in 1990 to 25.6% in 2020) terms, whilst its proportional share of the population decreased from 24.4% in 1990 to 15.9% in 2020. This represented an overall population density of 6,816 people per km^2 in 2020.

[1] Urban built-up extent defined and calculated based on individual pixel threshold values for 'urban extent' as outlined in Sect. 6.4within mesozone level typology classification.

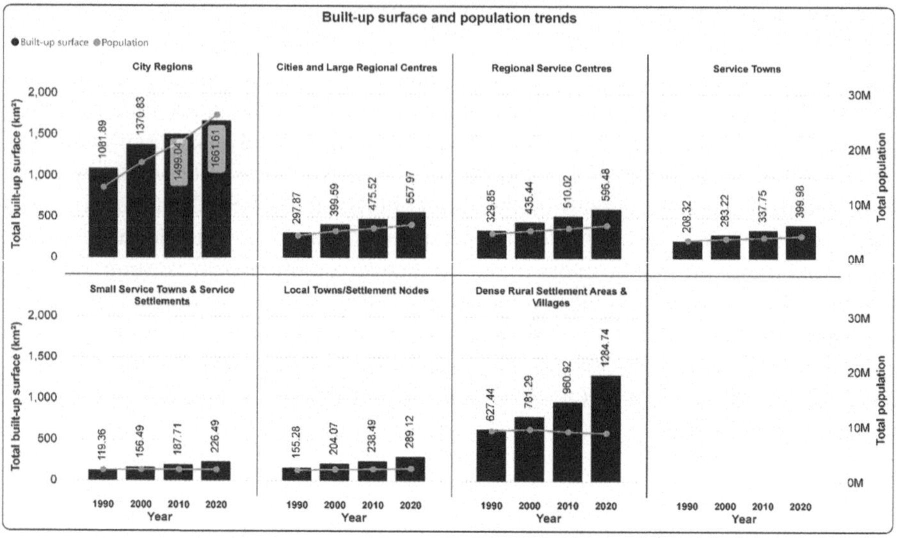

Fig. 6.2 Comparative built-up surface and population trends 1990–2020 (original to the author)

Table 6.1 Built-up area and population contribution per settlement type (1990–2020)

Settlement type	Proportion of built-up area				Proportion of population			
	1990	2000	2010	2020	1990	2000	2010	2020
City regions	38.4	37.8	35.6	33.1	34.4	39.4	43.4	47.3
Cities & large regional centres	10.6	11.0	11.3	11.1	10.6	10.9	11.0	10.9
Regional service centres	11.7	12.0	12.1	11.9	11.8	11.3	11.1	10.8
Service towns	7.4	7.8	8.0	8.0	8.7	8.0	7.7	7.4
Small service towns & service settlements	4.2	4.3	4.5	4.5	5.1	4.5	4.0	3.6
Local towns/settlement nodes	5.5	5.6	5.7	5.8	5.0	4.6	4.3	4.0
Dense rural settlement areas & villages	22.2	21.5	22.8	25.6	24.4	21.3	18.5	15.9

In contrast, although both the built-up footprint and population of the other settlement types increased between 1990 and 2020 (see Fig. 6.2), their proportional contribution to both the built-up footprint and population remained relatively constant between 1990 and 2020 (see Table 6.1) with Cities and Large Regional Centres and Regional Service Centres both contributing around 11% to built-up area and population, Service Towns approximately 8%, and Small Service Towns and Local Towns approximately 5%.

6.5.2 Land Consumption Rate (LCR), Population Growth Rate (PCR), and Land Consumption Rate per Population Growth Rate Index (LCRPGR)

The LCRPGR, or Land Consumption Rate per Population Growth Rate index, provides an indication of the LUE of a defined urban area. This index is determined by dividing the LCR by the PGR for a specific designated time frame. The first component of the LCRPGR indicator is LCR that measures the rate at which land is converted into built-up areas[2] over a specific period. Table 6.2 indicates that most of the settlement categories had similar LCR values ranging between 2.0% and 2.2% per annum over the period 1990–2020, while City Regions had the lowest LCR value of 1.4%. This implies that despite representing the largest physical increase in built-up footprint amongst the different settlement categories, City Regions grew at the lowest annual composite growth rate of 1.4% over this period. Conversely, the highest LCR was experienced in Dense Rural Settlement Areas and Villages at an annual rate of 2.4%. These figures imply possible less efficient land use patterns and lower densities and possibly urban sprawl tendencies in dense rural settlement areas, compared to the other settlement categories. LCR is however a single-dimensional indicator and cannot in isolation confirm land use inefficiencies that are influenced by multiple factors.

The results reflected in Fig. 6.3 highlight several important trends. Firstly, all settlement categories (except for Dense rural settlement areas and Villages) experienced an overall decrease in LCR between 1990 and 2020, indicating a general decrease in the LCR. Secondly, and in contrast to the other settlement categories, the LCR in Dense Rural Settlement Areas and Villages increased between 1990 and 2020. Thirdly, some settlement categories (including Cities and Large Regional Centres, Regional Service Centres, and Service Towns) experienced a continuous decrease across all three decades, whilst others (City regions, Small Service Towns,

Table 6.2 LCR per settlement type 1990–2020

Settlement type	Land consumption rate (LCR)	Population growth rate (PGR)
City regions	0.014	0.024
Cities & large regional centres	0.021	0.014
Regional service centres	0.020	0.010
Service towns	0.022	0.007
Small service towns & service settlements	0.021	0.002
Local towns/settlement nodes	0.021	0.006
Dense rural settlement areas & villages	0.024	−0.001

[2] Urban built-up extent defined and calculated based on individual pixel threshold values for 'urban extent' as outlined in Sect. 6.4within mesozone level typology classification.

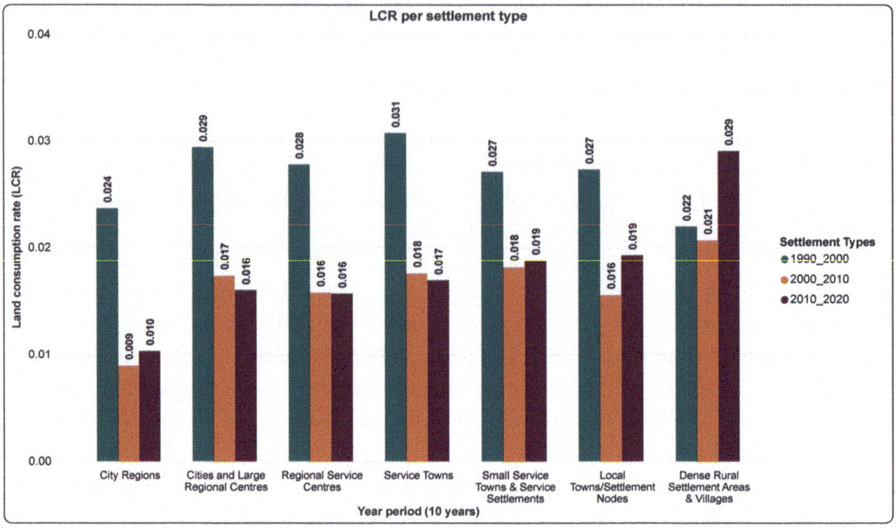

Fig. 6.3 LCR per settlement type 1990–2020 (original to the author)

and Local Towns) experienced a decrease in LCR during the 1990–2000 period and the 2000–2010 decade, followed by a higher rate in the 2010–2020 decade.

The second component of the LCRPGR indicator is PGR which measures the rate at which the population grows over a specified period. The figures in Table 6.2 reflect a differential urbanisation trend in the South African system between 1990 and 2020. The PGR as reflected in Table 6.2 depicts a consistent pattern with the higher order settlements reflecting the highest PGR (2.4% in the case of City Regions) and decreasing down the settlement hierarchy to 1.4% in Cities and Large Regional Centres, and only 0.2% in Small Service Towns & Service Settlements, and a negative PGR in the case of Dense Rural Settlement Areas and Villages. This decreasing PGR across the various settlement categories in the urban system is also clearly reflected in Fig. 6.4 which confirms the consistency of this pattern across all three decades between 1990 and 2020.

Figure 6.4 further shows that the highest PGR across all settlement categories were experienced over the 1990–2000 period in the immediate aftermath of the abolition of legislation governing the settlement of people based on racial lines, especially in City Regions (0.030) and Cities/Large Regional Centres (0.020). This dramatic change in the territorial governance approach allowed individuals the opportunity to respond rationally to economic and other opportunities within the context of population movements motivated by productionism to reside in settlements of their choice, and not constrained by territorial governance factors. This policy shift transformed City Regions into attractive destinations for many individuals seeking access to economic opportunities to enhance their quality of life. The overall trend reflects decreasing PGR values across the settlement categories and with negative PGR values in the case of Dense Rural Settlement Areas & Villages.

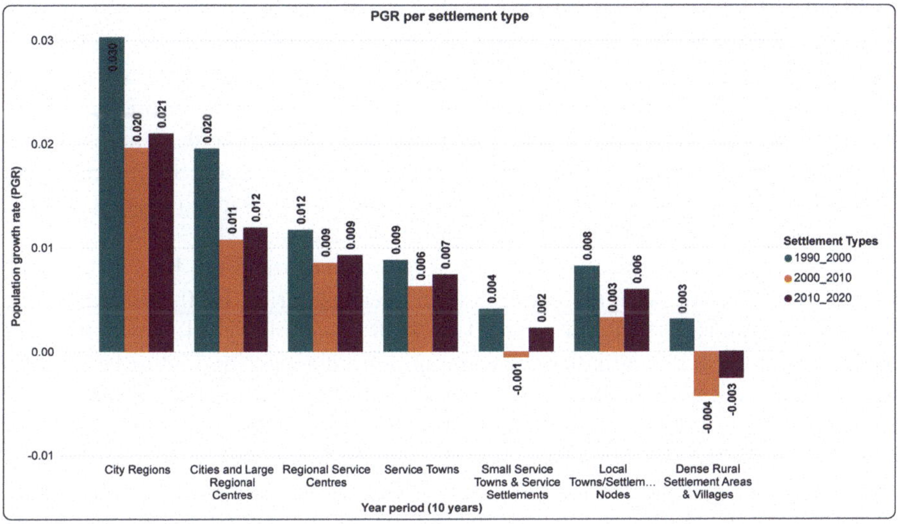

Fig. 6.4 PGR per settlement category (1990–2020) (original to the author)

This could partly be attributed to the fact that a significant proportion of the population pre-1994 had no choice but to reside in so-called 'self-governing territories' established in terms of the territorial governance policies of the apartheid era. The governance changes after 1994 allowed individuals the opportunity to respond to economic and quality of life considerations to reside in settlements of their choice and not be constrained by territorial governance factors in areas that were previously only accessible to specific population groups. This policy shift also contributed to migration from these Dense Rural Settlement Areas and Villages to higher-order urban settlements.

The LCRPGR (Land Consumption Rate per Population Growth Rate index) is the final indicator that combines LCR and PGR to provide an indication of the LUE of a defined urban area. It is calculated by dividing the LCR by the PGR for a specific designated time frame. LCRPGR values between zero and one are considered indicative of overall efficient land use with these settlements theoretically reflecting positive effects such as reduced urban sprawl, preserved green spaces, and potentially lower infrastructure costs. However, LCRPGR values between 0 and 1 will also have a lower threshold value beyond which it can lead to negative externalities, such as increased traffic congestion, overcrowding, rising housing prices, and a decrease in overall quality of life. LCRPGR values larger than 1 suggest less efficient use of land with LCR overshadowing PGR values. Negative LCRPGR values are the result of cases where either the PGR or LCR is negative values (negative PGR values will be indicative of settlements losing population over the study period, in the rare cases of negative LCR it will imply settlements with a decreasing built-up footprint over the study period). It is important to highlight that the measurement of LCRPGR is sensitive to small changes in PGR values and can lead to outlier values, resulting in LCRPGR figures that are either exceptionally small or exceptionally high. This sensitivity has been acknowledged by various studies (Calka et al. 2022; Jalilov et al. 2021; Jiang et al. 2021; Li et al. 2021; Wang et al. 2022), all of whom have noted that the LCRPGR can be significantly affected by minor fluctuations in PGR values. This is especially relevant in the case of small settlements with low population figures (such as Small Service Towns and Service Settlements, and Dense Rural Settlement Areas and Villages) where a small change in actual population numbers can represent a substantial proportional change.

Figure 6.5 clearly indicates that City Regions performed the best in terms of LCRPGR with values ranging between 0.78 over the period 1990–2000, and 0.49 between 2010 and 2020. In contrast, the LGRPGR values of Local Towns/Settlement Nodes fluctuated at much higher values ranging from a value of 3.34 in 1990–2000 to as high as 4.78 in the 2000–2010 period, and thereafter decreasing again to 3.25 between 2010 and 2020. These figures show a clear correlation between LUE (as measured by LCRPGR) and settlement type with the larger higher order settlements such as City Regions having the lowest LCRPGR and the comparative values much higher for Cities and Large Regional Service Centres, Service Towns, and Local Towns/Settlement Nodes. A further notable feature is that except for Local Towns/Settlement Nodes, the other settlement categories reflected in Fig. 6.5 experienced a continuous decrease in LCRPGR since 2000 suggesting an improving trend from a LUE perspective.

The detailed spatial analysis of LCRPGR between 1990 and 2020 depicted in Fig. 6.6 reveals several important patterns. Firstly, a significant proportion of settlements in the Free State, the Eastern Cape, KwaZulu-Natal and Limpopo have negative LCRPGR values because of negative population trends and continued migration

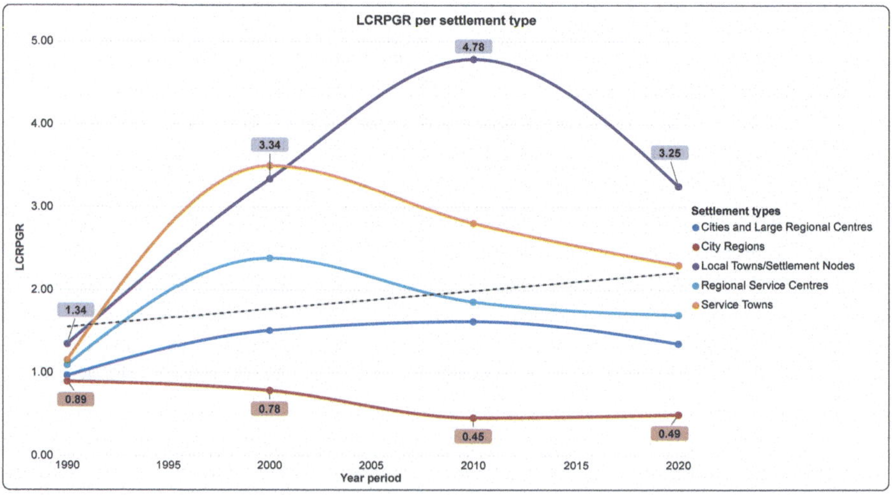

Fig. 6.5 LCRPGR per settlement type (Excluding dense rural settlement areas & villages and small service towns & service settlements due to occurrence of outlier values; original to the author)

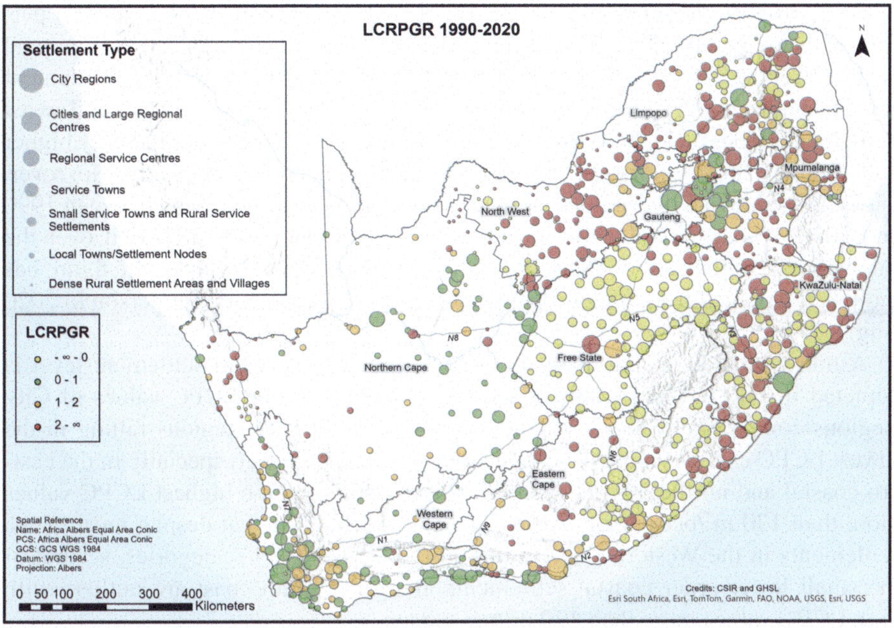

Fig. 6.6 LCRPGR 1990–2020 (original to the author)

from these areas to the larger city regions and cities in the country. A second notable trend is that settlements in the Western Cape, Gauteng and immediately surrounding parts (forming part of the Gauteng City Region), and to a lesser extent the Northern Cape have LCRPGR values between zero and one which is indicative of acceptable LUE. The surrounding lower-order settlements around the GCR and Cape Town also exhibit similar low LCRPGR index values, highlighting the interrelationship between major cities and their surrounding settlements. Thirdly, most settlements in the north-eastern part of the country in provinces such as Limpopo, the North West and KwaZulu-Natal have LCRPGR values in excess of two implying that the growth of the built-up footprint far outstrips the PGR and potentially indicative of unsustainable development trajectories from a LUE perspective.

6.5.3 Land Consumption Per Capita (LCPC)

UN-Habitat (2018) recommends the use of associated secondary indicators in addition to the primary indicators discussed in the previous section. These secondary indicators such as LCPC offer more nuanced insights into the multi-dimensional structures of human settlements (UN-Habitat 2018). The overall LCPC trend according to settlement category is broadly similar to that of LCR and LCRPGR, with larger higher-order settlements generally having lower LCPC values compared to lower-order settlements. In 2020 the LCPC in City Regions was 63.8 m^2/per person, with the comparative figures for Cities & Large Regional Centres, Regional Service Centres, Service Towns and Small Service Towns & Service Settlement ranging between 92.7 m^2/per person and 113.6 m^2/per person, and Local Towns/ Settlement Nodes and Dense Rural Settlement Areas & Villages significantly higher with figures of 130.1 m^2/per person and 146.7 m^2/per person respectively. Moreover, the LCPC value of City Regions showed a continuous declining trend between 1990 and 2020, whilst the values in all other settlement categories increased over the same period. In the case of Dense Rural Settlement Areas & Villages, the figure has more than doubled from 68.9 m^2/per person in 1990 to 146.7 m^2/per person in 2020 (Fig. 6.7).

A more detailed spatial analysis of LCPC at an individual settlement level is depicted in Fig. 6.8 and confirms the comparatively lower LCPC values of City Regions compared to other settlement types, with all City Regions falling in the lowest LCPC category. Conversely, smaller settlement types (especially in the eastern coastal and northern parts of South Africa) exhibited the highest LCPC values more than 170 m^2/per person. A further notable feature is that despite most of the settlements in the Western Cape falling in the lowest LCPC categories, several of the small lower-order coastal settlements along the south coast are outliers with high LCPC values more than 170 m^2/per person.

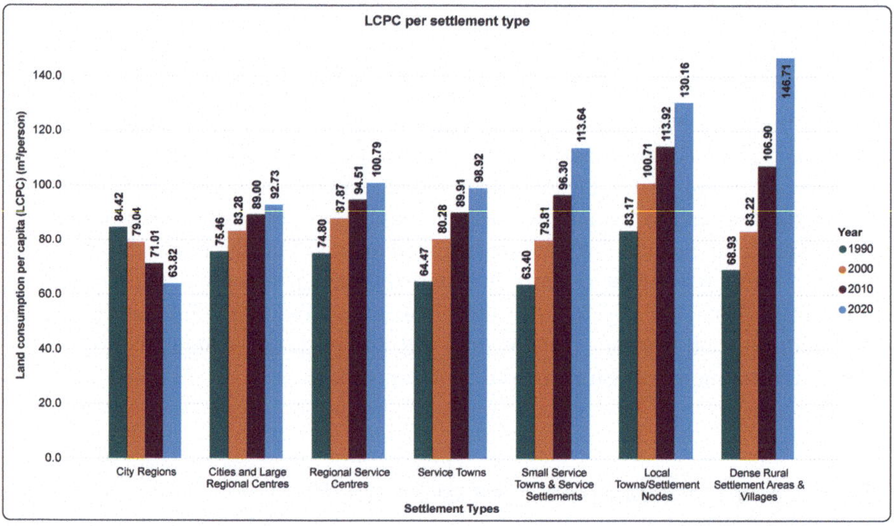

Fig. 6.7 LCPC per settlement type (original to the author)

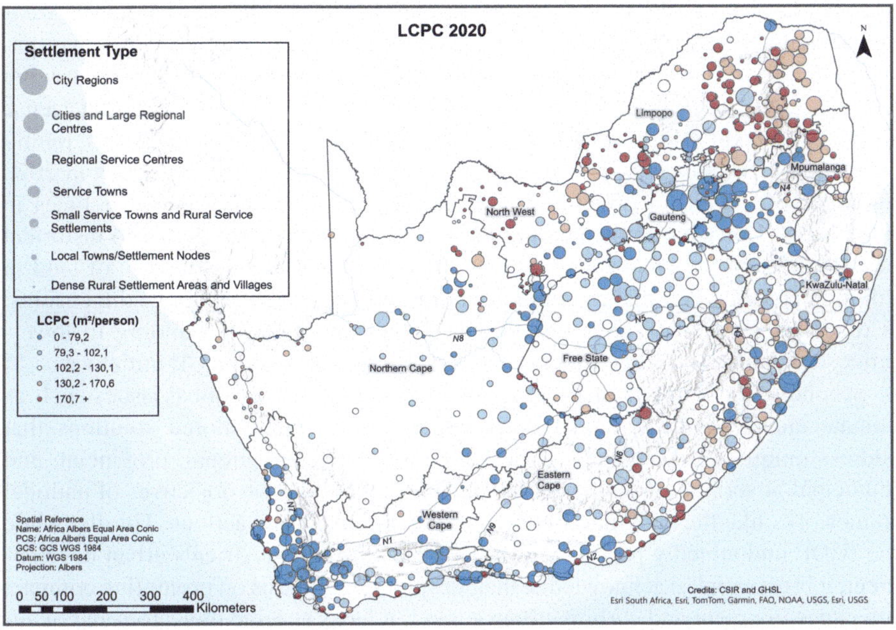

Fig. 6.8 Spatial distribution of land consumption per capita (2020) (original to the author)

6.6 Conclusions

The information presented in this chapter illustrated the applicability of the UN SDG 11.3.1 Indicator as an effective tool for assessing LUE in the South African context. Notably, the findings of this research illustrated metrics such as LCR, PGR, LCRPGR, and LCPC for diverse South African settlement types. Overall, the findings indicated that larger settlement types tend to outperform their smaller counterparts in terms of effectively managing both land consumption and population growth rates and are characterised by higher levels of LUE. Due to the complex nature of territorial governance policies and strategies and spatial plans, it is difficult to isolate the impacts of any one specific instrument as discussed in Sect. 6.3 on LUE. From the perspective of the LUE indicators used in this analysis, the territorial governance instruments as discussed in Sect. 6.3 however seem to have contributed to an overall improvement in LUE since 1990, albeit highly uneven amongst different settlement categories and different parts of the country. These results may be sobering for the future spatial planning and policy direction in South Africa and may indicate the need to further refine the local spatial plans and policies taking into consideration the full range of perspectives of LUE within varying local contexts. Several specific suggestions should be considered.

Firstly, these findings emphasise the need for targeted policy interventions in specific regions and specific settlement categories aimed at more efficient land use patterns. The higher LCR and LCRPGR values in Dense Rural Settlement Areas and Villages specifically emphasise the need for greater attention and effective spatial planning in these settlements. These lower-order settlements exhibit growing imbalances between built-up surface and population growth. The fact that Dense Rural Settlement Areas & Villages and Local Towns/Settlement Nodes jointly accounted for as much as 31.4% of the total built-up area of South Africa in 2020 highlights the need for effective spatial frameworks and development policies in these areas. However, implementing these policies is challenging due to historical inequalities and the unique social and cultural values regarding the use of land in different contexts. A holistic approach, addressing social and economic dimensions, is required, along with community engagement and capacity building in lower-order settlements as an investment in the long-term success of LUE initiatives.

Secondly, the concept of LUE is also directly related to global issues such as climate change and ecological preservation that require tailored solutions that address unique local challenges. Collaboration between national, provincial, and municipal governments must be strengthened to translate the objectives of national frameworks like the NDP and NSDF into local policies and actions. Thirdly, while the IUDF undoubtedly plays a vital role in shaping South Africa's urban development, it is essential to acknowledge that, despite its emphasis on promoting compact urban development and curbing urban sprawl, it lacks specific indicators and quantifiable targets for assessing these aspects. This implies increased emphasis on evaluation and monitoring at both provincial and municipal levels. This is essential to facilitate the pursuit of LUE in spatial development tailored to the unique characteristics of specific regions and settlement types. The results of the indicators such as those applied in this research can be very useful in these evaluation and monitoring processes and can also provide valuable baseline information in the preparation and evaluation of SDFs at various scales, specifically in quantifying possible future land requirements.

References

Auziņš A, Geipele I, Stāmure I (2013) Measuring land-use efficiency in land management. Adv Mater Res 804:205–210. https://doi.org/10.4028/WWW.SCIENTIFIC.NET/AMR.804.205

Berisha E, Caprioli C, Cotella G (2022) Unpacking SDG target 11.a: what is it about and how to measure its progress? City Environ Interact 14:100080. https://doi.org/10.1016/J.CACINT.2022.100080

Burton E (2002) Measuring urban compactness in UK towns and cities. Environ Plan B Plan Design 29:219–250

Calka B, Orych A, Bielecka E, Mozuriunaite S (2022) The ratio of the land consumption rate to the population growth rate: a framework for the achievement of the spatiotemporal pattern in Poland and Lithuania. Remote Sens 14(5):1074. https://doi.org/10.3390/RS14051074/S1

DALRRD & DPME (2022) National spatial development framework. www.gpwonline.co.za

Department: Cooperative Governance and Traditional Affairs (DCOGTA) (2016) Integrated Urban Development Framework. A new deal for South African cities and towns. DCOGTA, Pretoria

Ehrlich D, Freire S, Melchiorri M, Kemper T (2021) Open and consistent geospatial data on population density, built-up and settlements to analyse human presence, societal impact and sustainability: a review of GHSL applications. Sustainability (Switzerland) 13(14):7851. https://doi.org/10.3390/SU13147851/S1

European Commission (2020) Atlas of the human planet 2020—open geoinformation for research, policy, and action. 90. https://publications.jrc.ec.europa.eu/repository/handle/JRC122364

European Commission (2023) GHSL data package 2023

Gambe TR, Turok I, Visagie J (2023) The trajectories of urbanisation in Southern Africa: a comparative analysis. Habitat Int 132:102747. https://doi.org/10.1016/j.habitatint.2023.102747

Hubbell LL (2007) Quality, efficiency, and accountability: definitions and applications. N Dir High Educ 2007(140):5–13. https://doi.org/10.1002/HE.276

Hunter LM (2000) Population and environment: a complex relationship. https://www.rand.org/pubs/research_briefs/RB5045.html

Jalilov SM, Chen Y, Quang NH, Nguyen MN, Leighton B, Paget M, Lazarow N (2021) Estimation of urban land-use efficiency for sustainable development by integrating over 30-year Landsat imagery with population data: a case study of Ha Long, Vietnam. Sustainability 13(16):8848. https://doi.org/10.3390/SU13168848

Jiang H, Sun Z, Guo H, Weng Q, Du W, Xing Q, Cai G (2021) An assessment of urbanization sustainability in China between 1990 and 2015 using land use efficiency indicators. Npj Urban Sustain 1(34):1–13. https://doi.org/10.1038/s42949-021-00032-y

Kuo HF, Tsou KW (2015) Application of environmental change efficiency to the sustainability of urban development at the neighborhood level. Sustainability 7(8):10479–10498. https://doi.org/10.3390/SU70810479

Landman K (2006) Privatising public space in post-apartheid South African cities through neighbourhood enclosures. GeoJournal 66:133–146

Lemon A (1991) The apartheid city. In: Lemon A (ed) Homes apart. South Africa's segregated cities. Paul Chapman, London

Li C, Cai G, Du M (2021) Big data supported the identification of urban land efficiency in Eurasia by Indicator SDG 11.3.1. ISPRS Int J Geo Inf 10(2):64. https://doi.org/10.3390/IJGI10020064

Liu S, Xiao W, Li L, Ye Y, Song X (2020) Urban land use efficiency and improvement potential in China: a stochastic frontier analysis. Land Use Policy 99:105046. https://doi.org/10.1016/J.LANDUSEPOL.2020.105046

Ma Y, Zheng M, Zheng X, Huang Y, Xu F, Wang X, Liu J, Lv Y, Liu W (2023) Land use efficiency assessment under sustainable development goals: a systematic review. Land 12(4):894. https://doi.org/10.3390/LAND12040894

Melchiorri M (2022) The global human settlement layer sets a new standard for global urban data reporting with the urban centre database. Front Environ Sci 10:1003862. https://doi.org/10.3389/FENVS.2022.1003862/BIBTEX

Melchiorri M, Pesaresi M, Florczyk AJ, Corbane C, Kemper T (2019) Principles and applications of the global human settlement layer as baseline for the land use efficiency indicator—SDG 11.3.1. ISPRS Int J Geo Inf 8(2):96. https://doi.org/10.3390/IJGI8020096

Meng Y, Zhang FR, An PL, Dong ML, Wang ZY, Zhao T (2008) Industrial land-use efficiency and planning in Shunyi, Beijing. Landsc Urban Plan 85(1):40–48. https://doi.org/10.1016/J.LANDURBPLAN.2007.09.004

Mudau N, Mwaniki D, Tsoeleng L, Mashalane M, Beguy D, Ndugwa R (2020) Assessment of SDG Indicator 11.3.1 and urban growth trends of major and small cities in South Africa. Sustainability 12(17):7063. https://doi.org/10.3390/SU12177063

National Planning Commission (NPC) (2012) National development plan: vision for 2030. National Planning Commission, Pretoria

Nedd R, Light K, Owens M, James N, Johnson E, Anandhi A (2021) A synthesis of land use/land cover studies: definitions, classification systems, meta-studies, challenges and knowledge gaps on a global landscape. Land 10(9):994. https://doi.org/10.3390/LAND10090994

Schiavina M, Melchiorri M, Freire S, Florio P, Ehrlich D, Tommasi P, Pesaresi M, Kemper T (2022) Land use efficiency of functional urban areas: global pattern and evolution of development trajectories. Habitat Int 123:102543. https://doi.org/10.1016/j.habitatint.2022.102543
Seto KC, Pandey B (2019) Urban land use: central to building a sustainable future. One Earth 1(2):168–170. https://doi.org/10.1016/J.ONEEAR.2019.10.002
South Africa (Republic of) (1994) Reconstruction and development programme: a policy framework. The Presidency, Pretoria
South Africa (Republic of) (1995) Development Facilitation Act 67 of 1995. The Presidency, Pretoria
South Africa (Republic of) (1997) Urban development framework. The Department of Housing, Pretoria
South Africa (Republic of) (2001) Local government: municipal planning and performance management regulations. Government Gazette no. 7146, 24 August
South African Cities Network (SACN) (2016) State of South African Cities Report 2016. SACN, Johannesburg
StepSA (2015) Spatial temporal evidence for planning in South Africa. http://stepsa.co.za/socio_econ.htm
Todes A (2006) Urban spatial policy. In: Pillay U, Tomlinson R, du Toit J (eds) Democracy and delivery. Urban policy in South Africa. HSRC Press, Cape Town, pp 50–74
Turok I (2011) Deconstructing density: strategic dilemmas confronting the post-apartheid city. Cities 28:470–477
UN-Habitat (2018) SDG Indicator 11.3.1 training module: land use efficiency. United Nations Human Settlement Programme (UN-Habitat)
United Nations (2015) Transforming our world: the 2030 agenda for sustainable development
United Nations (2018) 68% of the world population projected to live in urban areas by 2050, says UN. https://www.un.org/development/desa/en/news/population/2018-revision-of-world-urbanization-prospects.html
van Huyssteen E, Mans G, le Roux A, Maritz J, Ngidi M, Maditse K (2015) Updated CSIR/SACN South African settlement typology. CSIR document. stepsa.org
van Huyssteen E, Green C, Sogoni C, Maritz Z, McKelly J (2018) South African functional town typology (CSIR 2018 v2). http://stepsa.org/socio_econ.html#Indicatorhttp://stepsa.org/socio_econ.html#Indicator
Varis O, Taka M, Tortajada C (2022) Global human exposure to urban riverine floods and storms. River 1:80–90. https://doi.org/10.1002/RVR2.1
Wang Y, Li B, Xu L (2022) Monitoring land-use efficiency in China's Yangtze River economic belt from 2000 to 2018. Land 11(7):1009. https://doi.org/10.3390/LAND11071009
Wineman A, Alia DY, Anderson CL (2020) Definitions of "rural" and "urban" and understandings of economic transformation: evidence from Tanzania. J Rural Stud 79:254. https://doi.org/10.1016/J.JRURSTUD.2020.08.014
Wu C, Dennis Wei Y, Huang X, Chen B (2017) Economic transition, spatial development and urban land use efficiency in the Yangtze River Delta, China. Habitat Int 63:67–78. https://doi.org/10.1016/j.habitatint.2017.03.012
Zhu X, Zhang P, Wei Y, Li Y, Zhao H (2019) Measuring the efficiency and driving factors of urban land use based on the DEA method and the PLS-SEM model—a case study of 35 large and medium-sized cities in China. Sustain Cities Soc 50:101646. https://doi.org/10.1016/j.scs.2019.101646
Zitti M, Ferrara C, Perini L, Carlucci M, Salvati L (2015) Long-term urban growth and land use efficiency in Southern Europe: implications for sustainable land management. Sustainability 7(3):3359–3385. https://doi.org/10.3390/SU7033359

Chapter 7
A Critical Evaluation of the Extent to Which Spatial Planning Integrate National Park Buffer Zones Within the Context of Territorial Governance and Spatial Transformation

Dirk Cilliers, Alexander Paterson, Francois Retief, Reece Alberts, and Claudine Roos

Abstract Amid the ongoing spatial transformation in post-apartheid South Africa, there's a growing urgency to understand and evaluate the nation's strategies towards creating a harmonious landscape, especially as it nears its 30th year of democracy. A prominent aspect of this landscape is the country's national parks, which are directly impacted by surrounding activities. To mitigate these impacts, buffer zones have been proposed. However, a challenge is that these buffer zones often fall outside the jurisdiction of protected areas, exposing them to external influences. Given the goals of reshaping South Africa's landscape, it's essential to integrate these buffer zones into spatial planning processes. The Strategy on Buffer Zones for National Parks in South Africa emphasises the importance of clearly defining these buffer zones and including them in municipal spatial development frameworks, known as MSDFs. This chapter examines how these MSDFs incorporate national park buffer zones, analysing ten MSDF case studies from nine national parks. The results show that many MSDFs don't adequately recognise or define these buffer zones, and they lack specific land use guidelines for them. This points to a gap in how MSDFs are being used as a key spatial planning tool in the context of mitigating the impact of land use change and development, especially within protected area buffer zones.

Keywords National parks · Governance · Buffer zones · Spatial planning · MSDF

D. Cilliers (✉) · F. Retief · R. Alberts · C. Roos
Unit for Environmental Sciences and Management, North-West University, Potchefstroom, South Africa
e-mail: dirk.cilliers@nwu.ac.za

A. Paterson
Institute of Marine and Environmental Law, University of Cape Town, Cape Town, South Africa

D. du Plessis (ed.), *Territorial Governance and Spatial Transformation in Post-apartheid South Africa*, Local and Urban Governance,
https://doi.org/10.1007/978-3-031-85489-7_7

7.1 Introduction

In the post-apartheid era, South Africa has been grappling with the challenges of spatial transformation, attempting to overcome historical patterns inherited from colonial and apartheid times. As the nation approaches 30 years of democracy, there's a pressing need to assess the strategies that have been employed towards achieving a harmonious and integrated landscape. A crucial element of this landscape is its national parks and their buffer zones. The extent to which protected areas can maintain ecosystem functions and achieve their conservation goals are directly influenced by activities within their surrounding landscapes (Perelló et al. 2012). Many of the problems affecting these protected areas are tied to land-use changes and activities along their boundaries or within their immediate surroundings (Du et al. 2015). The rate and scale of these changes and activities are consistently increasing (Wade and Theobald 2010; De la Fuente et al. 2020; Calderón et al. 2022).

Considering these challenges, the delineation of buffer zones around protected areas has been proposed as an effective strategy. However, buffer zones often fall outside the legal boundaries of the protected area and thus outside the jurisdiction of the protected area management authority. This complication emphasises the need for novel approaches, such as integrating buffer zone objectives into development planning tools like municipal spatial development frameworks (MSDFs), that align with South Africa's broader goals of spatial transformation.

In South Africa, buffer zones for national parks remain outside their official boundaries. As stipulated in the Strategy on Buffer Zones for National Parks (South Africa 2012), these zones play a dual role in both preserving the park and integrating it with its surroundings. Yet, despite the outlined strategy, to date, there are few buffer zones for national parks that have been officially recognised. It is important to emphasise that National Parks and their buffer zones not only serve conservation purposes but also significantly influence the broader spatial development patterns of municipalities, affecting urban and rural development planning goals and processes. MSDFs stand out as a pivotal instrument through which these zones can be implemented and managed towards achieving their objectives and the overarching spatial transformation goals of the country.

The objective of this chapter is twofold: to critically evaluate the extent to which MSDFs have integrated national park buffer zones, and to contextualize this within the broader narrative of territorial governance and spatial transformation in post-apartheid South Africa. As the nation reflects on the successes and shortcomings of its strategies adopted over the last three decades, this analysis provides valuable insights on the intersection of conservation goals and spatial transformation priorities.

The chapter begins by introducing the notion of buffer zones for protected areas, providing an overview of their nature, the relevant domestic law and policy framework and the general manner in which they have been implemented in the context of national parks to date. It then provides an overview of the form and nature of

MSDFs and the relevant domestic laws applicable to them. Having outlined this necessary context, it explains the methodology underpinning the assessment of the extent to which national park buffer zones are reflected in MSDFs. This is followed by a presentation of the results of this assessment. The chapter concludes with a discussion of the results and some recommendations regarding how best to recognise and give tangible effect to protected area buffer zones in the context of planning and decision-making.

7.2 Understanding Protected Area Buffer Zones

Understanding what protected area buffer zones are, what general purpose they serve and how they are given effect to in South Africa through domestic law, policy and practice, is naturally a necessary prerequisite to critically assessing the extent to which MSDFs successfully integrate them.

7.2.1 Nature and Purpose of Buffer Zones

Although views on what a buffer zone is and what it should achieve differ amongst scholars (Lynach and Urich 2002), two positions can broadly be distinguished: (1) buffer zones as extensions of protected areas; and (2) buffer zones as areas through which the protected area and communities around it can be integrated (Jusys 2016; Martino 2001). The preferred view affects the manner in which buffer zones are managed. Jusys (2016) argues that it is unrealistic to view buffer zones as formal extensions of protected areas as they will then effectively become protected areas themselves. Rather, buffer zones should be understood as multifunctional policy instruments aimed at minimising edge effects that might negatively impact on protected areas (Martino 2001; Perelló et al. 2012; Nadhira and Basuni 2021). This is achieved by reducing human impact on protected areas while also attending to the socio-economic needs of those residing near them by, for example, allowing but managing the use of land within the buffer zone (Robinson et al. 2013). Buffer zones are, therefore, understood here as designated areas on the periphery of a protected area in which activities are partially restricted to provide an extra layer of protection to the protected area (Sayer 1991; Wells 1992).

7.2.2 Domestic Legal Context

The *National Environmental Management: Protected Areas Act* (NEMPAA) (South Africa 2003), which regulates national parks, contains very few references to buffers. The only express reference relates to the purpose for which protected environments may be declared, which include regulating the area as a buffer zone for the conservation and protection of certain other types of protected areas, including national parks (South Africa 2003, s. 28(2)(a)). NEMPAA does not provide for the formal designation of a buffer to accompany the initial declaration of a national park. After declaration, the Minister of Forestry, Fisheries and the Environment (the Minister) must appoint a management authority for the area, with this always being South African National Parks (SANParks) in the context of all national parks (South Africa 2003, s. 31(1)(aA)). SANParks in turn must prepare and submit a management plan to the Minister for approval, with none of the prescribed or discretionary content outlined in NEMPAA for inclusion in the plan referring to buffer zones (South Africa 2003, s. 39–41). NEMPAA does empower the Minister to prescribe norms and standards (South Africa 2003, s. 11) and exercising this power, the Minister prescribed *Norms and Standards for the Management of Protected Areas* (South Africa 2016). Their application extends to national parks and one specific norm refers to the need for a protected area to have a determined buffer zone and its management authority being involved with 'outside planning structures', such as local municipalities, to ensure the integrity of the protected area (South Africa 2016: 13). The term 'buffer zone' is defined as including 'the immediate setting of the national park and attributes that are functionally important as a support to the national park and its protection' (South Africa 2016: 6). Standards relating to the above norm include establishing a buffer zone for the protected area and integrating the protected area into land-use and water-use planning instruments governing the area adjacent to its boundary. (South Africa 2016: 13–14). Key indicators associated with these standards against which management authorities must annually report, include: reflecting the buffer zone in the protected area's management plan; introducing mechanisms to implement the buffer zone; encouraging landowners surrounding the protected area to follow 'conservation-friendly land uses'; developing and implementing a policy for commenting on activities undertaken in the buffer zone which impact on the protected area; actively engaging with other organs of state for land-use and water-use planning affecting the protected area; playing an active role in land-use and water-use planning affecting the protected area; and ensuring alignment between land-use and water-use planning, and the protected area's management objectives. Examples of conservation-friendly land uses that could support the ecological health of buffer zones include eco-tourism related uses and low-impact agricultural practices such as managed grazing. Eco-tourism, when properly implemented, can serve as an economic incentive for conservation, and can also benefit local communities while uses such as managed grazing can help in maintaining the health of grassland ecosystems, preventing soil erosion, and supporting biodiversity.

Other standards relate to the content of a protected areas management plan, which must be developed in accordance with NEMPAA and the *Guidelines for Development of a Management Plan*. This content includes ensuring that any municipal integrated development plan (IDP) has 'considered the ecological sensitivity of the protected area, its buffer zones and any priorities areas for protected area expansion' (South Africa 2016: 15–16). When preparing the management plan,

SANParks is compelled by both NEMPAA (South Africa 2003, s. 39) and the *Regulations for the Proper Administration of Special Nature Reserves, National Parks and World Heritage* to consult with all other relevant organs of state (including municipalities) and consider their plans (including their IDPs and MSDFs) when preparing their management plan for the national park (South Africa 2005, reg. 57).

Outside of the remit of NEMPAA, protected area buffers for national parks are expressly identified in the context of the EIA regime prescribed under the *National Environmental Management Act* (NEMA) (South Africa 1998). *Listing Notice 3* defines 'buffer areas', 'unless specifically defined', as the 'area extending 10 kilometres from the proclaimed boundary of a .... national park' (South Africa 2014a: reg. 2). A range of activities outlined in *Listing Notice 3* proposed to be undertaken in these buffer areas trigger the need for an environmental authorisation granted under NEMA, which must be informed by a basic assessment undertaken in compliance with procedure prescribed in the *EIA Regulations* (South Africa 2014b).

Furthermore, under the *National Environmental Management: Biodiversity Act* (NEMBA) (South Africa 2004), ecosystems in buffers to protected areas are identified as 'ecosystems of special concern' potentially requiring protection through biodiversity management plans for ecosystems adopted under the Act and informed by the *Norms and Standards for Biodiversity Management Plans for Ecosystems* (South Africa 2014c). If included in a biodiversity management plan, provision is made in NEMBA for coordination and alignment between this plan and other plans, including MSDFs and IDPs developed by municipalities (South Africa 2004, s. 48). The biodiversity management plan would also constitute a relevant consideration in the context of any development proposals undertaken in the area it covers, which could include a buffer zone to the protected area.

7.2.3 Domestic Policy Context

The *Strategy on Buffer Zones for National Parks,* while introduced in 2012 prior to many of the above statutory interventions, informs their implementation with the definition of a buffer zone identical to that reflected in the *Norms and Standards for the Management of Protected Areas* (South Africa 2012). The Strategy's focus is, however, far narrower as it relates only to national parks. It identified purpose of a buffer zone is to support the management objectives, purpose and value of the national park, protect areas of high biodiversity value on the periphery of a park, while also promoting interaction between the park and adjacent communities with the aim of securing sustainable benefits for communities. Informed by a set of guiding principles and a range of relevant policy objectives drawn from the initial *White Paper on the Conservation and Sustainable Use of South Africa's Biodiversity* (South Africa 1997), it outlines the following six objectives of a buffer zone: ensure the persistence of important species and ecological processes; promote broad-based and sustainable economic activity; preserve, adapt, restore, and stabilise cultural heritage and secure the sustainable use thereof; preserve and improve the quantity and quality of water from catchments in park and the buffer zone; protect, enhance, and restore the unique and memorable character (the sense of place) that underpins the image of the national parks and their approaches; and protect and enhance the wilderness experience of park users (South Africa 2012).

Buffer zones for protected areas have also been mentioned in the revised *White Paper on the Conservation and Sustainable Use of South Africa's Biodiversity* (South Africa 2023). It identifies buffer zones as a tool to facilitate improved integration of protected areas (including national parks) within their broader ecological and social landscapes and advocates that activities undertaken in buffer zones must be compatible with the objectives underpinning the protected area (South Africa 2023).

It can therefore be argued that national park buffer zones are expected to operate as multifunctional policy instruments as described earlier in this article (Martino 2001; Perelló et al. 2012) as they should contribute to both the protection of sensitive features around the park and the integration of communities surrounding the park. Buffer zones are, therefore, established to provide an extra layer of protection to the national park and assist in the attainment of the six objectives mentioned above. In this regard, buffer zones should ideally offer complementary legal and management restrictions to ensure that pressures are managed, and the integrity of these areas is maintained.

The *Strategy on Buffer Zones for National Parks* identifies three types of areas for inclusion in a buffer zone, namely: priority natural areas; catchment protection areas; and viewshed protection areas (South Africa 2012). Priority natural areas contribute to the long-term protection of important biodiversity around the park, while catchment protection areas and viewshed protection areas consider the catchment areas affecting water resources inside the parks and the visual character of the park, respectively. It proceeds to identify nine goals to be achieved by the Strategy. These are reflected in Box 7.1 below. In respect of each, the Strategy outlines an array of specific and tangible implementation mechanisms available to government and these largely match those reflected in the *Norms and Standards for the Management of Protected Areas*. Relevant mechanisms include: identifying buffer zones for all national parks in management plans; establishing these buffer zones by publication in the Gazette; where necessary and appropriate, declaring the buffer zones or parts thereof as protected environments; subjecting activities undertaken within buffer zones to the requirements to obtain an environmental authorisation preceded by some form of EIA; integrating buffer zones into MSDFs; active participation by management authorities in EIA and land-use planning and decision-making processes relevant to buffer zones; encouraging and supporting sustainable land-use practices by landowners in buffer zones; and an array of mechanisms encouraging and supporting community-based biodiversity management initiatives within buffer zones and the flow of benefits to these communities (South Africa 2012).

Box 7.1 Strategy on Buffer Zones for National Parks Goals

- GOAL 1: Develop and introduce appropriate strategies, mechanisms, and incentives to integrate national parks areas within the broader ecological and social landscape and encourage conservation in adjacent private and communal areas.
- GOAL 2: Support and promote activities adjacent to protected areas that are compatible with, and which complement the objectives of the protected area.
- GOAL 3: Discourage development in areas in which biodiversity and ecological function· would be adversely affected.
- GOAL 4: Conserve and make sustainable use of biological resources in the buffer zone and avoid or minimise adverse impacts on the biodiversity of such areas.
- GOAL 5: Support the development of community-based biodiversity management initiatives as part of a broader set of approaches to land-use planning and developing local sustainable development strategies.
- GOAL 6: Promote the development of partnerships between the management authority, other conservation authorities, community organisations, non-governmental organisation (NGOs), and private entrepreneurs for purposes of planning and managing the use of resources within the· national park buffer zone, and optimising benefits for local people.
- GOAL 7: Enhance the capacity of communities residing in or adjacent to protected areas to participate in protected area management through providing appropriate training and education, and through recognising local expertise and traditional institutions.
- GOAL 8: Take steps to avoid or minimise damage caused to people and property by wildlife.
- GOAL 9: Improve benefit flows to people in and around protected areas.

If achieved, these goals would ensure that buffer zones are maintained and managed in line with the objectives of the Strategy. The manner and the extent to which the abovementioned mechanisms are implemented in buffer zones, however, directly affects the attainment of these goals.

7.2.4 Domestic Implementation and Management of Buffer Zones

Informed by the Strategy on Buffer Zones for National Parks, the Norms and Standards for the Management of Protected Areas and the Guidelines for Development of a Management Plan, SANParks has developed management plans for all national parks. Reflected in each of these management plans are maps, typically annexed to the management plans, depicting the buffers, with these generally matching the three types of areas for inclusion in buffer zones, namely: priority natural areas; catchment protection areas; and viewshed protection areas. The text of the management plans themselves also identify development guidelines relating to each of the different types of areas, which generally equate to careful consideration by SANParks of any development proposals undertaken in these areas and opposing those which would detrimentally impact on the national park.

Some studies have investigated the effectiveness of buffer zones in terms of certain socio-economic factors, such as tourism and community benefits (Atsri et al. 2020; Whitelaw et al. 2014; Gross-Camp et al. 2015). Few studies have investigated the effectiveness of buffer zone implementation and management, with identified challenges including weak policy; lack of capacity; boundary uncertainties; lack of legal status of buffer zones; lack of financial support; limited stakeholder buy-in; lack of awareness amongst landowners; and lack of clarity on management responsibilities (Heinen and Mehta 2000; Lynach and Urich 2002; Ahmad et al. 2012; Atsri et al. 2020). In addition, complexities associated with multiple decision makers and lack of harmonisation and co-operative governance between conservation agencies and local governments have been identified as compounding these challenges (Wallace et al. 2005; Garcia and Revah 2013).

In the South African context, the four most tangible mechanisms for implementing buffer zones for national parks once reflected in the management plan, amongst those identified in the relevant domestic policy frameworks, are seemingly: the formal establishment of buffer zones through gazetting; the declaration of buffer zones as formally protected environments; subjecting certain identified activities undertaken in buffer zones to the requirement to obtain an environmental authorisation preceded by some form of EIA; and integrating buffer zones within MSDFs. To date, no buffer zones have been formally gazetted. It is unclear under which provision in NEMPAA this would take place, with the only feasible option seemingly being under the broad powers accorded to the Minister to make regulations regarding anything which is 'necessary or expedient … for the proper implementation and administration of the Act' (South Africa 2003, s. 86(1)(g)). No buffer zones around national parks have been declared protected environments, possibly due to challenges associated with obtaining the necessary buy-in from private landowners, communities, and other role-players affected by the buffer zones. Some activities have been listed within buffer zones to national parks triggering the need for compliance with EIA procedures and the need to secure an environmental authorisation under NEMA. The rather discrepant use of the above mechanisms seems to place potentially great significance on the integration of buffer zones into MSDFs.

7.2.5 Influence of Protected Areas and Buffer Zones on Regional and Local Spatial Dynamics

Protected areas in South Africa have had a notable impact on spatial planning and development at both regional and local levels. Their presence has naturally influenced the trajectory of development projects, with activities like mining often forced to avoid these ecologically sensitive zones. This strategic avoidance over time has shaped the spatial configuration of numerous regions across the country. Similarly, these protected areas have acted as magnets for specific types of development, particularly those related to tourism. Businesses and services gravitate towards these areas to capitalise on the ecological and recreational value they offer, leading to a clustering effect that further defines the spatial character of these regions. The introduction and recognition of buffer zones, though a relatively newer concept, can be expected to further nuance this spatial dynamic and by incorporating buffer zones into Municipal Spatial Development Frameworks (MSDFs) and Integrated Development Plans (IDPs), there is an expectation for an even more pronounced influence on the spatial form of municipalities. This inclusion not only underscores the importance of these areas in environmental conservation but also highlights their role as catalysts for guided and sustainable local development. The evolving spatial landscape, shaped by the interplay of conservation and development, reflects a deepening awareness of the need to balance ecological integrity with human progress.

7.3 Municipal Spatial Development Frameworks

Before evaluating of the extent to which national park buffer zones are integrated into MSDFs, it is necessary to briefly outline their general nature, purpose and the legal framework applicable to them.

7.3.1 Nature and Purpose of MSDFs

MSDFs fall within the broad realm of planning law, with this concept perplexing many a scholar regarding its scope, nature and constituent parts. In the words of one scholar, 'no precise and permanent definition … exists' and while 'various attempts at a definition have been made', the definition invariably changes overtime (Van Wyk 2020: 10). A common distinction drawn by scholars to distinguish the range of components embedded in the broad notion of planning, is that between spatial planning components (referred to by some scholars as land management planning) and land use/development planning components (referred to by some scholars as land development planning and/or development control) (Van Wyk 2012, 2020; Viviers et al. 2021). Spatial planning is forward looking and creates 'a blueprint for future land use that will support sustainable development' (Van Wyk 2020: 11). Typical mechanisms associated with spatial planning include MSDFs and IDPs. Spatial planning informs land use/development planning, with the latter defined as 'processes and procedures to change and manage land use, invariably involving decision-making' relating to land use change (such as applications to rezone land or remove title deed restrictions attached to it) and land development (such as applications to establish townships, build structures, subdivide or consolidate land) (Van Wyk 2020: 11).

MSDFs fall within the realm of spatial planning. They are considered forward planning mechanisms aimed at guiding the spatial development of a region (Musvoto et al. 2016) and are the main instruments through which future development and growth are managed within South Africa's municipalities. MSDFs achieve this by indicating, amongst others: directions of growth; areas in which particular types of land uses should be encouraged and others discouraged; and areas in which the intensity of land development could either be increased or reduced.

7.3.2 Domestic Legal Context

The above delineation between spatial planning and land use/development planning is reflected in South Africa's contemporary national planning legislation, the *Spatial Planning and Land Use Management Act* (SPLUMA) (South Africa 2013), with Chap. 4 effectively dealing with the spatial planning component and Chap. 5 with the land use/development planning component. Prior to the introduction of SPLUMA, the *Local Government Municipal Systems Act* (LGMSysA) compelled all municipalities to adopt an IDP, which had to include a MSDF providing basic guidelines for land use management within the municipality (South Africa 2000, s. 25–26). These vague provisions relating to MSDFs have now been significantly elaborated on in SPLUMA, which requires all municipalities to develop MSDFs, with the Act clearly setting out the content to be included in the MSDF and the process to be followed in preparing and adopting it (South Africa 2013, s. 20–21). These provisions indicate that the MSDF must be prepared in terms of the LGMSysA, with it providing for consultation with the local community and relevant organs of state, and the identification of relevant plans for consideration when developing the MSDF (South Africa 2000, s. 27–34). Once developed, the MSDF informs the development of the municipality's land use scheme which sets out the zoning and associated land-use rights and restrictions for all properties falling with the jurisdictional boundary of the municipality (South Africa 2013, s. 24–26). Municipal authorities are also required, when making any land use/development decisions, to ensure that these are consistent with the relevant MSDF (South Africa 2013, s. 22). Therefore, MSDFs have huge potential to inform land use/development decisions within protected area buffer zones in cases where these zones and associated development guidelines have been adequately integrated into the MSDFs.

7.4 Methodology

To determine the extent to which national park buffer zones have been integrated into MSDFs, a sample of MSDF cases was evaluated. Similar case study approaches are commonly employed in studies dealing with the evaluation of policy implementation instruments (see for example Retief 2007; Sandham et al. 2013, 2020; Alberts et al. 2021) and was therefore deemed a valid approach for addressing the aim. The selection of cases and the methodological approach followed for their evaluation is discussed below.

7.4.1 Selection of Cases

A sample of four to ten cases is recommended for research aiming to provide an in-depth evaluation of an issue (Eisenhardt 2002). According to Eisenhardt (2002), it can be difficult to generate theory and results with much complexity and strong empirical grounding with fewer than four cases, while with more than ten cases, it starts to become difficult to deal with the complexity and volume of data. A sample of ten MSDF cases was, therefore, selected for this research based on the following criteria:

- Only MSDFs were considered as this is the sphere of government responsible for land use management.
- Only local municipalities whose jurisdictional boundary intersected with a national park buffer zone were considered.
- Buffer zones—and subsequently the selected local municipalities—had to be representative of distinct national parks.
- The demarcation of the buffer zone had to precede the review date of the MSDF to ensure that it was applicable when the MSDF was reviewed.

The ten selected cases are listed in Table 7.1 and the national parks to which they apply are shown Fig. 7.1. However, in line with the ethical requirements for the research, the data was analysed anonymously without reference to a particular park or municipal context.

In Table 7.2, the land cover composition of the buffer zones for the various protected areas is summarised. While urban development within these zones is notably low, with the highest occurrence in the buffer zone of Kruger National Park at 7.1%, this figure is substantial when considered in the context that only 1% of the global habitable land surface is used for 'urban' purposes (Ritchie and Roser 2019). This comparison underscores the relative significance of even minimal urban development within protected areas' buffer zones, highlighting the need for careful management and planning to balance conservation with the impacts of urbanisation. Cultivated lands are more prominent, particularly in the buffer zones of Bontebok National Park (42.4%) and Agulhas National Park (25.5%). Mining activities are

Table 7.1 Summary of selected MSDFs included in the evaluation

Local municipality	MSDF date	Applicable protected area/s	Buffer date
Beaufort West	2017	Karoo National Park	2013
Cederberg	2017	Tankwa Karoo National Park	2014
Dawid Kruiper	2018	Kalahari Gemsbok National Park	2016
Saldanha Bay	2019	West Coast National Park	2013
Ba-Phalaborwa	2019	Kruger National Park	2018
City of Mbombela	2019	Kruger National Park	2018
Overstrand	2020	Agulhas National Park	2013
Swellendam	2020	Bontebok National Park	2013
Witzenberg	2020	Tankwa Karoo National Park	2014
Dr. Beyers Naude	2021	Camdeboo & Addo National Parks	2015 & 2016

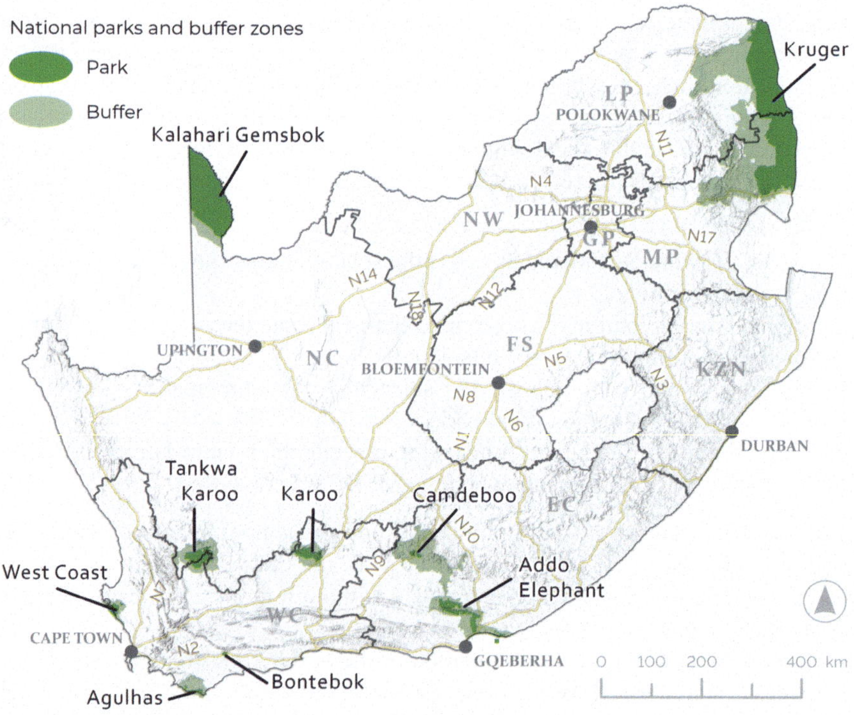

Fig. 7.1 National parks and Buffer zones included in the evaluation (original to the author)

almost non-existent in all buffer zones. When considering "near-natural" land cover, comprising forested, shrubland, grassland, and barren areas, a clear dominance of this category is evident. For instance, the buffer zone of the Kalahari Gemsbok National Park is almost entirely near-natural, with 99.9% of the area falling under this category, mostly as shrubland. Similarly, the buffer zones of Karoo National

Table 7.2 Land cover characteristics

	Land cover class as percentage (%) of buffer zone							
Protected area	Urban	Cultivated	Forested/ Woodland	Shrubland	Grassland	Mining	Barren	Water
Karoo National Park	0.3	0.6	1.1	57.8	7.1	0.0	32.8	0.2
Tankwa Karoo National Park	0.1	1.7	1.1	56.7	2.9	0.0	37.3	0.2
Kalahari Gemsbok National Park	0.0	0.0	0.0	98.4	1.0	0.0	0.5	0.0
West Coast National Park	2.2	12.3	0.7	71.7	0.3	0.2	1.6	11.1
Kruger National Park	7.1	11.4	69.5	0.0	10.3	0.1	0.4	1.2
Agulhas National Park	0.7	25.5	11.8	44.2	1.0	0.0	1.1	15.9
Bontebok National Park	2.5	42.4	11.6	38.7	1.5	0.1	0.9	2.3
Tankwa Karoo National Park	0.4	1.8	8.1	33.9	53.6	0.0	1.3	1.0
Camdeboo & Addo National Parks	1.0	17.7	27.2	23.5	24.4	0.1	5.6	0.6

Park and Tankwa Karoo National Park are overwhelmingly near-natural, with percentages of 98.7% and 97.9%, respectively. The Kruger National Park's buffer zone also has a significant near-natural cover, predominantly woodland, accounting for 81.2% of the land. This pattern underscores the preserved, less disturbed nature of lands surrounding these protected areas, highlighting their significant role in conservation and ecological stability.

7.4.2 Evaluation and Analysis

The following three key performance indicators (KPIs) were developed against which buffer zone and MSDF integration could be evaluated. The KPIs were informed by the requirements of the *Strategy on Buffer Zones for National Parks* (South Africa 2012).

- KPI 1—The extent to which the national park buffer zone is acknowledged and discussed in the MSDF.
- KPI 2—The extent to which the national park buffer zone is spatially reflected in the MSDF through maps and spatial datasets.
- KPI 3—The extent to which the MSDF provides guidance on land use management within the national park buffer zone.

Table 7.3 Evaluation scores and criteria

Evaluation Scores	Score Description		Colour Code
A	Adequately achieved	The score was awarded if the KPI was adequately achieved and if all or most requirements pertaining to the KPI were achieved.	Green
B	Partially achieved	The score was awarded if the KPI was only partially achieved and if some but not all requirements pertaining to the KPI were achieved.	Yellow
C	Not achieved	The score was awarded if the KPI was not achieved and if very few or none of requirements pertaining to the KPI were achieved.	Red

The MSDFs were reviewed against the three KPIs and each KPI was scored against a predefined set of evaluation scores (Table 7.3). Scores of A, B or C were awarded, where A was awarded when the KPI was adequately achieved, B where it was partially achieved and C where it was not achieved.

7.5 Results

The evaluation results are summarised in Table 7.4 and presented and discussed in the following sections structured around the KPIs. The cases in Table 7.4 are intentionally not presented in the same order as listed in Table 7.1 to maintain anonymity of the evaluated cases.

7.5.1 Extent to Which National Park Buffer Zones Are Acknowledged and Discussed in MSDFs (KPI 1)

For buffer zones to be integrated into MSDFs they must be acknowledged and adequately discussed. The applicable national park buffer zones were mentioned by name in only two of the ten MSDF cases (KPI 1 = A for these two cases), a finding

Table 7.4 Summary of results of evaluation of the ten cases against the three KPIs to determine the extent of national park buffer zone integration in MSDFs

KPI	Description	Cases									
KPI 1	The extent to which the national park buffer zone is acknowledged and discussed in the MSDF.	C	B	A	A	C	C	B	C	C	B
KPI 2	The extent to which the national park buffer zone is spatially reflected in the MSDF through maps and spatial datasets.	C	B	B	B	C	C	C	C	C	C
KPI 3	The extent to which the MSDF provides guidance on land use management within the national park buffer zone.	C	B	C	A	C	C	C	C	C	C

similar to that of Lima and Ranieri (2018) in the Brazilian context. The first of these MSDFs explained the purpose and importance of the buffer zone and highlighted its objectives within the context of the area. This understanding was further considered in later sections of the MSDF. The second MSDF, however, acknowledged and described the buffer zone, but did not consider it beyond the MSDF *status quo* analysis. Three of the ten MSDF cases referred to the concept of protected area buffer zones (KPI 1 = B for these three cases). However, they were only discussed in general terms and did not necessarily reflect the cadastral delineated buffer zones as contained in the protected area management plans. These MSDFs often also referred to buffer zones within the context of critical biodiversity areas (CBAs) or wetlands. The remaining five MSDFs made no mention of the protected area buffer zone (KPI 1 = C for five of the ten cases).

7.5.2 *Extent to Which National Park Buffer Zones Are Spatially Reflected in the MSDFs Through Maps and Spatial Datasets (KPI 2)*

Most of the buffer zones for the national parks included in the evaluation distinguished between priority natural areas, catchment protection areas, and viewshed protection areas. These distinctions were, however, not indicated on any of the MSDF maps and although the buffer zones were acknowledged in some of the MSDFs, they were not accurately reflected on MSDF maps. MSDFs generally consist of numerous maps and spatial datasets informing the spatial development vision of the municipality. These maps and datasets analyse and map different environmental attributes ranging from socio-economic to biophysical factors. It can therefore reasonably be expected that a national park buffer zone would at least be reflected on some of these maps, especially those dealing with the biophysical environment. Although this was found not to be the case, in three of the MSDFs parts of the buffer zone, for example sections of the 'priority natural areas' zone, did overlap with features such as critical biodiversity areas (CBAs) that were mapped as part of the biophysical analysis. While these areas still did not accurately reflect the buffer zones, they could contribute—at least in part—to their management (KPI 2 = B in these three cases).

None of the MSDFs made a clear distinction between the type of buffer zone being represented, which has a direct effect on management guidance as the three different types of buffer zones have different objectives. To give effect to its development vision, an MSDF must further include a spatial development plan depicting the spatial development of the municipality for a 5-year period. These plans are meant to indicate the areas that are deemed suitable for different types of land uses, such as residential development, industrial development, and open space systems. If acknowledged in these plans, MSDFs could potentially contribute to the management of buffer zones by guiding land use development decisions within them.

7.5.3 Extent to Which MSDFs Provide Guidance on Land Use Management Within National Park Buffer Zones (KPI 3)

In terms of the *SDF Guidelines* (South Africa 2017), land use management guidelines should accompany an MSDF. These guidelines should inform the land use management system (LUMS) through which land use is managed based on an understanding of land use desirability and compatibility, informing where certain land uses should be promoted, and others discouraged. Clear guidance on land use management within the national park buffer zone was provided by only one of the ten evaluated MSDF cases (KPI 3 = A for only one of the ten cases). This MSDF both recognised the buffer zone and provided guidance on the land uses that should be both encouraged and discouraged. The MSDF further promoted a desired state for the buffer zone that was aligned with both the objectives for buffer zones as discussed in the *Strategy on Buffer Zones for National Parks* (South Africa 2012), and those defined through the specific national park's management plan. This MSDF, therefore, successfully integrated the buffer zone and highlighted it as a priority area that must be actively managed. However, it still did not reflect accurate cadastral boundaries, which might have implications for effective management and decision making.

A second MSDF provided some land use guidance. However, the buffer zone it applied to was not representative of the entire national park buffer zone and represented mostly patches of CBAs that overlapped with some sections of the buffer zone (KPI 3 = B for this case). The remainder of MSDFs did not provide any guidance on the management of land uses within buffer zones (KPI 3 = C for the remaining eight cases), which meant very weak performance in relation to KPI 3.

7.6 Discussion and Conclusions

Within the frame of spatial transformation, MSDFs play a crucial role in reimagining and reshaping the South African landscape. This chapter aimed to assess the degree to which these transformative instruments integrate national park buffer zones, using ten representative MSDF cases tied to nine national parks, evaluated against three KPIs. The majority of MSDFs evaluated failed to integrate the applicable national park buffer zones into their development visions and land use guidelines. The fact that most MSDFs did not even acknowledge the existence of buffer zones is significant (KPI 1). This raises questions about whether local authorities are intentionally ignoring these buffer zones or simply not aware of them. Given the above, it is not surprising that buffer zones were, for the most part, not reflected in MSDFs maps (KPI 2). Although CBAs and other areas of environmental sensitivity mapped in some MSDFs did overlap with applicable protected area buffer zones, and indirectly promoted their management, these only represented small sections of the buffer zones. Guidance on land use objectives within national park buffer zones were, therefore, mostly absent from MSDFs (KPI 3). A direct effect of this is that the majority of MSDFs were neither promoting land uses compatible with the objectives of the buffer zone, nor discouraging ones that were deemed incompatible. To this extent, it would be challenging to achieve a desired state for these buffer zones as land use change and development applications would not be considered within the applicable buffer zone context.

Historically, post-apartheid spatial planning has grappled with inclusivity and equitable representation. The oversight of buffer zones might be symptomatic of larger challenges faced in actualising transformational objectives. The under-representation and omission of buffer zones might, unintentionally, perpetuate past spatial injustices by neglecting vital areas meant for environmental protection.

While exploring the underlying reasons for such omissions exceeds the purview of this chapter, it's a topic demanding attention in future studies. Assessing the status quo of national park buffer zones and gauging their practical viability against the backdrop of broader spatial transformation goals will be pivotal. It might be that some buffer zones are too expansive or poorly considered against MSDF objectives, and are subsequently being ignored by local authorities. Reviewing buffer zones through a consultative process with the relevant local authorities should provide insight into reasons as to why buffer zones have not been integrated into MSDFs; and should further also contribute to the buy-in needed to facilitate integration in future. This consultative process could take significant time. What other avenues could be explored in the short to medium term to contribute to the more effective implementation of national park buffer zones?

Firstly, the government could ensure that buffer zones are officially gazetted. This status should promote their uptake into spatial planning instruments such as MSDFs and their consideration in decision-making processes. The question in this regard is who is empowered to gazette them and in terms of what legislation. As highlighted previously in this chapter, one apparent option would be for the Minister to do so exercising her broad powers to make regulations relating to anything necessary or expedient to implement NEMPAA.

Secondly, where national parks are threatened by large scale developments or activities undertaken within the buffer zone or even further afield, and applying the risk averse and cautious approach, the Minister could be requested to exercise her powers in terms of NEMA to prohibit or restrict a list of specified activities taking place in a specified geographical area adjacent to the national park (South Africa 1998, s. 24(2A)). This would have the effect of nullifying any existing applications and precluding any similar ones.

Thirdly, the Minister of Rural Development and Land Reform could be encouraged to propose amendments to the prescribed content to be included in a MSDF as outlined in SPLUMA, to refer more expressly to protected area buffer zones, thereby hopefully ensuring their express reflection in MSDFs. Currently, the prescribed content for a MSDF includes a "strategic assessment of the environmental pressures and opportunities within the municipal area, including the spatial location of environmental sensitivities" (South Africa 2013, s. 21(j)). While this could naturally include protected area buffer zones, perhaps more express reference to the concept would preclude their omission. An alternative to this onerous legislative reform, could be to achieve the above improved clarity and integration by way of tweaking SPLUMA's regulations or the *SDF Guidelines* (South Africa 2017) to more expressly refer to the need to factor in protected are buffer zones within MSDFs.

Fourthly, perhaps the annual reporting by SANParks to the Minister on adherence to the *Norms and Standards for the Management of Protected Areas*, inclusive of their standards and indicators dealing with buffer zones, could be used as an opportunity to raise awareness among local authorities. This could feasibly be achieved by forwarding the annual reports through to relevant local authorities for consideration or meeting with them on an annual basis to discuss the content of these reports and associated implications.

Fifthly, when seeking to review national park buffer zones alongside the review of the management plan, SANParks could actively engage all relevant local authorities thereby potentially facilitating awareness and buy-in, and feasibly resolving possible conflicts with local development objectives in a pro-active manner. As previously mentioned, SANParks is under a statutory obligation to consult with all local authorities in the context of developing and reviewing management plans, and may well be doing so, but consultation and active and meaningful engagement can constitute very different things.

Sixthly, SANParks, as an interested and affected party, could more actively contribute to the EIA process and thereby feasibly influence decision making in respect of applications for environmental authorisations to undertake certain activities within national park "buffer areas" outlined in NEMA's *Listing Notice 3*. This approach, however, will not cover all buffer areas as some national park buffer zones extend beyond the threshold specified in Listing Notice 3 (>10 km). To further enhance the consideration of national park buffer zones during EIA screening processes, attempts could be made to have national park buffer zones (or at least some sections) reflected as areas of critical importance in conservation plans and bioregional plans. This would effectively mean that any development proposal within these areas would trigger an EIA. Furthermore, introducing biodiversity management plans for ecosystems falling within the national parks buffer zone, in terms of NEMBA, could add a layer of important relevant considerations in the context of any application for a land-use change, land development or other environmental approval.

Finally, the Government should task dedicated staff within SANParks with buffer zone management. These representatives should represent SANParks during IDP and MSDF development processes, engage with local authorities on a regular basis, monitor activities within buffer zones, act as interested and affected parties during EIA processes, and promote the integration of buffer zones into other decision support instruments such as bioregional plans, biodiversity management plans and environmental management frameworks. This should contribute to directing multiple policy instruments towards the protection of buffer zones.

Considering the focus on territorial governance and spatial transformation, it becomes essential at this juncture to explicitly underline the integral relationship between the recommendations provided above and broader issues of territorial governance. These suggestions, while diverse in nature, collectively aim to strengthen the governance frameworks surrounding national park buffer zones. They not only seek to address the immediate concerns of environmental conservation but also touch upon the larger canvas of territorial governance. This involves enhancing the legal and policy environment, fostering greater intergovernmental cooperation, and ensuring community engagement and participation in spatial planning processes.

Reflecting on the potential limitations of this research, it's important to acknowledge that the study's focus on a limited number of MSDFs and specific Key Performance Indicators (KPIs) may not capture the full spectrum of practices and attitudes towards buffer zones across South Africa. Furthermore, the analysis might be influenced by the evolving nature of spatial planning and policy frameworks. This research also does not delve deeply into the practical challenges faced by local authorities in integrating buffer zones into MSDFs, which could range from resource constraints to conflicting land use priorities.

In conclusion, the incorporation of national park buffer zones into MSDFs symbolises more than mere environmental conservation. It's an embodiment of South Africa's broader aspirations for spatial transformation and territorial governance, post-apartheid. The present shortfall highlights the urgency to reorient our strategies and ensure that these zones, representative of our commitment to the environment, don't get marginalised in the grand narrative of South African transformation.

References

Ahmad CB, Hashim IHM, Abdullah J, Jaafar J (2012) Stakeholders' perceptions on buffer zone potential implantation: a preliminary study of Tasek Bera, M'sia. Proc Soc Behav Sci 50:582–590

Alberts RC, Retief FP, Cilliers DP, Roos C, Hauptfleisch M (2021) Environmental impact assessment (EIA) effectiveness in protected areas. Impact Assess Proj Apprais 39(4):290–303

Atsri KH, Abotsi KE, Kokou K, Dendi D, Segniagbeto GH, Fa JE, Luiselli L (2020) Ecological challenges for the buffer zone management of a West African National Park. J Environ Plan Manag 63(4):689–709
Calderón MM, Ávila EB, Chaves AL, Jiménez EC, Lozano CB (2022) Anthropization in buffer zones of protected areas. Ann Tour Res Emp Insights 3:100072
De la Fuente B, Bertzky B, Giacomo D, Mandrici A, Conti M, Florczyk AJ, Freire S, Schiavina M, Bastin L, Dubois G (2020) Built-up areas within and around protected areas: global patterns and 40-year trends. Glob Ecol Conserv 24:e01291
Du W, Penabaz-Wiley SM, Njeru AM, Kinoshita I (2015) Models and approaches for integrating protected areas with their surroundings: a review of the literature. Sustainability 7(7):8151–8177
Eisenhardt K (2002) Building theories from case study research. In: Huberman A, Miles M (eds) The qualitative researcher's companion, 1st edn. Sage Publications, pp 5–36
Garcia DAD, Revah IO (2013) La reserve de la biosfera alto golfo de California y delta del Rio: planceación territorial. Región y Sociedad 25:57–85
Gross-Camp ND, Martin A, McGuire S, Kebede B (2015) The privatization of the Nyungwe National Park buffer zone and implications for adjacent communities. Soc Nat Resour 28(3):296–311
Heinen JT, Mehta JN (2000) Emerging issues in legal and procedural aspects of buffer zone management with case studies from Nepal. J Environ Dev 9(1):45–67
Jusys T (2016) Quantifying avoided deforestation in pará: protected areas, buffer zones and edge effects. J Nat Conserv 33:10–17
Lima EACF, Ranieri VEM (2018) Land use planning around protected areas: case studies in four state parks in the Atlantic Forest region of south-eastern Brazil. Land Use Policy 71:453–458
Lynach FM, Urich PB (2002) A critical review of buffer zone theory and practice: a Philippine case study. Soc Nat Resour 15(2):129–145
Martino D (2001) Buffer zones around protected areas: a brief literature review. Electr Green J 1(15):1–19
Musvoto G, Lincoln G, Hansmann R (2016) The role of spatial development frameworks in transformation of the eThekwini Municipality, KwaZulu-Natal, South Africa: reflecting on 20 years of planning. Urban Forum 27:187–210
Nadhira S, Basuni S (2021) Implementation of the concept of conservation area buffer zone in Indonesia. J Manajemen Hutan Trop 27(1):32–41
Perelló LFC, Guadagnin DL, Maltchik L, Dos Santos JE (2012) Ecological, legal, and methodological principles for planning buffer zones. Natureza Conservação 10(1):1–9
Retief F (2007) A performance evaluation of strategic environmental assessment (SEA) processes within the South African context. Environ Impact Assess Rev 27(1):84–100
Ritchie H, Roser M (2019) "Half of the world's habitable land is used for agriculture" Published online at OurWorldinData.org. Retrieved from: 'https://ourworldindata.org/globalland-for-agriculture' [Online Resource]
Robinson EJZ, Albers HJ, Busby GM (2013) The impact of buffer zone size and management on illegal extraction, park protection, and enforcement. Ecol Econ 92:96–103
Sandham LA, Van Heerden AJ, Jones CE, Retief FP, Morrison-Saunders AN (2013) Does enhanced regulation improve EIA report quality? Lessons from South Africa. Environ Impact Assess Rev 38:155–162
Sandham LA, Huysamen C, Retief FP, Morrison-Saunders A, Bond AJ, Pope J, Alberts RC (2020) Evaluating environmental impact assessment report quality in South African national parks. Koedoe Afr Prot Area Conserv Sci 62(1):1–9
Sayer J (1991) Rainforest buffer zones: guidelines for protected areas manager. Forest Conservation Program, IUCN
South Africa (1997) White paper on the conservation and sustainable use of South Africa's biodiversity. Government Notice No. 1095, Government Gazette 18163, July 1997
South Africa (1998) National Environmental Management Act (Act no. 108 of 1998). Government Printer, Pretoria
South Africa (2000) Municipal Systems Act (Act no. 32 of 2000). Government Printer, Pretoria

South Africa (2003) National Environmental Management: Protected Areas Act (Act no. 57 of 2003). Government Printer, Pretoria

South Africa (2004) National Environmental Management: Biodiversity Act (Act no. 10 of 2004). Government Printer, Pretoria

South Africa (2005) Regulations for the proper administration of special nature reserves, national parks and world heritage sites, Government Notice No. 1060. Government Gazette 28181, October 2005

South Africa (2012) Biodiversity policy and strategy for South Africa: strategy on buffer zones for national parks. Government Notice No. 106, Government Gazette 35020, February 2012

South Africa (2013) Spatial Planning and Land Use Management Act (Act no. 16 of 2013). Government Printer, Pretoria

South Africa (2014a) Listing Notice 3: List of activities and competent authorities identified in terms of Sections 24(2) and 24D. Government Notice No. 985, Government Gazette 38282, December 2014

South Africa (2014b) Environmental impact assessment regulations. Government Notice No. 982, Government Gazette 38282, December 2014

South Africa (2014c) Norms and standards for biodiversity management plans for ecosystems. Government Notice No. 83, Government Gazette 37302, February 2014

South Africa (2016) Norms and standards for the management of protected areas in South Africa. Government Notice No. 382, Government Gazette 39878, March 2016

South Africa (2017) SDF guidelines: guidelines for the development of provincial, regional and municipal spatial development frameworks and precinct plans. Department of Rural Development and Land Reform

South Africa (2023) White paper on the conservation and sustainable use of South Africa's Biodiversity. Government Notice No. 3537, Government Gazette 48785, June 2023

Van Wyk J (2012) Planning law, 2nd edn. Juta, Cape Town

Van Wyk J (2020) Planning law, 3rd edn. Juta, Cape Town

Viviers J, Retief F, Cilliers D (2021) Municipal integrated development planning. In: Du Plessis A (ed) Environmental law and local government in South Africa, 2nd edn. Juta, Cape Town

Wade AA, Theobald DM (2010) Residential development encroachment on U.S. protected areas. Conserv Biol 24(1):151–161

Wallace GN, Barborak JR, Mac Farland C (2005) Land use planning and regulation in and around protected areas: a study of legal frameworks, best practice and capacity building needs in Mexico and Central America. Brazil J Nat Conserv 3:147–167

Wells M (1992) Biodiversity conservation, affluence and poverty: mismatched costs and benefits and efforts to remedy them. Ambio 21(3):237–224

Whitelaw PA, King BE, Tolkach D (2014) Protected areas, conservation and tourism–financing the sustainable dream. J Sustain Tour 22(4):584–603

Chapter 8
Territorial Governance and Transformation of South Africa's Urban Spatial Economy

Cecil Madell

Abstract This chapter explores and reflects on the role of territorial governance and public policies in transforming the spatial economy of urban areas in South Africa during the post-apartheid era. Contemporary South African cities and towns are reproducing the legacy of colonial and apartheid patterns of economic exclusion and inequality, albeit at an industrial scale, perpetuating the entrapment of the majority of urban inhabitants in dormitory townships, devoid of meaningful economic opportunities, underserved in terms of basic utility services, public transport and critical social services and remotely located from established urban centres. This chapter reflects on why spatial and economic policy interventions by the state, inadvertently obstruct the poorest residents of towns and cities from accessing economic and other opportunities, even though South Africa's constitution and plethora of legislation are geared at building a socially, economically and spatially just and inclusive society. This critique aims to unearth the implications and contradictions in the state's policy and legislation over the last 30-years and reflection on the required changes to the state's approach to urban development and territorial governance to advance economic transformation of South Africa's urban spatial economy.

Keywords Territorial governance · Spatial transformation · Urban spatial economy · Economic development

---

C. Madell (✉)
City and Regional Planning, University of Cape Town, Cape Town, South Africa
e-mail: cecil.madell@uct.ac.za

D. du Plessis (ed.), *Territorial Governance and Spatial Transformation in Post-apartheid South Africa*, Local and Urban Governance,
https://doi.org/10.1007/978-3-031-85489-7_8

8.1 Introduction

Africa's population was estimated at 1.2 billion by 2018 and projected to double to 2.5 billion by 2030. South Africa, Africa's most industrialised country, had an estimated population of 62 million by 2022, with 68% living in urban areas with this figure predicted to increase to 66 million, with an urbanisation rate of 71.3% by 2030 (RSA 2022b). This pattern of rapid urbanisation in Africa and South Africa is inevitable, more so in that cities and towns are the epicentres and drivers of economic growth and productivity as they benefit from economies of scale and agglomeration (Un-Habitat et al. 2008; UN-Habitat and Rotterdam 2018).

South Africa is facing daunting and catastrophic problems of unemployment, poverty and inequality. Over the period 2010–2017, real GDP growth averaged 2.0% per year, significantly short of the targeted rate of 5% required on a sustained annual basis to meaningfully address these challenges. At 0.6%, GDP growth in 2016 was the lowest since the 2009 recession (RSA 2018a). Unemployment in South Africa has become chronic (i.e. structural unemployment), rather than acute or cyclical (UNDP 2005). The official unemployment rate reached an all-time high of 36.4% in 1999, dipping below 25% between 2006 and 2015, and reaching a low of 21.5% at the onset of the global financial crisis during 2008. Over the past decade the overall unemployment rate has risen continuously, moving towards 27.7% during 2017, the highest in 14 years, reflecting the inability of the economy to create sufficient jobs for an expanding labour force. There was a modest decline to 26.7% during 2017, however, unemployment levels remain unacceptably high, at 29% during 2019. Since then, it has been on the increase for at least a decade, back to the high of 27.1% in 2018. On average over the 25 years, it has hovered consistently around 25% (RSA 2019). The official unemployment rate was 31.9% in the third quarter of 2023 (RSA 2023).

South Africa is the most unequal country in the world, ranking first among 164 countries (World-Bank 2022). Contemporary South African metropolitan areas, secondary cities, small towns, rural areas and former Bantustans[1] reflect patterns of spatial development and economic exclusion emanating from colonialism and apartheid, forms of development inadvertently reinforced by the post-1994 democratic state. South Africa's urban development challenges are explicit spatially, with most inhabitants trapped in dormitory townships, located on the outskirts of towns and cities, devoid of meaningful economic opportunities, remotely located from established economic centres and underserved in terms of basic utility services and public transport. In contrast, wealthy residents live closer to historic urban centre, characterised by the clustering of economic activity, leafy suburbs and gated communities; and are serviced with better education, health, transport and other public facilities and services. Although significant strides have been achieved by the state

[1] Bantustans (also known as Bantu homeland, black homeland) were territories, mainly economically marginal rural areas, set aside by the apartheid government for different ethnic groups as part of their divide, rule and oppression strategy.

over the last three decade, in terms of delivering subsidised 'free' housing, health care, basic education and social grants; spatial and economic apartheid stubbornly persists. These patterns of underdevelopment and duality restrict economic growth and development, entrench social and economic exclusion, and reproduce poverty traps, while the immense inequality remains a source of discontent amongst South Africans.

South Africa's history of political oppression, spatial segregation and economic exclusion of the majority of black[2] inhabitants can be traced back to its colonial and apartheid eras. 'Segregation in housing, education, health, welfare, transport and employment left deep scars of inequality and economic inefficiency' (RSA 1994):7). The UN declared apartheid as crime against humanity during 1966 and this was followed by decades of international economic sanctions and isolation and rising unrest in townships throughout the country. During this period, the Freedom Charter, drawn up by anti-apartheid forces, was seen as a blueprint for a future democratic state '...there was sufficient "political hubris" within these organisations ... to believe that the post-apartheid dispensation would be at least be strongly Keynesian[3] in its leanings, if not even socialist' (Mc Donald 2008: 76). The period 1990–1994 was critical for South Africa, with the unbanning of liberation organisations and the onset of constitutional negotiations. In terms of the urban spatial economy, racial, economic, and spatial segregation were entrenched countrywide in towns and cities, the urban areas were seen as discrete elements that can be separated and delinked. Non-whites were forcefully removed from urban centres and relocated to the edge of towns and cities. Infrastructure and businesses in established urban centres were dominated by those classified as white, with urban employment opportunities for the black labour force centred on decentralised industrial areas and domestic work in suburbs.

Central to addressing South Africa's multi-dimensional and complex urban problems, is the need to spatially and economically transform colonial and apartheid created cities and towns. Internationally, spatial transformation is often associated with the physical conversion of neighbourhoods, towns and cities through processes such as urban regeneration and renewal. However, unlike other countries, spatial apartheid was formally adopted and enforced in South Africa with dire consequences for most of its population. Harrison and Todes (2015: 160) observed that in post-apartheid South Africa there is a deliberate '...attempt to create a new spatial order through restructuring cities towards a more compact and integrated urban form, offering better access to urban opportunities for those previously excluded, and greater efficiencies, as well as reducing the economic marginality of township'.

Oranje and Van Huyssteen (2011: 6) argued that 'transformation', is "concerned with restructuring the spatial economy from the national to the local level, with the

[2] The term 'black' in this article refers to all races not classified as 'white' in terms of apartheid legislation.

[3] John Maynard Keynes (1883–1946) advocated an active role for the state in the economy through monetary policy i.e. lowering interest rates to stimulate spending and deficit spending in order to increase demand and boost the economy (Peet and Hartwick 2009)

aim of ensuring a very different economic and social outcome, through the pursuit of shared, sustainable, equitable and inclusive growth". The National Spatial Development Framework (NSDF) (RSA 2022a: 20) described 'spatial transformation' as "the carefully planned and well-managed process of placing infrastructure, social services and economic activities in settlements in such a way that the segregated spatial patterns inherited from colonial and apartheid times are broken down, and the spatial inefficiencies, injustices and inequalities in access to opportunities resulting from these past patterns are corrected". The South African Local Government Association defines spatial transformation as "…the overhaul of an inherited segregated spatial dispensation and preventing it from re-establishing itself in new forms of post-apartheid class and income-based spatial segregation and spatial inequality" (SALGA 2012).

In response to these problems the state has formulated a wide range of urban development and economic legislation, policies and strategies aimed at creating more sustainable and economically inclusive towns and cities. Notably amongst these include the Reconstruction and Development Programme (RSA 1994), the Growth, Employment and Redistribution Macro-economic Strategy (GEAR) (RSA 1996b), the Accelerated and Shared Growth Initiative for South Africa (AsgiSA) (RSA 2006a), the New Growth Path (NGP) (RSA 2010), the National Development Plan (RSA 2011b) and the National Spatial Development Framework (RSA 2022a). However, the 10, 15, 20, 25 and 30-year nation Review Reports by the state conceded that meaningful spatial and economic transformation has not been happening within cities, towns, regions and rural areas in South Africa (RSA 2003, 2009b, 2014, 2019, 2024).

This chapter considers the intent and outcomes of state legislation, policies and strategies aimed at spatial economic transformation in South African towns and cities from the perspective of territorial developmental governance. It aims to respond to the research question, why state interventions, have seemingly no meaningful impact on reversing patterns of economic exclusion of post-apartheid South African towns and cities. The purpose is to identify reasons for the disconnection between the machinery of the state, its policies, the impact of private property developments, and the persistence of economic exclusion. The need to transform the spatial economy is inextricably linked to transforming the urban form and structure of cities and towns in South Africa and therefore the spatial policy interventions are also considered in terms of their impact on economic inclusion. A critical review of policies and legislation relevant to spatial and economic transformation is conceptualised within the context of the urban space, framed in relation to the state and territorial governance; market forces and property development; and the impact on society at large. More emphasis is placed on the economic dimension of spatial transformation, as the latter is covered more substantially in the other chapters. This critique draws on urban planning and economic development theory and practice. This review and analysis undertook a chronological assessment of landmark spatial and economic legislative and policy shifts by the post-1994 South African Government.

The review showed that the emphasis during the first decade post-1994 was on 'how' to achieve 'sustainable human settlements' through 'developmental local government'. Subsequently, in response to significant change not happening, the contemporary question is rather 'why spatial transformation has not been achieved?' Comparably, the initial RDP period focussed on reconstruction and poverty reduction, with a shift to neoliberal macroeconomic growth, then to 'radical economic transformation' and more recently, back to market driven economic policies with social welfare support. The review focusses on the implications of state interventions at city and town level, the arena where the spatial and economic impact have the most relevance. The landmark national and international policies reviewed, explored the nature and scope of how the legacy of spatial and economic apartheid have been framed, the search for an alternative spatial and economic trajectory for cities and towns, proposals to realise the interventions; and the implementation track record of these over the last three decades. This critique aims to unearth the implications and contradictions in the state's policy and legislation over the last 30-years and reflection on the required changes to the state's approach to urban development and territorial governance to advance economic transformation of South Africa's urban spatial economy.

8.2 Global Rise of Neoliberalism, Nation Building and Reconstruction in South Africa

The period is characterised by tectonic global events that had enormous impact on the collapse of the apartheid system in South Africa. The Global rise of neoliberalism, nation building and unset of reconstruction in South Africa for the period 1990s to 2006 is captured in Table 8.1.

8.2.1 Global Rise of Neoliberalism (1990s–2006)

The election of Margaret Thatcher as Prime Minister of the UK (1979–1990) and Ronald Reagan as US President (1981–1989) presented the rise of the 'new right' in the west. This period of neoliberalism was characterised by accelerated globalization, 'market fundamentalism' and 'cosmopolitan consumerism'. Harvey observed that neoliberalism in political and economic practices, result in deregulation, privatisation, and withdrawal of the state from many areas of social provision (Harvey 2007). Critics such as Lovering (2009), pointed out the consequence of the height of neoliberalism economic policies, including economic instability spearheaded by the rise of the value of, property as asset value, driven by a global wave of speculative investment in financial assets and land in the global north and the subsequent crash of financial institutions. He argued that the recession pushed up to 200 million people into extreme poverty, with the sharpest rise in unemployment in advanced economies (Lovering 2009). However, despite this consequence, many governments, including those in the global south, have continued to respond with

Table 8.1 Global rise of neoliberalism, nation building and reconstruction in South Africa

Global rise of neoliberalism—1990s–2006	RSA nation building, reconstruction and development (Mandela Presidency: 1994–1999)	
Global events and economic influences	RSA governance, legislative and policy shifts	Impact on urban spatial economy
• *"The New Right" (Thatcher 1979–1990 and Reagan 1981–1989).* • *Accelerated globalization-neoliberalism—based on 'market fundamentalism' and cosmopolitan consumerism.* • *Revival of the concept of the 'city-region' as an urban economic hub.* • *Competitive Advantage Theory (1980s–2007)* • *Michael Porter's Competitive Advantage Theory. Clusters and Agglomeration* (Porter 1998) • *Competitive Place Theory* • *'Modes of organization' of production, shift from Fordist mass production to flexible production systems.* (Storper and Walker 1989) • *Exclusion of the poor, growing inequalities.* • *Resistance to Washington Consensus—economic structural reform to increase the role of the market and the recoil of the state.*	Constitution of RSA (1996) *Governance* • First National Election, 1994 • Est. Government of National Unity (GNU), 1994. • The White Paper on Reconstruction and Development (RDP), 1994. *Spatial development* • Development Facilitation Act (DFA), 1995. • White Paper on Local Government, 1998. *Economic development* • Growth, Employment and Redistribution Strategy (GEAR). Macro-economic Strategy, 1996. • Local Economic Development Discussion Document, 1995. • White Paper on National Strategy on the Development and Promotion of Small Business in SA, 1995.	• Onset of large-scale, rapid migration from rural areas to towns and cities following abolishment of apartheid legislation. • Fragmented governance institutionally, racially and spatially. • No significant impact of DFA principle to promote residential and employment opportunities in close proximity to each. • Critique and resistance against industrial decentralisation and growth points—viewed as 'lame ducks' and 'castles in the desert'. • Focus on restructuring major cities, e.g., metropolitan spatial development plans for Cape Town, Joburg and Durban. • Growing informal economic activities in townships and acceptance in former 'white' CBDs.

conservative economic development policies, with emphasis on the central role of the market.

The concept of the 'city-region' as an urban economic hub was revived, fuelled by approaches such as a shift from 'Fordist' production systems to 'flexible' production systems and Michael Porter's theory on competitive advantage, economic clusters, and agglomeration (Porter 1998). Other related theories include competitive place theory, viewing places as market-based commodities (a shift from urban design emphasis on public spaces) and modes of organisation' of production (Storper and Walker 1989). However, this era was at the same time typified by growing inequalities and the exclusion of the poor in inner city areas in the west and capitals in the former colonies, and resistance to the 'Washington Consensus', with its recommendations on economic structural reform to increase the role of the market and the recoil of the state. This period eclipsed both the Mandela and Mbeki presidency in South Africa.

8.2.2 *Nation Building and Reconstruction in South Africa (Mandela Presidency, 1994–1999)*

Following the conclusion of the multi-party negotiations at Kempton Park, Johannesburg, the first post-apartheid national election was held in 1994 and won by the African National Congress in alliance with the COSATU and the SACP.

8.2.2.1 Governance, Legislation and Policy Shifts

The Mandela Presidency was characterised by reconsolidation, nation building and a government of national unity. The new Constitution of the Republic of South Africa (RSA 1996a) states that the objective of local government is 'to promote social and economic development'. The *Reconstruction and Development Programme* (RDP) (RSA 1994) was successful in establishing the developmental and transformation agenda of the post-apartheid state. (Marais 1998):179) noted that the RDP was a "…mixed 'neo-liberal' prescription with some residual Keynesian regulation, corporatist processes with a 'people-driven' approach, ostensibly from commitments to redistribution with stern macro-economic strictures". No specific reference was made to 'spatial intervention nor spatial transformation' in the RDP. Competition and conflict between the RDP office in the Presidency and line function ministry contributed to its eventual closing down during 1996 and the resultant legacy of national departments working in silos, continues up to today.

The Development Facilitation Act (DFA) (RSA 1995a) included the principles that residential and employment opportunities should be in close proximity to or integrated with each other; and the development of more compact towns and cities. *The White Paper on Local Government* (RSA 1998b) was aimed at establishing spheres of government, however, this presented challenges for the state to drive much needed change such as spatial and economic transformation in response to decades of colonial and apartheid development. For example, land use change can't merely be "authorised" from national government (as was to case when apartheid was enforced through centrally controlled, top-down 'hierarchy' of governance), but rather must be 'negotiated and agreed to' through often lengthy planning, environmental, heritage and other approval processes. *The White Paper on Local Government* noted (in accordance with neoliberalism) that local government is not directly responsible for creating jobs, but should rather ensure a conducive environment for the creation of employment opportunities.

During the period 1996–2000, the ANC's ideological shift towards neoliberalism was confirmed with the introduction of the *Growth, Employment and Redistribution* (GEAR) (RSA 1996b) macro-economic strategy (Turok 2008). Watson (2002) noted that critics immediately labelled GEAR as 'neoliberal' and 'cautious Thatcherism'. Bond (2005) argued that GEAR adhered to the prescriptions of the 'Washington Consensus', i.e. rigid 'neoliberal', while Hart (2008: 678, 691) commented that GEAR was regarded as "…non-negotiable, because of globalisation there was no alternative" and shift could be regarded as a "home-grown version of structural adjustment". The outcomes of GEAR were mixed, with improved financial discipline and macroeconomic stability, however, formal employment continued to decline, poverty remained massive, and wealth remained unevenly distributed along racial lines. The *White Paper on the National Strategy on the Development and Promotion of Small Business in SA* (RSA 1995c) identified constraints facing small enterprises and these include the legal and regulatory environment, access to markets, finance and affordable and quality business premises and associated infrastructure in poor areas.

With the *Local Economic Development Discussion Document* (RSA 1995b) (DPLG) the RDP office, DPLG ministries and organisations such as SANCO viewing local economic development from a state led poverty reduction perspective, while other such as the Department of Trade and Industry (DTI) and the Urban Foundation argued for a market-led approach. A reflection during this period was "…if transformation can be achieved either radically through measures such as nationalisation where political control would be transposed into economic power, or via a more liberal route involving the promotion of macro-economic development and the growth of company earnings?" (Mosala et al. 2017: 327). Initial local economic theorists such as Nel, Pieterse, Rogerson and Tomlinson argued for strategies that will lead to poverty reduction and economic inclusion (Nel 1997; Pieterse 1998). Early academic writings on local economic development made no reference to a substantive the body of international theory and knowledge that already existed, and it appeared that South Africa was developing its own theory, as an extension of theories on poverty reduction.

8.2.2.2 Impact on Urban Spatial Economy

The immediate post-1994 period in South Africa was characterised, by the onset of large-scale, rapid migration from rural areas to major urban centres. Although GEAR as macro-economic strategy improved macro-economic stability, following the crisis at the end of apartheid, no significant attention was given to restructuring the spatial economy. While the industrial decentralisation points associated with the homeland system were criticised as 'lame ducks' and 'castles in the desert', state support remained for some of the industrial areas closer to established major urban centres (Todes and Turok 2017). The DFA principle to promote residential and employment opportunities in close proximity to each had no significant impact on the spatial economy. Some major cities, notably, Cape Town, Johannesburg and Durban initiated spatial restructuring plans while the Kempton Park talks were underway.

8.2.3 Consolidation and Developmental Government (Mbeki Presidency. 1999–2004 and 2004–2008)

The legacy of the Mbeki Presidency is embodied in continuation of neoliberal macro economic policies, consolidating, transforming, and establishing the legislative and administrative governance of the post-apartheid state (See Table 8.2).

Table 8.2 Consolidation and developmental government (Mbeki Presidency. 1999–2004 & 2004–2008)

Reconstruction, consolidation and developmental government (First 1999–2004 and Second Mbeki Presidency: 2004–2008)	
RSA governance, legislative and policy shifts	Impact on urban spatial economy
Governance • Second National Election, 1999 • Local Government Municipal Systems Act, 2000. • Towards a Ten-Year Review, 2003. • Third National Election, 2004. *Spatial development* • National Urban Renewal Programme (URP) (2001). • Breaking New Ground: Integrated Sustainable Human Settlements, 2004. • National Spatial Development Perspective (NSDP), 2006. *Economic development* • Endorsed of GEAR in 1st Mbeki term. • Accelerated and Shared Growth Initiative for South Africa (AsgiSA), 2006. • Local Economic Development: Refocusing Development on the Poor, 2001. • National Policy and Strategy for Local Economic Development, 2004. • National Framework for Local Economic Development in South Africa. 2006–2011, 2006. • Extended Public Works Programme (EPWP), 2002 • The Integrated Small Business Strategy, 2005 • A National Industrial Policy Framework, 2007	• Focus on urban renewal strategies for selected large townships, no significant economic strategies included. • Continuation of unstructured city, irrespective of new forward plans. • Private sector driven new towns, housing estates and freeway driven malls, following the logic of the distorted urban land market. • Informal and illegal viewed as transitional—clampdown on 'informal and illegal'. • Flight of private sector investments from inner city to 'edge cities', Century City, Sandton. • Public sector 'created' jobs, temporary. • Unemployment became chronic (i.e. structural unemployment), rather than acute or cyclical. • Shift in local economic development from a focus on 'poverty reduction in townships' to 'pro-market' approach. • More in line with the state's market macro-economic policy and trade and investment small business and sector policies. • Onset of electricity 'loadshedding' and negative impact on the economy and jobs creation.

8.2.3.1 Governance, Legislation and Policy Shifts

This period was characterised by the reconstruction of national government departments, consolidation of apartheid created municipalities into a system of wall-to-wall municipalities and a focus on establishing developmental local government. Landmark legislation and policies aimed at driving a transformation agenda included *the Local Government: Municipal Structures Act* (RSA 1998a) and the *Local Government Municipal Systems Act* (RSA 2000). *The National Urban Renewal Programme* (URP) was launched in 2001 and focussed on 8 large townships located in metropolitan cities. Although National Government made funding available for planning, implementation of the URP was dependent on line-function departments re-prioritising their budgets, consequently no significant outcomes are evident from this initiative. *Breaking New Ground (BNG)* (RSA 2004a: 16 policy) advocated "…pursuing a more compact urban form, facilitating higher densities, mixed land use development, ...adequate access to economic opportunities, and integrating land use and public transport planning, (and) reducing travelling distances". BNG noted that the inequalities and inefficiencies of the apartheid spatial economy persisted, evident by the continuous provision of low-income housing on the periphery of cities and towns.

The *National Spatial Development Perspective* (NSDP) (RSA 2006c) argued that government spending on fixed investment should be focused on localities of economic growth and/or economic potential. In localities where there are both high levels of poverty and economic potential, the focus should be on fixed capital investment, while in localities with low economic potential the focus should concentrate on human capital development, social grants, and poverty-relief programmes. Future settlement and economic development opportunities should be channelled into activity corridors and nodes that are adjacent to or that link the main growth centres. The NSDP was "...never used as envisaged and fell out of favour, ...it was criticised as being neo-liberal, urban-focused and anti-rural, and taking a too narrow a view of development potential" (RSA 2022a : 28). This reflected the rural biasness of the political elite, a recurring pattern when policies are seen to be urban focussed. GEAR and its pro-market macroeconomic approach was retained during the first Mbeki presidency and was replaced by the *Accelerated and Shared Growth Initiative for South Africa* (AsgiSA) (RSA 2006a), with the mandate to halve poverty and unemployment by 2014. AsgiSA included policies such as inflation targeting, floating exchange rate and a growth target of 6%, however, growth averaged only at 3% between 1994 and 2004.

The Local Economic Development Policy Paper: Refocusing Development on the Poor (RSA 2001) argued for both a poverty relief and an economic growth strategy. This approach was criticised as '…akin to social work rather than the building of the competitiveness of localities' (Meyer-Stamer 2002). Similarly, the *Extended Public Works Programme* (EPWP), 2002 was aimed at creating temporary jobs and not significant skill development required by the targeted growth sectors. *The National Policy and Strategy for Local Economic Development* (RSA 2004b: ii) noted that "…within the cities, the majority of people were not allowed to live and trade in the city centre, were restricted to the townships that had limited economic opportunities… and had to travel long distances to work… spatial economy was characterised by pockets of deprivation and poverty". The 2004 policy argued that local economic development is both a poverty relief and an economic growth strategy. The document recognised that the private sector has shown limited willingness to invest in poorer areas, including townships. A shift occurred with the *National Framework for Local Economic Development in South Africa.* 2006–2011 (RSA 2006b). It argued that it represents a move towards 'new institutionalism' and was based on a market driven economy.

The *Integrated Small Business Strategy,* 2005 (RSA 2005) promoted entrepreneurship, more flexible regulations, better access to finance and markets and business support. The *National Industrial Policy Framework* (NIPF) in 2007 (RSA 2007) acknowledged the crisis in manufacturing as many industries found themselves on the brink of collapse (TIPS 2016). The policy promoted more labour-absorbing industrialisation, favoured areas of traditional industrial clustering, IDZs, industrial parks, high-tech and science parks. The primary aim of the industrial policy was radical economic transformation of the racially skewed ownership, management and employment profile of the economy (RSA 2018a).

8.2.3.2 Impact on Urban Spatial Economy

Although significant effort was placed on urban renewal programmes for the identified townships in major cities, including funding for business plans and urban planning, no significant outcomes can be claimed due to these strategies. Typically, local municipalities were required to raise matching capital funding for major projects, and this did not materialise. None of the urban renewal programmes had significant economic strategies, even though National Treasury was involved. Consequently, the apartheid patterns of development continued unabated into the democratic area, irrespective of new spatial plans that envisaged futures that were economically, socially and environmentally more sustainable. Private sector driven gated housing estates and freeway driven malls continued to develop as the 'first economy' followed the logic of the distorted urban land market. Although former 'white' suburbs become more multi-racial as the black middle class grew, townships stagnated economically. The Urban Development Zone (UDZ) tax incentive was introduced by National Treasury in order to stimulate economic revitalisation and urban regeneration in declining inner-city areas. However, it resulted in "...accelerating gentrification, i.e., the exclusion and eviction of long-term residents and businesses that are unable to afford the property prices and rentals of the redevelop and post-upgrade neighbourhood" (Madell 2024)

The flight of private sector investments from inner city areas cities to 'edge cities' continued, such as Sandton (City of Johannesburg) and Century City (Cape Town). Street based informal economic activities continued to be viewed as transitional by the state with limited tolerance in formal, established CBDs, while jobs 'created' by the state were temporary (Todes and Turok 2017). The shift in local economic development from a focus on 'poverty reduction in townships' to 'pro-market' approach was more in line with the state's market macro-economic policy and trade policies. South Africa's energy crisis, or 'loadshedding' began during 2007, towards the end of the Mbeki presidency, an outcome of the failure of Eskom to generate sufficient electricity due to ageing coal power stations, endemic corruption and 'state capture', and delayed response to activate independent (largely renewable energy) power suppliers. The national shortage of electricity over the last 18 years have impacted significantly on the performance and growth of the South African economy and its ability to generate jobs.

8.3 The Great Recession and Radical Economic Transformation in South Africa

The period of the Great Recession and the onset of 'radical economic transformation' is captured in Table 8.3.

8.3.1 The Great Recession

The Great Recession was triggered by the world-wide financial crisis (December 2007 to June 2009), due the collapse of the US banking system, accompanied by the disintegration of the housing market and downturn of the stock market. The 2007 recession has been compared with the Great Depression of the 1930s and resulted in closure of businesses, decline in consumer wealth, and general decline in economic activity not only in the US, but also worldwide. Lending agencies such as the IMF and World Bank advocated a recoil of public sector spending as precondition for bailouts.

Local economic theory that evolved during this period include collaborative advantage theory (post 2009) and later the quality of life and sustainability theories (2010 onwards) partially in response to the recession (Blakely 1994; Blakely and Leigh 2010). Collaborative advantage theory placed emphasis on aspects such as business development, human resource development and community development. Quality of life and sustainability theory argued that recovery from the global recession requires a shift from 'business as usual' towards an economy focussed on reinventing itself through new technologies, innovations, and renewed commitments to ethical leadership. This theory contended that economic development focussing on reinvestment in community productivity, is more sustainable. They challenge the conventional wisdom that flexible markets, low regulation, low taxes, weak unions, low workers' compensation and low unemployment insurance rates are good for businesses. Although this may be appealing for many businesses, low bargaining power of workers often means low wages, low benefits, and job insecurity in the community. Low salaries, means that the multiplier effect on money spent locally is significantly reduced and less tax revenue per worker to pay for public services. The

Table 8.3 The great recession and radical economic transformation in South Africa

	Radical economic transformation: Zuma Presidency (2009–2014) & (2014–2018)	
Global events and economic influences	RSA governance, legislative and policy shifts	Impact on urban spatial economy
• Great Recession Dec. 2007–June 2009. • IMF & World Bank, recoil of public sector spending as precondition for bailouts • Collaborative Advantage Theory (Post 2009) • Quality of Life and Sustainability Theories (2010 onwards). • FIFA World Cup, 2010 hosted by South Africa • Post-recession, resurgence of neo-liberalism interrupted with economic recessions and pandemics.	*Governance* • Polokwane Conference, 2007: Mbeki replaced with Zuma, 2008 • Fourth National Election, 2009 • Towards a 15 Year Review, 2009 • City Support Programme, 2011 onwards. • Sustainable Development Goals (SDGs), 2015. • Fifth National Election, 2014 • Twenty Year Review. South Africa. 1994–2014 *Spatial development* • National Development Plan. Diagnostic Overview, 2010 • National Development Plan. Vision 2030, 2011 • National Infrastructure Plan (NIP), 2012 • The Spatial Planning and Land Use Management Act (SPLUMA), Act 16 of 2013. • International Guidelines on Urban & Territorial Planning (IGUTP), 2015 • Integrated Urban Development Framework (IUDF), 2016 • New Urban Agenda (NUA), 2017 *Economic development* • New Growth Path (NGP), 2010. • National Tourism Sector Strategy. 2016–2026. 2017. • The Industrial Policy Action Plan IPAP 2018/19-20/21. 2018.	• 'Radical economic transformation' (RET) in response to outcome of pro-market macro-economic policies. • Progressive NDP & IUDF policy proposals. • RET translated to individual businesses and malfeasant activities. • No meaningful change for majority, spatial and economic exclusion. • Exclusionary, high-income housing, retail and office developments. • 'Sustainable, integrated cities' in rhetoric and active resistance from the property elite. • Continuation of attempts to formalise 'illegal' settlements and informal trading. • Separated high, medium and low value property investments. • Private investment based on non-public sharing of 'community' facilities. • Public investment prioritising the needs of the wealthy (e.g., private road infrastructure). • Local economic development debased to 'back to basics' utility services. • IDZs underperformed in terms of jobs and investment. • SEZ success not apparent in relation to infrastructure investment and intended outcomes. • Industrial Parks Revitalisation Programme not successful.

externalities of the cost of urban sprawl, bulk services and public transport are often not considered when the focus is only on economic growth, i.e. the economy and the environment are separate entities that compete with each other. However, post-recession, a resurgence of neo-liberalism has occurred, interrupted with further economic recessions and the Covid-19 pandemic (Blakely and Leigh 2010).

8.3.2 'Radical Economic Transformation' and Territorial Governance (Zuma 2009–2018)

The Zuma presidency was preceded with his election as leader of the ANC at its 52nd National Conference in Polokwane in 2007 on a *radical economic transformation* (RET) agenda. A notable feature at the conference was the concern raised by the ANC's alliance partners, Cosatu, and SACP of the impact of neoliberal macroeconomic policies on the poor and the marginalised. The Zuma presidency was characterised by 'state capture', i.e. a form of highly organised corruption, costing the taxpayer an estimated at R1.5 trillion (US$ 100 billion) and the collapse of state-owned enterprises' such as Eskom, Transnet, SAA as well as SARS. Zuma was replaced by Ramaphosa as President of the ANC during 2017 and resigned as President of RSA on 14 February 2018.

8.3.2.1 Governance, Legislation and Policy Shifts

The Fourth National Election, 2009, extended the ruling party's office in government and a number of landmark policies and legislation were implemented during this period (See Table 8.3). *The New Growth Path* (NGP) (RSA 2010) identified the mass joblessness, poverty and inequality as the core challenges. This macroeconomic policy was a continuation of neo-liberalism and consistent with previous GEAR policy. The NGP had a goal of creating five million new jobs by 2020, argued for re-industrialisation through value adding activities and support for SMMEs, however, none of these ideas were new. *The National Development Plan* (NDP). *Diagnostic Overview* (RSA 2011a) remarked that "…in the cities, the poorest live far from places of work and economic activity, the situation has been aggravated since 1994" (RSA 2011a: 235). *The National Development Plan. Vision for 2013* (NDP) (RSA 2011b) observed that due to weak spatial governance, spatial planning tends to follow patterns set up by private-sector investment and not by the long-term public interest. The NDP argued that the capability of the state to engage with the private sector must be improved. The NDP also argued that spatial transformation will increase mobility for poor households to access better jobs and education opportunities. The NDP prioritised development in inner cities and in other areas of economic opportunity such as around transport hubs and corridors within metropolitan areas, cities and towns. A radical revision of the housing finance regime is proposed, including shifting state investment support to incentivise the acquisition of well-located land; and the provision of affordable, gap housing by the private sector.

National Treasury launched the *Cities Support Programme* during 2011 and (CSP) included reference to a conceptual plan linking established urban core areas (with economic opportunities) with underdeveloped townships via activity corridors (Duminy et al. 2020). CSP required the preparation of Built Environment Performance Plans (BEPPs), inclusive of projects and budgets. Other initiatives included the *Intermediate City Municipal Support Programme* and the *Small-Town Regeneration Programme. The Spatial Planning and Land Use Management Act* (SPLUMA) (RSA 2013) development principles, do not specifically refer to ensuring residential and employment or economic opportunities are in close proximity to or integrated with each other, a principle that was central to the DFA. Local economic development at municipal level from 2014 onwards was replaced by "back to basics", focussing on providing utility services at a local level, as the minimum requirement for businesses, as the collapse of municipalities unfolded. This period was characterised by less focus on economic support for local businesses.

The Integrated Urban Development Framework (IUDF) (RSA 2016) build on the NDP and highlighted the importance of cities as places of concentrated economic activity, driving force for economic development, developing a dynamic competitive advantage, and as engines of growth and employment opportunity. Territorial governance challenges identified by the IUDF (RSA 2016) include weak long-term and forward planning and coordination within government and with the private sector; private sector investments frequently fail to align with public sector plans; and spatial planning following patterns set by private-sector investment instead of the long-term public interest. International policies impacting on spatial economic development in South Africa include the United Nations 17 Sustainable Development Goals (UN-SDGs) (UN-Habitat 2015b), the International Guidelines on Urban & Territorial Planning (IGUTP) (UN-Habitat 2015a) and the New Urban Agenda (NUA) (UN-Habitat 2017).

8.3.2.2 Impact on Urban Spatial Economy

The overriding political agenda during this period was 'radical economic transformation' (RET) in response to the exclusionary outcome of pro-market GEAR and AsgiSA macroeconomic policies pursued by the state. The NDP and IUDF spatial and economic proposals can indeed by regarded as 'radical', as major changes to the urban land market, urban development and business support interventions were proposed. However, at a political level RET was translated to individual 'black' business empowerment through state procurement, however, the outcome was no meaningful change for the majority of the population in terms of jobs and economic inclusion. Exclusionary, high-income housing, retail and office developments investment continued in established areas and 'edge cities'. 'Sustainable, integrated cities' rhetoric were incorporated in spatial planning policies, however, encountered active resistance from the property elite, when goals of social justice or environmental sustainability are pursued. This is evident with inner city urban renewal programmes and associated Urban Development Zone (UDZ) tax incentives and the development of shopping centres (Madell and Karam 2022; Madell 2024). In terms of manufacturing activities, the state continued to focus on Industrial Development Zones (IDZs) (2000–2014), Special Economic Zones (SEZs 2014-current) and Industrial Parks Revitalisation (2016 to current). These are intended to attract foreign direct investment and export value added commodities, however, underperformed in terms of investment, economic development and jobs (Todes and Turok 2017).

8.4 Climate Change, COVID-19 Pandemic and the 'New Dawn' in South Africa (Ramaphosa Presidency: 2018—Current)

This period, capturing the impact of climate change, the COVID'19 pandemic and the 'New Dawn' in South Africa (Ramaphosa Presidency: 2018—Current) is reflected in Table 8.4.

Table 8.4 Climate change, COVID'19 pandemic and the 'New Dawn' in South Africa

	The New Dawn: Ramaphosa Presidency: 2018–2024	
Global events and economic influences	RSA governance, legislative and policy shifts	Impact on urban spatial economy
• Global Covid-19 Pandemic crisis (2020–2023) and impact of climate change on communities and economies. • Global multilateralism & multipolarity (e.g. Brexit – 2020, BRICS Plus 2023) • Conflicts in Central and West Africa, Ukraine and Russia, the Middle East and Asia. • Trade wars between West, Europe, China • UNHabitat SDGs as guiding development goals at a global scale. • Election of Donald Trump as US President	*Governance* • Sixth National Election, 2019. • Introduction of District Development Coordination Model (DDM) during 2020. • Seventh National Election, 2024 • Towards a 25 Year Review: 1994–2019. *Spatial and Economic Development* • National Spatial Development Framework, 2019 • National Framework for Local Economic Development. Creating Innovative Driven Economies, 2018–2028, 2018.	• Identification of urban areas, regions and development corridors as engines of national transformation, innovation and inclusive economic growth. • National initiatives aimed at attracting inward investment, in particular for major transport and logistic infrastructure projects. • Local economic development focussing on pro-market approaches. • Continued slow economic growth.

8.4.1 *Climate Change, COVID-19 Pandemic, Global Conflicts and Trade Wars*

The contemporary period, globally, has witnessed rapid increase in the ongoing crisis and impact of climate change, the COVID-19 global pandemic, global multilateralism versus multipolarity (e.g., Brexit-2020, BRICS Plus 2023); the conflicts in Africa, eastern Europe, the middle East and Asia and the unfolding trade wars between the West and China and more recently the between the US and its traditional trading partners, such as Europe, Canada and Mexico. These post 2007 recession events have stifled the global economic recovery and have deepened the inequality between the wealthy global north and struggling global south, while the SDGs remain the guiding development goals at a global scale.

8.4.2 *The New Dawn in South Africa*

Ramaphosa was elected president of the ruling party, at the ANC's 54th National Conference in December 2017, and was elected as president of South Africa on 14 February 2018 following the resignation of Zuma and the sixth national election of 2019. He promised a 'new dawn' and the end of state capture and corruption. The Judicial Commission of Inquiry into allegations of state capture, with Chief Justice Raymond Zondo as chair was appointed in 2018 and the final report was submitted to the Presidency in June 2022. The seventh national election took place in 2024.

8.4.2.1 Governance, Legislation and Policy Shifts

Landmark national policies approved during this period include the *National Spatial Development Framework* (RSA 2022a) and the *National Framework for Local Economic Development. Creating Innovative Driven Economies*, 2018–2028 (RSA 2018b).

The *National Spatial Development Framework* (NSDF) (RSA 2019) proposed a national scheme for spatial targeting that includes national competitiveness corridors; nodes of competitiveness; and special intervention areas (jobs, growth management zones; green economy zones), amongst others. The NSDF includes a wide range of spatial targeting levers, namely urban areas and regions as engines of national transformation, innovation, and inclusive economic growth; national spatial development corridors as incubators and drivers of new economies and quality human settlements and a national transport, communications and energy infrastructure network to ensure a shared, inclusive and sustainable economy (RSA 2022a: 90). The NSDF recommends that "relevant municipalities should establish transformation zones, to act as catalysts and demonstration projects for urban integration and densification. Specific projects could regenerate run-down inner-city areas, develop growth magnets on large vacant sites that have the potential to accommodate job- creating investment and large residential populations in accessible locations; and cement links between peripheral townships and urban cores" (RSA 2011a, b, c: 249–250).

The *National Framework for Local Economic Development* 2018–2028 (RSA 2018b) views local economic development as an integral part of the Back to Basics (B2B) Programme (RSA 2018b). This underscores the declining role of local economic development within the public space in governance. The main challenges identified by the framework facing local economic development policy and implementation include a lack of a shared conceptual understanding of what Local Economic Development is; a lack of integration of sector plans as part of local economic development planning and implementation; limited funding and financing for municipal local economic development programmes; uncoordinated project approach to developing local economies; a skills deficit and human resource challenges.

8.4.2.2 Impact on Urban Spatial Economy

The *National Framework for Local Economic Development* (NFLED) policy focus areas refer to established urban centres and inner-city revitalisation and small towns' regeneration. Other references to economic spaces include business attraction, retention and expansion, image enhancement, beautification and marketing of local spaces. This framework encourages municipalities to leverage on the Strategic Infrastructure Programmes (SIPs), Special Economic Zones (SEZ), industrial parks, logistics parks, industrial estates, agri-parks and innovation hubs. The policy emphasizes the 'transformative' potential of local economic development, in marginalised localities and economically distressed district municipalities and cities. Government, partners, and stakeholders are "urged to revitalise township economies, and to invest in factories, workshops, technical hubs and locally owned retail operations in townships. The driving force behind job creation must be the realisation that high-value economic activity in townships must also be encouraged" (RSA 2018b: 14). Proposals include business incubation, development of informal enterprises to formalised small businesses, access to funding and finance, and strengthening local economic development agencies. Despite the expansive NSDF and NFLED strategies, actual impact have not materialised.

8.5 Synthesis of Key Urban Spatial Economic Challenges

The state's 10-, 15-, 25- and 30-year reviews provide some sombre reflection of its performance and the country's development challenges. Ironically, there are significant consistency on what the challenges are, however, it appears that each review had no impact on subsequent new policy introduced. Interestingly, with the looming

2024 game changing elections, the 30-Year review (1994–2024) focus more on progress achieved and less on development challenges. From this overview, it is apparent that there are components of past and existing legislation, policies and strategies that offer opportunities to respond constructively (albeit in a disjointed, incoherent and uncoordinated manner) to the challenges of spatial and economic exclusion and marginalisation of the majority of inhabitants of towns and cities in South Africa.

The following recurring themes appear from the critical reflection of legislation and policies, and highlighted by the state reviews:

- South Africa remains one of the most unequal societies in the world, and the "... majority of black people still commute between their residential townships and the cities where they are employed, paying up to a third of their income for transportation" (RSA 2011a, 2016, 2019).
- Many new housing settlements are located far from economic activities and work opportunities. Many of the state's urban settlement interventions and other affordable housing projects remain on the peripheries of cities (RSA 2009b, 2011a, 2016, 2019, 2022a).
- The state has not adequately utilised the levers at its disposal, such as legislation, regulations, policies, procurement systems and incentive schemes to reverse the legacy of apartheid through integrated development and densification of well-located land (RSA 2011b, 2016, 2019).
- Economic development challenges include structural constraints to potential growth (e.g., energy security, transport logistics efficiency, infrastructure backlogs, low R&D and technological innovation, etc), vulnerability to global shocks and resilience, the country's risk premium, declining manufacturing, and the slow pace of broad based economic empowerment and transformation (RSA 2024).

Some proposals and recommendations, emanating mainly from the state reviews, that could contribute to transformation of the urban spatial economy include the following:

- The need for improved spatial planning capabilities of local government; and closer alignment, co-operation and co-ordination between all speres of government (RSA 2004c).
- Every municipality should have an explicit spatial restructuring strategy; prioritising development in inner cities areas; providing support for economic development hubs, nodes and linkages with historically black townships (RSA 2004c).
- The private sector should be incentivised to service the market in better-located areas, including revitalisation of neglected inner-city areas and inclusionary housing (RSA 2019; Madell 2024).
- Devolving economic opportunities to the places and localities where most poor people live in townships and rural areas (RSA 2019).
- Implementation of the NSDF should be accelerated, through partnerships with the private sector and financial institutions to increase the supply of capital and the provision of housing (RSA 2024).

Significant congruence exists across the legislation policies over the last three decades on alternative and future development visions, highlighting the importance of cities as the drivers for economic development, striving for sustainable human settlements, the desire for a more compact urban form, mixed land use development, locating affordable housing in areas with access education, health and other public facilities and services, and integrating land use and public transportation in order to reduce travel distances and generate economic thresholds for local businesses.

Most municipalities do not have significant planning nor economic development capacity beyond delivering basic services and land use management. This lack of capacity is exacerbated with limited ability to manage economic activities in urban centres and neighbourhood localities. Some of the well-resourced municipalities, such as metropolitan cities and secondary towns do not have the internal capacity to deliver nor strategically engage with major catalytic projects, including collaborating with private sector driven commercial initiatives. Those that do have development agencies are ineffective in mobilising the private sector when delivering on capital intensive projects.

A review of SPLUMA principles, norms and standards is required, as these are ineffective tools for spatial and economic transformation. Property developers have understandably, argued that the planning and building plan approval processes as required by SPLUMA, NEMA and the National Heritage Act are arduous, costly, and significantly increase the uncertainty and risks for investors. The NDP concurred that the planning approval processes requires to be reviewed and integrated.

8.6 Leverage the Tools of Territorial Governance to Transform the Urban Spatial Economy

The state has ploughed through a plethora of national legislation, policies and programmes impacting on the urban spatial economy over the last 30 years. It is apparent that the enormous spectrum of often diverse and contradictory approaches and policies have not resulted in structurally changing the entrenched spatial form of the apartheid city (Harrison and Todes 2015; Madell 2012; 2024). The siloed independent state line function departments, contradictory and weak political leadership from developmental perspective, have certainly muddied the waters, leaving officials, businesses and private sector role players somewhat confused about the coherence of national economic developmental approaches and strategies. The National Developmental Plan does provide some direction, but more importantly, the conundrum of sectoral economic and urban policies remains contradictory and the major obstacle for meaningful reform.

The following section critiques the state's approach to these challenges, highlighting aspects of its policies that are contradictory, while some components could be included in a more coherent and sensible approach to transforming the urban spatial economy. It is argued that the state's approach to development and territorial governance of the urban spatial economy hinders effective and meaningful transformation:

- The state's unquestionable subservience to neoliberalism and the inability to exploit its contradictions.
- The state's dogmatic approach to economic growth vis à vis economic development, viewing these as opposing extremes of political philosophies, the extremities of the unhindered market vis à vis social welfare.
- Incoherent understanding of the notion of developmental government,
- Contradictory understanding and policies on the role of cities in the national economy, weakened by municipal capacity and competence to develop local economies.
- Inability to implement the meaning of sustainable human settlements, beyond mere rhetoric.
- Anti inner-city stance and practice, contradicting its own policies and practices.

The following are recommendations in response to the spatial economic development challenges facing South African towns and cities. Many of these interventions are included in policy documents prepared by the state over the last three decades, but have not been realised nor implemented, either because they appear in piecemeal in these documents or they have been ignored and therefore not carried forward. The recommendations are embedded in a critique of the state's approach to development in general, and the urban spatial economy specifically. It aims to unearth the barriers, obstacles as well as the unused inducements that state has at its disposal to advance urban economic transformation in South Africa. Urban spatial economic patterns reflect, to a large extent the historical and current power relations and major changes to spatial and economic arrangements have a track record of provoking resistance from those who have a vested interest in maintaining the status quo (RSA 2011b).

8.6.1 The State's Unquestionable Subservience to Neoliberalism and the Inability to Exploit Its Contradictions

Neoliberalism, power relations and vested interests are intertwined in global economies reflected locally, more so in complex urban spaces such as towns and cities. Neoliberalism remains the politically and economically 'hegemonic as mode of discourse' in the global north and much of global south. The post-apartheid South African state macro-economic policies over the last three decades have demonstrated unquestionable subservience to neoliberalism as well as its inability to exploit its contradictions. National Treasury and the Reserve Bank have championed market orientated macro-economic policies on behalf of the ruling party, while some line function departments have in tandem pursued poverty reduction and social interventions through social welfare grants and subsidised 'free' housing, targeting the poorest. Consequently, the state has managed to keep extreme levels of poverty at bay with this dualistic approach, mainly through increasing tax on a limited middle-income base, while meaningful participation in the market economy remains impossible for most citizens. Unemployment, inequality, and the realities of poverty have continued unabated, especially in urban areas in the global south and more so in South Africa (Madell 2012).

Those with economic power, including investors and owners of higher valued residential, retail, commercial and office properties, have a significant impact on how the city is structured, with the state generally serving this interest, or at least does not confront the forces at play (Madell and Karam 2022). Often, resistance to change occurs through a wide range of measures, for example when social housing is located within well-resourced urban centres or nodes, or in other cases when micro and small businesses are excluded from top-end malls through high rentals. However, a fallacy of the state and sometimes urban planning is conceptualising the private sector as a homogenous entity, cemented with a pointless political notion of 'white monopoly capital' vis à vis the disempowered urban poor. Those investing in urban economies, through infrastructure or real estate have rather a narrower view of their interests. In many cases opportunities for economic inclusion can be successfully negotiated with the private sector through meaningful technical policies, dealing with reducing risk and increasing returns. These could include how land is made available, how development rights are determined or how supporting infrastructure and financial incentives are devised. These are acceptable practices to enhance economic and social inclusion, even for the most capitalist cities of the global north (Madell 2024).

8.6.2 The State's Dogmatic Approach to Economic Growth vis à vis Economic Development

The South African state tends to view *economic growth vis à vis economic development* as opposing political-economic philosophies, the extremities of the unhindered market vis à vis social welfare. The state's economic development policy position accepts the assumptions that economic growth must be the primary priority and achieving high levels of GNP growth will unquestionably lead to trickle down benefits for the economically marginalised. Further, the state tends to conflate its policies dealing with local economic development and business support with poverty reduction. Given that the state's macro-economic policies are essentially neoliberal and pro-market, while local economic policies remain grounded in poverty reduction, is problematic. Consequently, publicly owned property (land and buildings) in highly competitive locations (inner city and other urban centres) remain unused by the state as leveraging tools to further social and economic inclusion.

Economic development policies at local level cannot contradict national policies and to do so negates the ability of entrepreneurs and proprietors to establish and sustain competitive businesses. Greenwood and Holt (2010) differentiates between economic growth and economic development. They challenge the assumption that markets respond well to changing consumer demand and will always create upward mobility and reduces poverty. While the benefits of economic growth could include higher quality of goods and services; more leisure, vitality and innovation, higher taxes and increase GNP per capita; disadvantages include an increase in urban sprawl and associated infrastructure costs; increase in economic exclusion, degradation of the environment, an increase in the carbon footprint of society, more congestion and commuting times and overall decrease standard of living. In contrast, economic development is achieved when there is a sustained improvement in a community's overall standard of living and quality of life, through a process of human and physical development, based on principles of equity and social inclusion. Therefore, the issues of growth, development, quality of life and sustainability must be tackled jointly, and this require thinking and acting locally, as well as globally.

8.6.3 Incoherent Understanding of the Notion of Developmental Government

The state and its various organs understanding of the notion of developmental government is incoherent and rudimentary, weakened by declining municipal capacity and competence with regard to comprehending how to develop local economies. The notion of developmental government was recognised with the Constitution, the RDP, White Paper on Local Government and further elaborated upon with subsequent legislation and policies. The NDP argued that spatial planning does not have significant institutional force within the governmental system because this responsibility is institutionally fragmented across different line function departments, the same argument applies to economic development. Frequently, state policies argue that there is a need for strategic co-ordination across national government departments, however, mere alignment with new public management practices such as performance management, are more concerned with measurable outcomes such as process and procedure than actual transformation of the urban spatial economy. There is perhaps a case to separate the administration of government from the more developmental arm. This implies that one dimension of the state focusses on ensuring that the regulatory and services delivery aspect of government is safeguarded and massively improved. This will enable the second dimension of the state to focus on programme and project delivery, in essence on spatial and economic development and transformation. Closer cooperation and alignment would then be more crucial across spheres, from national to provincial to local municipalities. Partnerships and collaboration between state, local development entities and businesses would be critical for such an approach.

8.6.4 Contradictory Understanding and Policies on the Role of Cities in the National Economy

A common attribute of managing urbanisation is the apparent contradiction between two main alternative policy approaches; one a desire to achieve regional 'balance' across a national space (i.e. lowered disparities between wealthy and poor regions) through direct assistance to poor or lagging regions. Alternatively, targeted state assistance to regions or urban areas that have potential for economic growth (on the assumption that in the long term there might be 'trickle down' to poorer regions). Todes and Turok (2017) discusses the three contrasting approaches, i.e., spatial rebalancing, space-neutral and place-based aimed at addressing spatial inequalities within nations and regions. They conclude that it is risky spending high levels of capital in economically marginal locations (for example the ruling party has strong support in rural areas) with strong political pressures and weak public institutions. Although place-based policies have potential, stronger vertical and horizontal policy alignment is required to effectively address spatial inequalities.

"A key feature of managing urbanisation in Africa, including South Africa, is the dualistic focus on rural areas. This often contradicts efforts by the state to mobilise resources for urban development and to implement strategies aimed at creating compact and denser cities" (Madell and Karam 2022: 27). This can be observed with the earlier reflection on the NSDP approach and NSDF critique. The critical role of cities in national economic growth and development has been acknowledged by international agencies such as UN- Habitat and the World Bank (UN-Habitat 2022; The-World-Bank 2013). Although state policies (such as the NSDP, IUDF and NSDF) acknowledge the economic importance of cities, this understanding has not been transferred to state action, for example in how it views publicly owned property in areas with intensive economic activity or how it engages with private sector real estate developments with no or limited concern to capture value for the public good (Turok et al. 2020; Madell, 2024).

8.6.5 Inability to Implement the Meaning of Sustainable Human Settlements

The state's inability to implement the meaning of 'sustainable human settlements', beyond mere rhetoric is core to the lack of spatial and economic transformation. It appears that the state reference to 'sustainable settlements' only implies to low-income, affordable housing and townships. Often the critical requirement for neighbourhoods to have a wide range and mixture of income groups, is not included in the conceptualisation of 'sustainable settlements.' BNG (2004) remains a landmark policy on defining 'sustainable human settlements', in terms of its description of what this means, but more so its implication for housing. However, in spite of BNG's critique that the implementation of the National Housing Code (RSA 2009a) has resulted in continuation of apartheid settlement developments, it remains in place. National policies acknowledge the need for the public sector to be active participants in the urban land and property market, with the objective to achieve spatial and economic transformation and sustainable human settlements. The NDP and BNG identified several critical actions and proposals, including the assemblage of public and privately owned land and buildings as well as fiscal incentives (and disincentives) to support the development of sustainable human settlements.

8.6.6 Anti-Inner-City and Urban Core Policies and Practices

National policies acknowledge that historic urban centres, inner city areas, decentralised urban nodes and centres in established neighbourhoods have the highest concentrations of economic opportunities, jobs and higher order public facilities. However, despite this policy rhetoric, anti-urban core and inner-city policies is evident by the lack of strategies dealing with inner city public property, the lack of inclusionary housing policy and no significant support for emerging businesses and SMMEs to find a foothold in these highly competitive environments. It should be obvious that overcoming the legacy of apartheid requires that these areas should be the highest priority for spatial and economic transformation. A common avoidance strategy by many municipalities is to argue for economic and other developments within envisaged future urban activity corridors and nodes, linking townships with established urban centres. This strategy might have long term merit but require massive capital investment over decades to become viable. It becomes expedient to focus on these areas, but in reality, is a way of avoiding considering more viable, better located options, that may be politically more challenging. Todes (2008) caution the use of nodes and corridors became a standard form of planning in the post-apartheid era spatial frameworks. Todes argued that planning should engage with the complexity of the socio-spatial dynamics of the city, develop a stronger understanding of urban economic space and a stronger link to infrastructure development.

Linked to optimising the well -resourced urban centres, is the use of planning policies, streamlining and fast tracking development approvals, land-based financing and value capture, the use of public sector infrastructure as well as financial incentive to entice private sector real estate investors and developers to undertake more inclusionary economic and housing developments (Madell 2024). National policies such as the IUDF, recommended the development of measures to counter the displacement impact of gentrification on poor households. For example, National Treasury's urban development zone (UDZ) tax incentive to revitalise inner city areas, has result in gentrification and forced removals of households and businesses that have a lengthy historic connection with these areas (Madell and Karam 2022). There is a requirement for municipal planning departments to understand the basic principles of property development and engage with developers in order to crowd in private investment in projects advancing economic restructuring and inclusion.

8.7 Conclusion

South Africa's public policy formulating institutions responsible for spatial planning, economic development, transport, housing, infrastructure development, finance, and others, all have a critical role in enabling spatial and economic transformation of town and cities in South Africa. It is apparent from the policies and strategies formulated by the different state departments that there is common understanding of the nature, scope and impact of development challenges facing South African urban settlements. The reviews by the state over the last three decades, acknowledge that the desired post-apartheid settlements that are economically, socially and environmentally sustainable, are far from being realised. This is most obvious with the shift from the initial optimistic view during the 1990s that sustainable human settlements can be achieved within a decade or two; followed by the call for radical economic transformation and more recently the question; why spatial and economic transformation is not occurring.

A review of state policies reveals that the lack of meaningful change cannot be attributed to a lack of ideas, proposals and recommendations. This in particular evident with policies such as the BNG, NDP, IUDF and NSDF. At the same time there are major contradictions and uncertainly in the policies when engaging with the market. It is argued that challenges relating to territorial governance and urban spatial transformation cannot effectively be addressed unless the state reconceptualise and adjust its approach to key areas of policy and practice dealing with economic and spatial development. This includes the state's unquestionable subservience to neoliberalism and the inability to exploit its contradictions; the dogmatic approach to economic growth vis à vis economic development; the incoherent understanding of the notion of developmental government, weakened by municipal capacity and competence about comprehending the meaning of developing local economies. Other areas of concern include the state's contradictory understanding and policies on the role of cities in the national economy; the inability to implement the meaning of sustainable human settlements; and its anti-inner-city and urban core policies and practices.

References

Blakely EJ (1994) Planning local economic development. Theory and practice, 2nd edn. SAGE Publications, London

Blakely EJ, Leigh NG (2010) Planning local economic development. Theory and practice. SAGE Publications, London

Bond P (2005) Elite transition. From apartheid to neoliberalism in South Africa. University of KwaZulu-Natal Press, Pietermaritzburg

Duminy J, Parnel S, Brown-Luthango M (2020) Supporting city futures. The cities support programme and urban challenges in South Africa. African Centre for Cities

Harrison P, Todes A (2015) Spatial transformations in a "loosening state": South Africa in a comparative perspective. Geoforum 61:148–162

Hart G (2008) The provocations of neoliberalism: contesting the nation and liberation after apartheid. Antipode 40(4):678–705

Harvey D (2007) A brief history of neoliberalism. Oxford University Press, New York

Lovering J (2009) The recession and the end of planning as we have known it. Int Plan Stud 14(1):1–6

Madell C (2012) Local economic development, business stimulation and poverty reduction in poor urban areas: the disjuncture between state policies and household survival strategies in the case of Heideveld. University of Cape Town, Cape Town

Madell C, Karam A (2022) African cities, urban growth management, and property development. In: Karam A, Viruly F, Kariuki C, Akujuru V (eds) Understanding African real estate markets. Routledge. Taylor & Francis Group, London/New York. https://doi.org/10.1201/9780429279256

Madell, C (2024). Inclusionary housing policy and practice in South African cities: planning as a spatial transformation tool. Journal of Housing and the Built Environment. https://doi.org/10.1007/s10901-024-10149-y

Marais H (1998) South Africa: limits to change: the political economy of transition

Mc Donald DA (2008) World city syndrome: neoliberalism and inequality in Cape Town. Routledge, London

Meyer-Stamer J (2002) Current challenges for local economic development in South Africa: experience from Mpumalanga. Report for GTZ (German Technical Cooperation), Pretoria

Mosala S, Venter J, Bain E (2017) South Africa's economic transformation since 1994: what influence has the National Democratic Revolution (NDR) had? Rev Black Polit Econ 44(3–4):327–340. https://doi.org/10.1007/s12114-017-9260-2

Nel E (1997) Evolving local economic development policy in South Africa. Reg Stud 31(18):67–72

Oranje M, Van Huyssteen E (2011) Nestling national 'transformation' imperatives in local 'servicing' space: critical reflections on an intergovernmental planning and implementation project. Town Region Plan 58:6–16. http://www.ajol.info/index.php/trp/article/viewFile/77204/67655

Peet R, Hartwick E (2009) Theories of development. Contentions, arguments, alternatives. Guilford Press, London

Pieterse E (1998) A framework to link local economic development and poverty alleviation. Isandla Institute and Department of Constitutional Development, Cape Town

Porter ME (1998) Competitive advantage: creating and sustaining superior performance. The Free Press, New York

RSA (1994) The white paper on reconstruction and development

RSA (1995a) Development Facilitation Act No 67 of 1995

RSA (1995b) Local economic development. A Discussion Document'. Department of Provincial and Local Government (DPLG)

RSA (1995c) White paper on the National strategy on the development and promotion of small business in SA, 1995. Department of Trade and Industry

RSA (1996a) Constitution of the Republic of South Africa No 108 of 1996

RSA (1996b) Growth, Employment and Redistribution—a macro-economic strategy (GEAR). Department of Finance, Government Printer, Pretoria

RSA (1998a) Local Government: Municipal Structures Act

RSA (1998b) The white paper on local government

RSA (2000) Local Government: Municipal Systems Act, 2000. (Act 32 of 2000)

RSA (2001) Refocusing development on the poor and LED manual series. Department of Provincial and Local Government (DPLG)

RSA 2003. Towards a Ten-Year Review. Pretoria, South Africa.

RSA (2004a) Breaking new ground. A comprehensive plan for the development of integrated sustainable human settlements. Department of Human Settlements

RSA (2004b) National Policy and Strategy Framework for Local Economic Development. Department of Provincial and Local Government (DPLG)

RSA (2004c) Towards a ten-year review. The Presidency

RSA (2005) Integrated small-enterprise development strategy. Unlocking the potential of South African entrepreneurs. Department of Trade and Industry

RSA (2006a) Accelerated and Shared Growth Initiative for South Africa (AsgiSA). The Presidency

RSA (2006b) National framework for Local Economic Development in South Africa (2006–2011). Stimulating and developing sustainable local economies. Department of Provincial and Local Government (DPLG)
RSA (2006c) National spatial development perspective. The Presidency
RSA (2007) A National Industrial Policy Framework. Department of Trade and Industry
RSA (2009a) The national housing code. A simplified guide. Department of Human Settlements
RSA (2009b) Towards a fifteen year review. The Presidency
RSA (2010) The New Growth Path Framework (NGP). Department of Economic Development
RSA (2011a) National Development Plan. Diagnostic report. National Planning Commission
RSA (2011b) The National Development Plan. Vision for 2030, November 2011. National Planning Commission
RSA (2011c) The National Development Plan. Our future make it work. National Planning Commission
RSA (2013) The Spatial Planning and Land Use Management Act, Act 16 of 2013
RSA (2014) Twenty year review. South Africa. 1994–2014. Department Planning, Monitoring and Evaluation
RSA (2016) Integrated Urban Development Framework. A new deal for South African cities and towns. Department Co-operative Government and Traditional Affairs
RSA (2018a) Industrial Policy Action Plan IPAP 2018/19-20/21. Department of Trade and Industry
RSA (2018b) National Framework for Local Economic Development. Creating innovative driven economies, 2018–2028. Department Co-operative Government and Traditional Affairs
RSA (2019) Towards a 25 year review (1994–2019). Department Planning, Monitoring and Evaluation
RSA (2022a) National Spatial Development Framework. Department of Rural Development and Land Reform
RSA (2022b) South African National Census. Statistics South Africa
RSA (2023) Quarterly Labour Force Survey quarter three 2023. Statistics South Africa
RSA (2024) Towards a 30-year review of South Africa's democracy. 1994–2024. Department Planning, Monitoring and Evaluation
SALGA (2012) Spatial transformation barometer: towards refinement of indicators for municipalities, Pretoria
Storper M, Walker R (1989) The capitalist imperative: territory, technology and industrial growth. Blackwell
The-World-Bank (2013) Planning, connecting & financing cities-now. Priorities for city leaders
Todes A (2008) Rethinking spatial planning. KovsieJournals
Todes A, Turok I (2017) Spatial inequalities and policies in South Africa: place-based or people centred? Prog Plan 413:31
Trade and Industry Policy Strategy (TIPS) (2016) Analysis of existing industrial policies and the state of implementation in South Africa. Pretoria
Turok B (2008) The evolution of ANC economic policy. From the Freedom Charter to Polokwane. New Agenda, Cape Town
Turok I, Scheba A, Visagie J (2020) Cape Town: a city still divided by race and class. Human Science Research Council, RSA
UNDP (2005) United Nations Development Programme. Annual Report. A Time for Bold Ambition. Together we can cut poverty in half
UN-Habitat (2015a) International guidelines on urban and territorial planning
UN-Habitat (2015b) Sustainable development goals
UN-Habitat (2017) New urban agenda
UN-Habitat (2022) World cities report 2022. Envisaging the future of cities
Un-Habitat, ECA-CEA, UNEP, IDRC, & CRDI (2008) The State of African cities 2008, a framework for addressing urban challenges in Africa
UN-Habitat, Rotterdam I-EU (2018) The State of African cities 2018: the geography of African investment. United Nations Human Settlements Programme (UN-Habitat)
Watson V (2002) Change and continuity in spatial planning. Cities and regions: planning, policy and management. Routledge, London
World-Bank (2022) Inequality in southern Africa. An assessment of the Southern African Customs Union

Chapter 9
Negotiated Privilege and Territorial Governance: Exploring the Implications of Gated Developments on Public Coastal Access in South Africa

Tshepiso Monnakgotla and Karina Landman

Abstract Gated developments have proliferated in South Africa as a response to security and local management. However, these manifestations of territorialisation have numerous implications for the country's aspirations towards spatial transformation and justice. This chapter focuses on gated developments within the notion of public coastal access in South Africa. It draws from a study that dealt with negotiated privilege by exploring the relationship between luxury gated estates, municipalities, and fishermen communities on the coast. Public coastal access is often restricted through gated developments, which has implications for spatial transformation related to coastal access for poor communities. The chapter examines how municipalities deal (or fail to deal) with public coastal access restrictions in the broader context of territorial governance and spatial transformation. Case studies comprise three local municipalities within three districts spread across KwaZulu-Natal and Western Cape Provinces. The data reveals how some gated developments negatively impact spatial transformation and highlights the role of local and district governments in relation to territorial governance. The chapter is framed within the larger policy debate in South Africa to contextualise the findings with the current relevant planning policies and legislation related to spatial transformation and justice.

Keywords Gated developments · Public coastal access · Spatial transformation · Spatial justice · South Africa

---

T. Monnakgotla (✉) · K. Landman
Department Town and Regional Planning, University of Pretoria, Pretoria, South Africa
e-mail: tmonnakgotla@dffe.gov.za

D. du Plessis (ed.), *Territorial Governance and Spatial Transformation in Post-apartheid South Africa*, Local and Urban Governance,
https://doi.org/10.1007/978-3-031-85489-7_9

9.1 Introduction

In 1994, South Africa transitioned to a democracy, built on a Constitution (1996) that promotes equality, justice and access to public spaces and resources for all. The National Development Plan (NDP) (2012) reiterates the calls for spatial transformation in urban and rural communities and expanding infrastructure. At the same time, the country is committed to implementing the Sustainable Development Goals as outlined in the Agenda 2030 (Fourie 2018). The key to achieving this is to ensure that all people in the country have access to resources and opportunities to pursue socioeconomic goals, including sustainable livelihoods. Aligned with these goals, the Integrated Coastal Management (ICM) Act (2008) emphasise the importance of public coastal access in South Africa. However, several gated developments along the coast have asserted control over beaches and access routes to these beaches, which reflects trends in the USA, where affluent communities have restricted access to many beaches (Kranz 2009; McKeon 1970).

Restricting access can be seen as part of a more significant trend of the territorialisation of public and private property. Several authors have indicated how certain types of gated developments in the country have privatised public space or restricted access to public space (Low 2003; Lemanski et al. 2008; Madanipour 2010) or given rise to patterns of socio-spatial exclusion (Lemanski 2004; Durington 2009) and tensions related to planning and development applications (Makhale and Landman 2018). In addition, gated developments have also contributed to the privatisation of governance through the establishment of micro-governments (Lemanski et al. 2008). Privatised governance raises several questions about the impact of gated communities on public coastal access and its implications for territorial governance. According to Arango Espinal et al. (2021), territorial governance in the academic spectrum is progressively becoming a unit of analysis and a topic of interest, especially in regional economic development strategies. They further deduce that there are opportunities for future research that aims to promote territorial sustainability through land use planning and resource management. Davoudi et al. (2008) also expressed the need for further spatial planning reflections on different experiences through the lens of territorial governance action and processes. In addition, Jerzak et al. (2019) argue that sea or maritime space has undergone significant transformation and should be considered from a multi-disciplinary perspective, including urban planning as a source of satisfying human needs.

This chapter focuses on the nature and impact of gated developments along the coast in South Africa and its relation to territorial governance. The discussion spotlights gated developments in three areas in Kwa-Zulu Natal and the Western Cape to explore the territorialisation of space and its influence on public coastal access. The chapter reports on the location and development of the gated developments, the role of district and local municipalities in coastal management, and the impact on local fishermen from the surrounding communities. The authors argue that the notion of democracy should include public coastal access and that any restriction thereof impedes notions of spatial justice and sustainable development. Depending on the reading of territorial governance, this challenges notions of territorial cohesion and governance.

9.2 Gated Developments, Territorial Governance and Public Coastal Access

9.2.1 Gated Developments and Territorial Governance

Several types of gated developments have proliferated worldwide and in South Africa. The most common reason for the growth of gated communities is a need for safety and security, including protection from crime (Blakely and Snyder 1997; Landman and Schönteich 2002; Atkinson et al. 2004; Durington 2009). However, other prominent motivations include the need for privacy, segregation, and a culture of control (Garland 2001; Low 2003). Crime control manifests through many dimensions, including spatial or territorial control through various initiatives such as gated communities. Gated developments tend to reroute movement, distribute power and sustain social inequalities by reinforcing race and class boundary lines (Dinzey-Flores 2013).

Studies of gated developments have often focused on an understanding of the social and economic forces that drive the production of gated developments, their management and impact on residents and those outside gated developments (Atkinson and Blandy 2004). Pow (2009) interrogated the moral politics of urban planning in gated communities by critically demonstrating that they are entrenched in the politics of a life of pleasure, material comfort, and luxury. He noted that gated developments are protected and sustained through the territorialisation of private property and privilege amid a highly segregated landscape.

Territorial governance is becoming more prevalent in European policy discourse and, more generally, spatial planning discourse (Stead 2013). In some circles, it has even been used as a synonym for spatial planning (Faludi 2012). Despite this, there is a lack of consensus on its meaning (Stead 2014; Faludi 2012). Territorial governments refer to policy-making bodies of institutions that command the ability to impose rules on outsiders within a given geographic territory. The emergence of territorial governance is usually linked to prosperous communities in territories that were initially peaceful (Congleton 2011: 78). Davoudi et al. (2008: 35) defines territorial governance as "an organisational mode of territorial collective action, based on openness and transparency of the process itself, cooperation/coordination among actors (horizontally and vertically) and in a framework of a more or less explicit subsidiarity". They continue to note that territorial development at different scales aims to ensure the spatial coherence of different actions. Given this, territorial governance is concerned with managing territorial dynamics, monitoring territorial impacts, and delineating boundaries (Stead 2014: 1380).

As a process, territorial governance relates to the territorial organisation of many relations and interactions among different but non-conflictual, interests. Such a process aims to maintain and reproduce specific territorial capital, which is the territorial dimensions of various forms of capital—intellectual, social and political. In this sense, territorial governance is organising and coordinating actors to develop territorial capital in a non-destructive way to enhance territorial cohesion (Davoudi et al. 2008). Territorial cohesion has been referred to as spatial cohesion—a crucial part of spatial development—and related to connectivity improvements, as well as addressing territorial imbalances, spatial disparities and the differences in the potential for development (Zaucha and Böhme 2020: 629). Territorial governance actions are based on a shared vision and three characteristics: (1) territory as a common good, (2) the identification and valorisation of territorial capital and (3) the definition of the territory as part of the participation process. Participation in this context should ensure involvement by organised groups capable of making their voices heard and those who tend to be excluded from such processes (Davoudi et al. 2008). Given this, these actions should be inclusive to ensure territorial cohesion.

Territorial governance manifests at different territorial scales, including macro, meso and local. For example, the EU and national levels could be regarded as the macro level, with regional and polycentric urban networks as the meso level. The micro level could refer to functional urban or metropolitan regions (Davoudi et al. 2008). At the meso or micro level, territorial governance can take the form of a coastal conservancy comprising an organised community-based group of people dedicated to caring for the coastal environment and thus managing or controlling access to coastal resources in a given territory. Such actions align with the debates concerned with the apparent shift towards territorial governance from territorial government in the claim that such undergoing will increase flexibility and responsiveness within the function and structure of territorial government. Territorial governance formations operate within various factors at local and regional levels, comprising voluntary organisations, non-elected state-run agencies, and private businesses, which gives rise to scenarios of diffusion of responsibility as a result of ongoing collaborations, negotiations, and exchange of resources amongst network partners which in turn, constrain democratic accountability (Dostál 2002).

The debate advances issues related to territoriality, territorialisation, and territorialised actions. Territoriality is considered "a spatial strategy to affect, influence or control resources and people, by controlling an area" (Sack 1986: 1). Territoriality can be applied to many levels and geographical areas or functional zones, including river basins, vegetation zones, population concentrations and other possible regionalisation (Faludi 2012). Jerzak et al. (2019) highlights the prominence of sea or maritime space as a specific functional zone or, in the context of this discussion, a territory. Policy territorialisation is concerned with adjusting policies to the territorial context and integrating them to become more efficient (Zaucha and Böhme 2020). Davoudi et al. (2008) argue that this is necessary to ensure that the actions of local actors are considered, and local resources valorised. Territorialised actions are directly influenced by the role of the specific territory at various levels. While the higher levels are generally less problematic, the lower levels could cause concern due to incorrect delineation of the territory and actions within due to negative attitudes and a failure to see it as a common good with potential (Davoudi et al. 2008).

In summary, territoriality is the territorialisation of a geographical area or territory through shared actions to manage this area and its relation to other territories at the same or other (higher or lower) levels. In this context, gated developments can be regarded as a type of spatial development aimed at territorialising a specific territory through actions to manage behaviour in the specific area. However, territorialising or delineating space through rigid boundaries has implications for access in and around these developments.

9.2.2 Public Coastal Access

South Africa is endowed with a rich and diverse coastline in terms of biodiversity, history, culture, and socio-economic opportunities. Its coastline spans more than 3000 km and interfaces three oceans namely the Indian Ocean, the Southern Ocean and the Atlantic Ocean making it the third largest coastline after Madagascar and Somalia respectively. The coastline stretches along four provinces and 46 local municipalities mandated by the ICM Act to secure public access to the coast and Part B Schedule 5 of the constitution to manage beach facilities. The White Paper for Sustainable Coastal Development in South Africa was promulgated in the year 2000 and subsequently presented to Parliament in 2006 as the Integrated Coastal Management Bill. This signalled an important paradigm shift in coastal governance. Sowman and Malan (2018), reviewed the progress made in terms of ICM in South Africa since the dawn of democracy and note some significant challenges in relation to challenges to implementing what would be a progressive ICM agenda. Amongst those challenges, they highlight inadequate institutional capacity, lack of political support and clarity on the application of the act on private and communal land. Further suggesting that urgent action should be taken on the revitalisation of public interest about the coast, declaring coastal access land and improving cooperation across the relevant government role players.

Public Coastal Access deals with the ability of any person to enter, use, and enjoy public coastal spaces and resources. The "public" in the Public Coastal Access concept demonstrates openness and commonness or concerns all people and therefore 'public interest' or actions for the common good. Public interest serves to justify and defend development planning interventions and reforms. Thus, planning relies on recognising public interest in the practice to legitimise itself (Chettiparamb 2016).

The second aspect of PCA is "coastal", which many (Coccossis 1985; Naar and Mahenge 2014; Bavinck et al. 2017) have described in its spatial or territorial form as a land-sea interface except for Jeans (1990), who has defined it as a space of culture-nature interface. According to Phoenix et al. (2021), health and well-being benefits can also be derived from being or living near the coast. The coast comprises beaches, estuaries, cliffs, and an extent of land above the high-water mark and the sea below the watermark. In more recent debates, maritime space covers both the sea and terrestrial gateways servicing the sea or offering access. Jerzak et al. (2019) further point out that humans are the critical factor influencing the development of coastal space through their economic, social, and emotional bond to the sea and hence have a direct impact on the sustainable development of this space. The Natal Town and Regional Planning Commission (1984) has declared that the coastal area is where pressure and competition on land is greatest, subsequently leading to the potential for conflict between differing interests.

The third aspect of PCA is "Access", which, according to Ribot and Peluso (2003), is a term most popular in studies concerned with property and natural resource management (NRM). For them, "access" is primarily the ability and the right to benefit from things. However, from a property management perspective, it can also be associated with "power" rather than "rights". The work of Ribot and Peluso (2003) is valuable in strengthening knowledge and understanding of "access," particularly by exploring notions of power relations and actors (Myers and Hansen 2020). Likewise, access to public space governance is built from the power derived from social networks based on socioeconomic status. When access to public spaces and resources is analysed through a power relations framework, results show that the elite groups can claim and dictate access. In contrast, the general public is left with the option to negotiate for access.

The maintenance and control of access, along with webs of powers, hierarchies and social differentiation, can also have a spatial and territorial bearing (Ribot and Peluso 2003). García and van Dijk (2020) propose a framework to understand better typologies of access, particularly concerning claims to natural resources and land. The framework includes (1) grounding claims, which are about how the landscape is inscribed to visibly mark a territory through, for example, fencing and signage display; (2) talking claims, which refer to speech to contest and justify access; and (3) the practice of representing claims, that is claims as represented or reflected on material objects separate from the actual resource for example depicted on a map or plan and title deed documents. This methodological approach to analyse access includes the mapping and identification of the mechanisms deployed by different actors to control, distribute, and sustain the benefits flow, as well as to analyse the power relations at play (Ribot and Peluso 2003: 161). It has been recorded in the literature that in places such as Florida and California in the United States of America, there is evidence of territories where rich people often close off access to beaches for the public to create private beaches for exclusive enjoyment, which in many instances are overturned by court judgments in favour of the public (Kranz 2009; McKeon 1970). Similar actions have seldom been recorded in academic writings in South Africa, except in some official documents and online newspaper articles. For example, an official document mentioned the following:

> "Access to the coastal zone and its assets has historically been inequitable. There are currently large tracts of private land through which the public may not traverse and even may not approach along the seashore. This has resulted in restrictions to access to the coast and its resources in conflict with the South African Constitution in addition to conflicting with the ICMA". (Western Cape Provincial Government 2017).

9.2.3 *Public Coastal Access and Spatial Justice*

In democratic South Africa, Public Coastal Access should be recognised as a spatial transformation tool to support spatial visions related to integration, justice and inclusiveness in access to economic opportunities. The legacy of spatial exclusions in South Africa did not spare access to beaches and other coastal areas. Public Coastal Access is a concept born in an attempt to address spatial injustices of the past and to seek democracy, equity and integration amongst all members of the South African society. Such recognition is important because the country has had a history of dispossession, exclusions and discrimination in coastal space and resource distribution. Public coastal access restrictions remain a challenge for many previously marginalised communities in South Africa owing to the history of Apartheid. Mafumbu et al. (2022) add that there is a need for drastic shift in spatial planning to address these ongoing spatial imbalances and while they recognise that the Integrated Coastal Management Act (2008) which is the second law after the Constitution (1996) that advocate for public coastal access, lacks the capacity and boldness to implement its provisions in the interest of the general public. Bearing this in mind, it raises questions about the impact of gated development as a form of territorialisation on territorial governance and cohesion at a broader level.

9.3 Methodology

This chapter focuses on gated coastal estates and their relationship to public coastal access in South Africa within the context of territorial governance. Research completed by Monnakgotla (2023) which dealt with negotiated privilege, presents empirical data that was collected at macro, meso and micro levels, including district and local municipalities, luxury gated estates fronting the coast and fishermen communities spread across the provinces of KwaZulu-Natal and Western Cape. The study examined how public coastal access is restricted in South Africa, how municipalities responded to those restrictions and what the implications of public coastal access restrictions were for poor communities, mainly local fishermen. Appropriate research methods and designs were selected based on these three research questions (Leedy and Ormrod 2014), including three case studies utilising documentation review, spatial analysis and semi-structured interviews. Social science research in the Global South deploys innovative methods in response to environments characterised by tensions, diversity and inequalities (Laher et al. 2019). Furthermore, studies that seek to unpack the interaction between socioeconomic, environmental, cultural and political factors that shape spatial planning require context-based research (Tesner-Smith 2019).

The study focussed on three case studies—one in Kwa-Zulu Natal province and two in the Western Cape (Fig. 9.1). The case studies included various District and Local Municipalities and some gated developments along the coast (Table 9.1). The

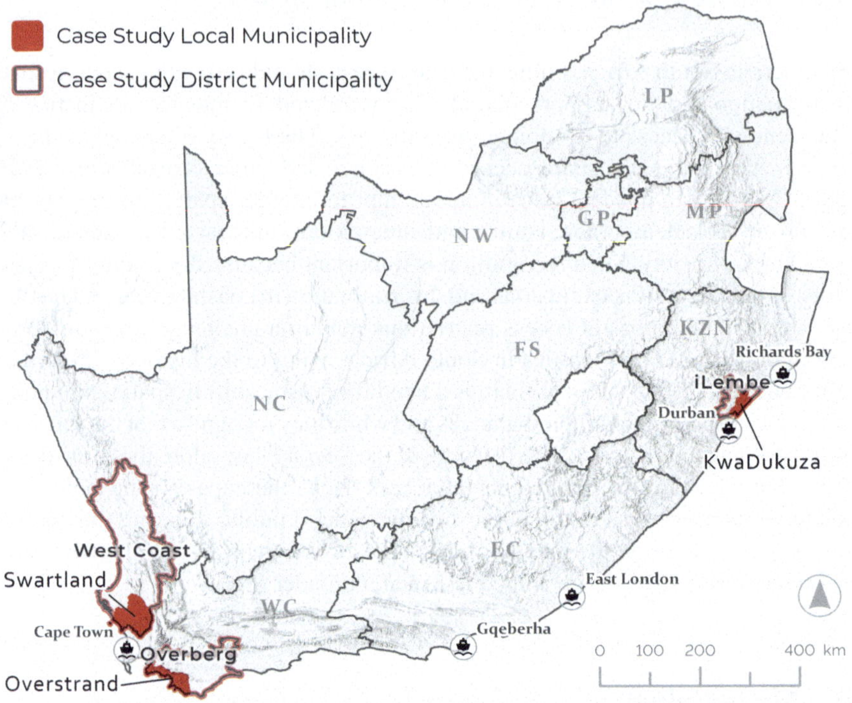

Fig. 9.1 Map of the distribution of case studies (original to the author)

Table 9.1 Case study samples

Samples	Case study 1	Case study 2	Case study 3
Province	KwaZulu-Natal	Western Cape	Western Cape
District Municipality	Ilembe	Overberg	West Coast
Local Municipality	Kwadukuza	Overstrand	Swartland
Coastal estate	Zimbali	Middlevlei/Sonesta	Grotto Bay
Focus groups with local fishermen	Groutville, Thembeni and Gledhow	Kassiesbaai and Hawston	Mamre, Atlantis and Darling

data collection incorporated the review of official documentation, which included planning permissions. The documentation analysis method supported the qualitative research by providing multiple sources of evidence in pursuit of validation and convergence (Bowen 2009; Yin 2009). Data was also collected through site observations at gated coastal estates which included observing the security and access control features of the estates. In-depth interviews with planners at district and local municipalities were conducted to gather information on how they manage coastal access in terms of planning laws and how they respond to coastal access restrictions. Focus groups with small-scale fishers in each case study were centered around their experiences in interacting with government policies on the one hand and coastal estates residents in areas where they fish. Innovative methods were used to combine various sources of evidence to illustrate public coastal access restrictions in a schematic drawing by overlaying nonspatial data such as planning approvals, observations in real-time, and data from interviews and focus groups to explain how territorial governance physically plays itself out at coastal territories.

9.3.1 *The Case Studies at Meso-Level*

The following sub-subsections present the case studies at meso-level to provide a local spatial context of each estate and its surrounding setting to briefly explain the relation between the estates and the fishermen, both spatially and functionally.

9.3.1.1 Case Study 1: Kwadukuza

Figure 9.2 shows the spatial context of the Kwadukuza case study with focus on the coastal gated estates. There are three significant coastal estates, including Zimbali coastal estate (Fig. 9.2). The Many fishers (participants in the focus groups) reside in the three neighbourhoods of Groutville, Thembeni and Gledhow located east of the coast and traditionally used the coastal area next to the gated estates.

9.3.1.2 Case Study 2: Overstrand

Hawston—a small village where fishers reside—is located closely to the Middlevlei coastal estate. The Middlevlei estate privatised the beach and a fishing spot that was used by many fishers from Hawston. Middlevlei estate used to be called Sonesta and a coloured people designated beach under the Separate Amenities Act 49 of 1953 during the apartheid racial segregation system in South Africa. Official records show that the property was sold off in the early 1990s as a public asset to private individuals, challenging the rights of public to access coastal spaces. A further challenge for the Hawston community is the restriction of boat launch access at the Hawston slipway in terms of the Management of Public Launch Sites in the Coastal Zone regulations under the ICM Act (Fig. 9.3).

9.3.1.3 Case Study 3: Swartland

The third study area is situated in two municipalities namely Swartland and the City of Cape Town. The areas of fisher communities' historical coastal access land are located in Swartland, and the fisher communities who formed the focus group are residing in Mamre and Atlantis in the City of Cape Town metropolitan municipality

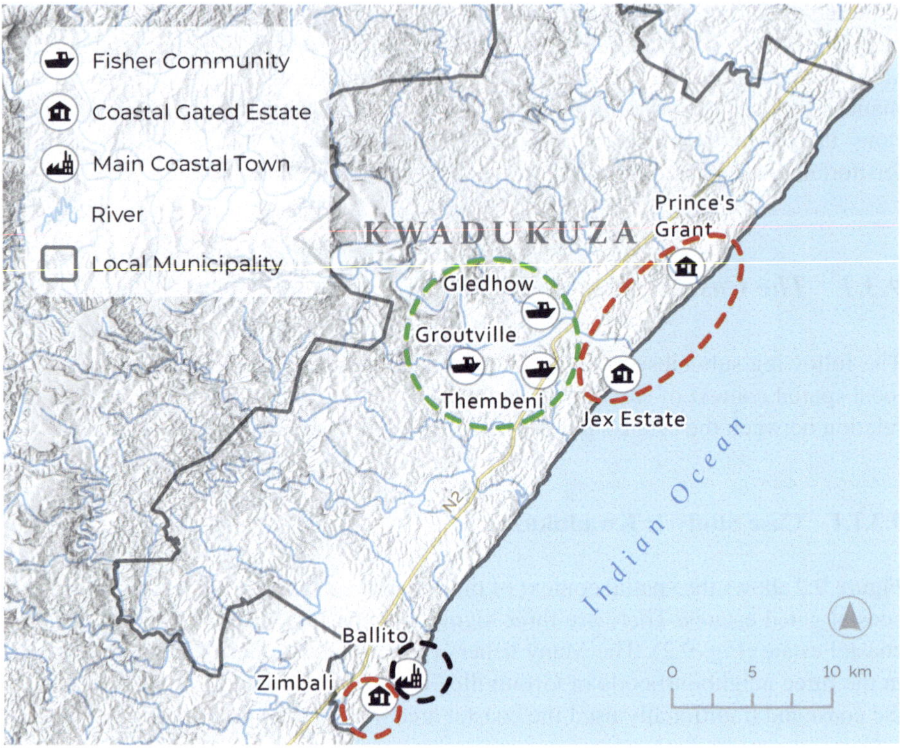

Fig. 9.2 Local spatial context of case study 1 to illustrate the coastal estate and the surrounding settings which include fisher communities (original to the author)

and Darling located in Swartland. The two gated estates that were assessed included Grotto Bay luxury estate, and Ganzekraal coastal resort. These estates restricted the use of historical launching sites comprising a slipway and natural beach surf launch which is in close reach of the fisher communities (Fig. 9.4).

9.4 Public Coastal Access Restriction in South Africa

This section discusses the location and nature of the gated developments, the role of district and local municipalities related to coastal management, and the impact on local fishermen from the surrounding communities to demonstrate the relationship between territorialisation and a lack of territorial governance.

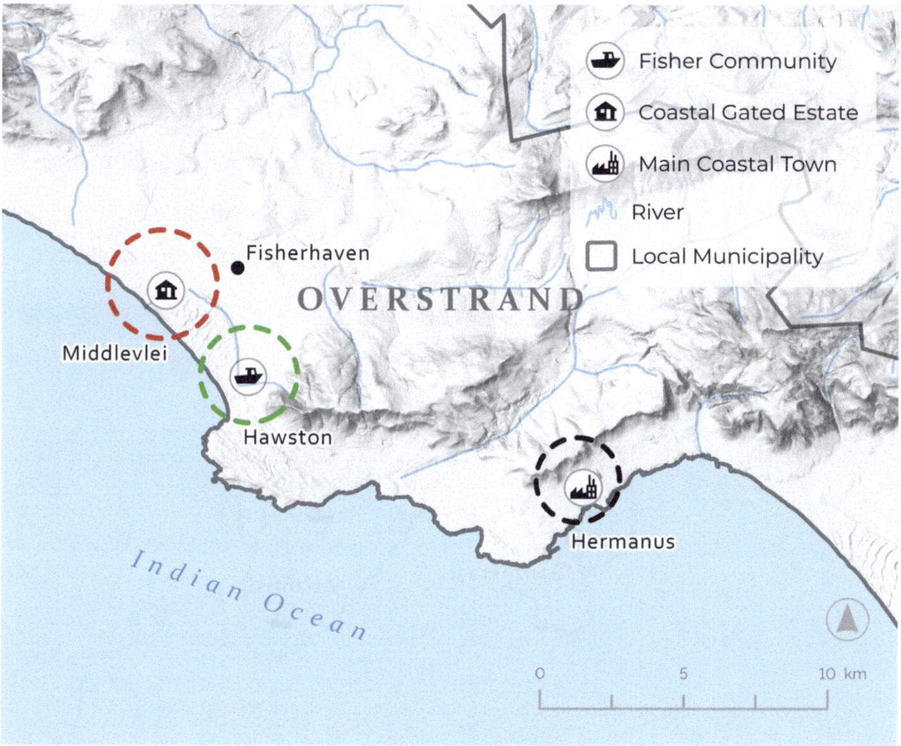

Fig. 9.3 Local spatial context of case study 2 to illustrate the coastal estate and the surrounding settings which include fisher communities (original to the author)

9.4.1 Territorial Control and Micro-Governance

The coastal developments considered in the study include Zimbali, Middlevlei and Grotto Bay coastal estates, spread across the case study municipalities. They are all affluent gated neighbourhoods with manned security access points, managed through Home Owners Associations (HOA) to oversee the security and maintenance of the estate as well as to enforce compliance of the estate rules by all residents and visitors. The interaction between the estates and local municipalities is also facilitated through the HOAs. The estates are set up to offer private and easy access to the coastline for their residents and guests (Fig. 9.6). In addition, the proximity and accessibility to the beaches and coastlines are vital features of these developments. These coastal estates do not permit free and unrestricted access to the public coast for general members of the public. However, public members may access the beaches in front of these estates by hiking on foot from either side through lateral access instead of the vertical access enjoyed by estate residents and guests. It must, however, be acknowledged that the Middlevlei coastal estate had, in recent times, started to allow access to members of the public who must present

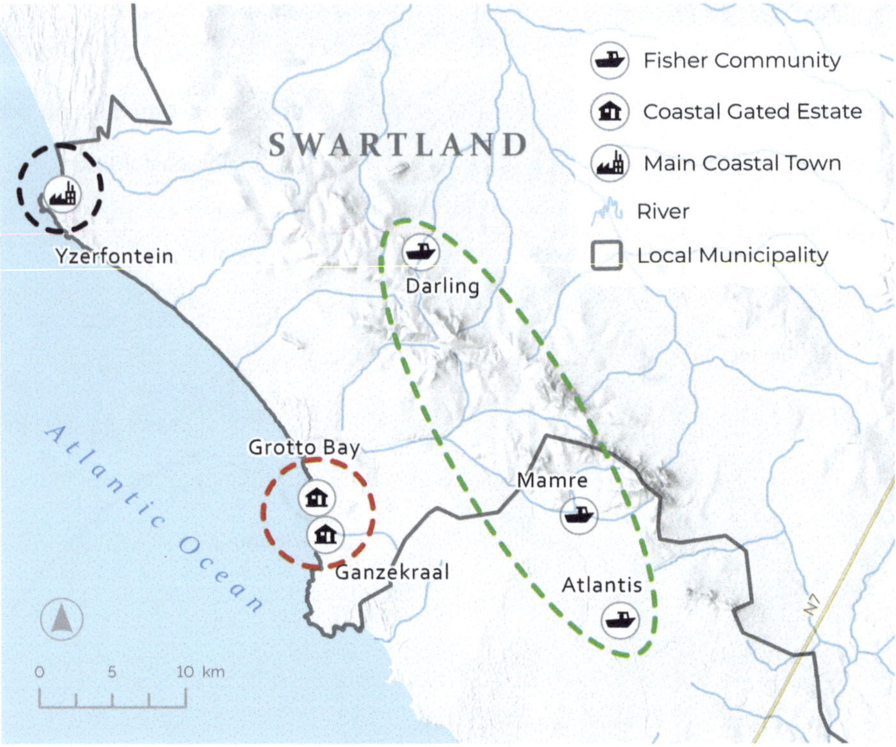

Fig. 9.4 Local spatial context of case study 3 to illustrate the coastal estate and the surrounding settings which include fisher communities (original to the author)

themselves at the security entrance for screening registration and to be issued with permit slips to access the coast via a private road within the estate. These are small-scale, non-commercial fishers from small, poor, marginalised communities.

The physical designs of the gated coastal estates restrict access to coastal spaces abutting them in several ways. These estates are manned by security guards and access control mechanisms like a guardhouse and a boom gate at the access point. Although this access control point is meant to service the estate, residents and the public must pass many control points to access the beach.

Alternative public coastal access is available but may require a long walk and it is unsafe. It would entail going around a large coastal estate—typically on foot—as the paths may be narrow and unofficial. In all the three case studies assessed, it can be presented that the coastal estates reviewed were all developed on or near historical access land and have undertaken to either deregister proclaimed public roads to make them private by using road ordinances, for example in Middlevlei. In the case of Zimbali, the highway was rerouted to restrict public road thoroughfare within the estate. In Grotto Bay, there is a closed off-road vehicle access by means of a steel boom on the admiralty reserve—a strip of coastal land registered under government ownership.

Coastal estates are often fenced on all sides of the property except on the coastal front. They further enhance access for their residents and paying guests. People within coastal estates enjoy free and easy access to coastal resources with minimal or no oversight from authorities. The design and securitisation of gated coastal estates make it difficult for environmental enforcement authorities to come in unannounced for inspections or law enforcement operations. This also implies that gated coastal estate residents and guests govern themselves. They can access the coast via secured and private access to swim, launch boats and fish at any time, including at night, without the presence of authorities to ensure that they comply with the country's strict natural resource management laws. As micro-governments, they can also establish estate rules that regulate use of coastal resources and access to the coastal space through the estate (Fig. 9.5).

Figure 9.6 offers a spatial schematic constructed from the various coastal estates observed in the study to demonstrate how public coastal access is typically restricted at such territories and how territorial governance takes shape in gated coastal estates. The restrictions are initiated by obtaining development approvals and environmental authorisations from respective authorities like municipalities and provincial departments. It must be noted that it is standard practice for authorities issuing approvals to such developments to include approval conditions, which clearly state that such developments must not impede on the rights of the public to access public coastal spaces. However, the said authorities rarely enforce such conditions. Figure 9.6 demonstrates how approved development designs are implemented to enable private actors to take control of coastal territories and common goods.

Fig. 9.5 A gated coastal estate outside of Hermanus. The road within the private property was previously a proclaimed public road, but now it is registered as a private road (original to the author)

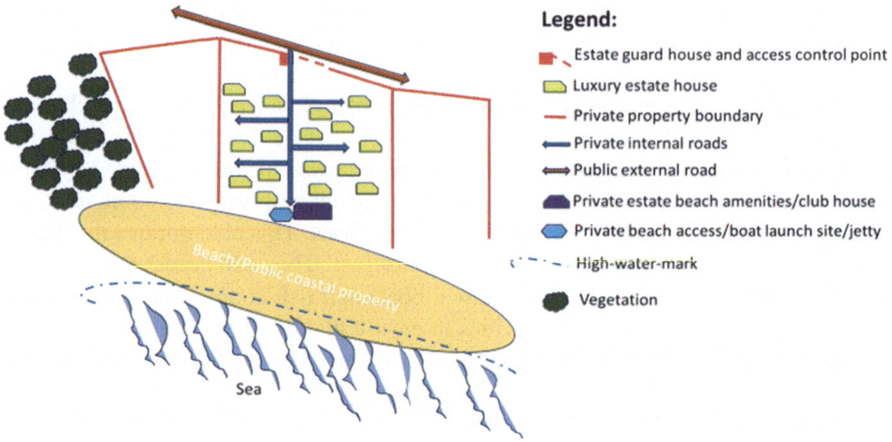

Fig. 9.6 A typical design of a luxury coastal estate to demonstrate physical public coastal access closures (original to the author)

9.4.2 The Role of Municipalities Related to Territorial Governance in the Coastal Regions

Local and district governments' responses to these new forms of territorialisation have been piecemeal. Both district and local municipalities have yet to act and enforce development conditions and legislation that proactively promote equitable access and distribution of public coastal spaces and resources. Prior to the formulation of the ICMA in 2000, District municipalities used to be the delegated authority and actively managed the coast. Among these duties were issuing boat licenses; ORV (beach driving) permits and law enforcement, which generally dealt with public coastal access. The district municipalities indicated that this created employment for many people, and municipalities were well-resourced, including having equipment like radios, boats, and vehicles for law enforcement. Approximately R1.8 million could be generated for municipal revenue through this function (Overberg District, Interview, 6 August 2018). The district's influence is limited to coastal management in the current environment. Although Spatial Development Frameworks (SDFs) at the district level influence local municipal SDFs to some extent, little detail can be provided on dealing with public coastal access or getting involved in broad coastal management issues generally. Thus, the district's involvement in issues of public coastal access and broad coastal management has been piecemeal rather than proactive.

District municipalities have developed Coastal Management Plans (CMP) to assist local municipalities. However, local municipalities currently lack the resources needed, such as human and financial capacity, which hampers the implementation of some of the planned initiatives. A District CMP's objectives include obtaining sustainable coastal development and must align with the provincial and national

CMPs visions, reflecting notions of vertical alignment required for effective territorial governance. Districts have highlighted issues of capacity and funding as the main barriers to actively performing coastal management actions.

Planners from the various municipalities also conveyed a reluctance to deal with the issue of public coastal access restriction. Officials from KwaDukuza Municipality indicated that the municipality was busy with a project to map and audit public coastal access points to establish which are public and which are private, which will assist in identifying those coastal access strips that are negatively impacting the environment. The municipality received several public coastal access complaints that the Council considered and delegated to the environmental management unit to resolve—it is unclear if these complaints were resolved or whether it was covered in the broader project of mapping access strips. The environmental management units handle matters of public coastal access within KwaDukuza and the Overstrand local municipalities. In contrast, the Swartland municipality still needs a dedicated official as they rely on the environmental management unit in the West Coast District for that purpose. The Kwadukuza municipality town planning units' involvement in public coastal access has been limited and only related to stipulating conditions like preserving public coastal access in new land development applications for properties abutting the coast and requiring developers to deliver beach facilities as part of their contributions. However, concerning older developments like Zimbali in KwaDukuza, there is no legal standing for the municipality to impose public coastal access according to the interviewees (KwaDukuza Local Municipality interview, 8 July 2019), as the applications were submitted under the Development Facilitation Act (1995), which did not make provision for public coastal management stipulated under the ICM Act (2008).

The Overstrand Municipality initiatives on public coastal access have been to hand over names and locations of contentious coastal access points to the District Council. Meanwhile, new coastal developments are handled through instruments like the SDF, Environmental Management Frameworks (EMF), Coastal Management lines (CMLs) and GIS overlays to guide them on coastal development decisions and ensure public coastal access preservation. The municipality considers its coastline as a shared space and denies that there are any private beach access developments. For them, the coast is open for the public to use. However, they acknowledged that there may be claims of closed-off historical access points that the municipality must be made aware of and would investigate if this comes to their attention.

A planner from the Swartland municipality stated that they do not perceive themselves as a coastal but an agricultural municipality, and that the municipality could be more active regarding coastal management or public coastal access. Coastal development is managed similar to any other land development in terms of the town planning scheme. Their reluctance to address coastal development from an explicit coastal zone management approach could be due to the lack of clarity of the ICMA, specifically regarding roles. Secondly, the municipality relies on the district and province in such matters. Similarly, the planning unit in this municipality uses conditions in the development planning approvals, i.e. conditions that state that development should not block access to the coast for members of the public. The municipality lacks capacity or interest to enforce those conditions. This discussion highlights the need for more cooperation and agreed actions between the district and local municipalities related to coastal resource management.

9.4.3 The Impact of the Public Coastal Restriction on Local Fishermen

Territorialising parts of the coastline resulted in privatising coastal space and resources for the consumption of private actors as demonstrated in the Zimbali, Middlevlei and Grotto Bay gated coastal estates. The boat launch sites practices have proven to be unfavourable to fishers in the small, poor and marginalised areas. For example, in the case of Kwadukuza, boat launch sites that are permissible for use are in affluent areas such as Ballito. In the Overstrand case, the boat launch site in Hawston has been removed from the list of permissible use launch sites even though it has a good and solid ramp for launching of boats. Meanwhile, in the Swartland case, there are several launch sites, some being natural surf launch sites and others being hard infrastructure such as in the Ganzekraal resort. All the launch sites which are within a short reach for the Mamre fishing community have been declared non-permissible for use. The next closest launch sites for these communities are also in affluent areas like Yzerfontein and Blouberg. Fishers in Mamre have expressed concern that the authorised launch sites are too far away and have cost implications to travel, while the closer ones also have significant historical value. The fishermen also indicated feeling intimidated to launch in affluent areas among people that have lavish equipment such as boats, jet skis and vehicles.

Fisher communities in the three different case study sites all indicated that the coastal access restrictions influence their lives in many adverse ways. Many of them mentioned that the restrictions have increased poverty levels as fishing is their primary source of livelihood. For them, fishing is a culture and heritage they learned from their forefathers and hoped to transfer to their children. However, with the advent of the ever-increasing restrictions on access to coastal resources, the fishing culture and heritage are rapidly eroding. Lastly, most fishermen highlited/indicated the need to organise themselves into social movements in search for justice and rights of access to coastal resources, which raises issues related to participation in the quest for territorial governance. Although in Kwadukuza, the Gledhow fisher community has made some attempts to protest against access restrictions, these have not been enough to have their voices heard by authorities effectively.

9.5 The Impact and Implications of Gated Developments on Territorial Governance

The discussion indicated that these gated developments territorialised parts of the coastline and that these actions restrict coastal access for local fishermen. In this way, the gated estates represent notions of territoriality, which, as noted by Sack (1986), is a spatial strategy to influence or control recourses and people through the control of an area. Like other gated developments in the country, these coastal estates also reflect notions of spatial or territorial control and private micro-governance that may constrain democratic accountability in some cases. The gated developments that were studied have all obtained planning approvals and environmental authorisations from respective authorities. Some patterns are evident in the assessments of documentation submitted by developers to authorities. Amongst these patterns are that all the land sites undertaken to develop luxury coastal estates were previously recreational and fishermen's coastal access areas. Those areas were either not managed or burdensome for authorities to manage. Developers took advantage of that situation by claiming that developments would ensure such land parcels' effective and sustainable use.

However, although these gated developments can be considered an example of territorialisation, they do not reflect the typical characteristics of territorial governance. As discussed, territorial governance encompasses collective action based on openness and cooperation of actors. While there may be consensus among the residents of the gated developments, there is a lack of cooperation with the surrounding fishermen communities who need access to the beach and boat launching sites. Secondly, territorial governance should relate to the territorial organisation of many relations and interactions among different, non-conflicting interests to manage territorial capital (Davoudi et al. 2008). In these cases, the beach and the maritime resources are the territorial capital. However, the public coastal restrictions highlight tensions related to conflicting interests about perceived safety for residents of the estates and the livelihoods of local fishermen. These tensions impede processes of addressing territorial imbalances and spatial disputes, which, as Zaucha and Böhme (2020) noted, is a critical requirement for territorial cohesion.

Furthermore, restricting coastal access has direct implications for achieving territorial governance actions, as there is no shared vision at a regional level related to regional spatial development. In these cases, territorialisation hinders the actualisation of the coast and the maritime resources as a common good. It does not valorise territorial capital for the benefit of all the stakeholders, two critical aspects of territorial governance actions (Davoudi et al. 2008). The third critical aspect, defining territory as part of a participation process, also suffers due to privatised spatial control through territorialisation. The broader region is not defined or governed as a holistic territory, as the voices of the fishermen are rarely considered and do not have an influence. Even though organisations representing fisher rights objected to these developments, the developers established their rights to develop. It must also be noted that planning approvals consistently included strict conditions that coastal developments must not restrict public coastal access. It appears that municipalities would need more capacity and resources to enforce such conditions and left the onus to developers. It also shows that authorities have ceded the powers to facilitate public coastal access and management of public coastal resources and space to large gated coastal estates.

Territorial governance also manifests at different scales and should facilitate multi-level governance (Davoudi et al. 2008; Stead 2014). The discussion indicated that the district and local municipalities tend to avoid their responsibilities specifically regarding coastal access management. While the Constitution and the ICMA call for justice and access to public space and resources, there is little evidence that the district and local municipalities enforce this. This may be due to a need for more clarity related to the roles of different spheres of governance or a lack of capacity and funding. Even though some municipalities have responded to some complaints from the fishermen, they have not boldly undertaken any measures that are in the public interest. Some municipalities have taken the stance that they cannot interfere with private developments, while others have attempted to bring poor communities closer to the coast using social housing. However, this was opposed by the existing coastal estates. In another instance, one municipality pointed out that it would be costly to service new coastal sites with ablution, parking, signage and other related beach access amenities. The municipality is not able to provide infrastructure and support it with security to curb potential vandalism. The evidence indicates municipalities need to prepare or be able to undertake their constitutional obligations, despite Schedule 5, Part B of the Constitution of South Africa (1996) stating that beaches and amusement facilities are local government matters. In addition, the ICM Act (2008) specifies that municipalities must designate public coastal land over private property and assign supporting beach access infrastructure. There is also a notion of side-stepping access responsibilities between the District and Local municipalities by claiming that there needs to be more clarity about which tier the ICM Act refers to when it refers to a municipality. Not addressing the issues reflects a reluctance from municipalities to support public coastal access, which impedes spatial transformation and justice. These actions also indicate that municipalities prefer the option of allowing private actors such as gated coastal estates to manage public coastal access on their behalf, or as Dostál (2002) notes; territorialisation gives rise to the diffusion of responsibilities and democratic accountability constraints as a result of negotiations, collaborations and the exchange of resources amongst a group of affiliates. The question then becomes, what resources can be exchanged between a gated estate development and a municipality within a planning and coastal management environment, while they collect revenue from the property taxes?

Consequently, fishers and other previously marginalised communities will endure access restrictions and have to access the coast via rugged terrain. Another option for the fishers is to negotiate access to private property, but the private property owner in a coastal estate reserves the right to grant access to a public asset. Reserving access relates to negotiating privilege and the notion of "access," as conceptualised by Ribot and Peluso (2003), where it demonstrates the role of power rather than rights in public coastal access relations (Myers and Hansen 2020). In the case where access to the coast over private property is granted to a fisherman, that fisherman will still need to abide by the rules of the estate owner; for example, time limits may be imposed despite the expertise and desire of the fisherman to target certain times and weather conditions. Public coastal access restrictions for fishermen imply that fisher communities lose their heritage, livelihoods and, to some extent, food sources for their families.

9.6 Conclusion

The discussion focussed on three gated developments in Kwa-Zulu Natal and the Western Cape to explore the territorialisation of space and its influence on public coastal access. The chapter indicated that the location and development of these gated developments restrict direct access to the coast through territorialisation and the spatial control of territories next to the coast. The restriction of public coastal access has a negative impact on the sustainable livelihoods of the fishermen from local communities in the surrounding areas. Nevertheless, the district and local municipalities still need to deal with public coastal access restrictions in the broader context of territorial governance. Firstly, the gated developments are examples of negotiated privilege, where they are allowed to control and restrict access to a demarcated spatial area along the coast.

However, these acts of territorialisation extend beyond the rigid boundaries of the estates. By closing off direct access to the beaches and traditional access routes, they make it difficult for local fishermen to access the beach conveniently and affordably. Secondly, the district and local municipalities have not managed to address the spatial imbalances, disparities and differences in the potential for development, which impedes the possibilities for territorial cohesion and challenges notions of spatial justice. The coast's territorialisation in these instances also challenges the directives set out in the Constitution of South Africa and the ICMA, which calls for access to public spaces and resources and the management of these areas through the provision of infrastructure. Thirdly, territorial governance actions are neglected as there are (1) restricted access to the beach and maritime resources as a common good, (2) limited or no valorisation of territorial capital (maritime resources) for everyone in the region and (3) no integrated definition of territory as part of a broader participation process. The negotiated privileges by the gated developments, therefore, have negative implications for territorial governance in these coastal regions.

There are several implications for spatial transformation in South Africa. In most cases, the infrastructure for public coastal access is either lacking or existing within restricted areas, hindering the ability of local fishers to access livelihoods opportunities. Secondly, due to segregated spatial patterns, many lower income communities are located further away from opportunities, in this case, the coast and specifically accessible launching sites. Consequently, they must travel further than those residing on the coast, implying greater physical effort to walk or additional costs to secure transport. Unfortunately, privatising beaches to secure safety is perpetuating the inefficiencies, injustices and inequalities of the past.

In 1994, South Africa became a democracy, promising equality, justice and access to public spaces and resources for everyone. The NDP also emphasises the need for spatial transformation in urban and rural communities and expanding infrastructure. At the same time, the SDGs highlight the importance of sustainable development, including sustainable communities, justice and partnerships. While territorial governance has the potential to facilitate these goals through the organisation of territorial actions, openness and cooperation, this is challenged through actions of territorialisation and spatial control in some coastal regions. Given this, democracy and spatial justice must include public coastal access for everyone in the region to ensure territorial cohesion.

References

Arango Espinal E, Ceballos Molano R, Patiño Giraldo LV (2021) Territorial governance: a bibliometric analysis. Cuadernos de Administración 36(68):14–27. https://doi.org/10.25100/cdea.v36i68.9702

Atkinson R, Blandy S, Flint J, Lister D (2004) Gated communities in England. Office of the Deputy Prime Minister, London

Bavinck M, Berkes F, Charles A (2017) The impact of coastal grabbing on community conservation—a global reconnaissance. Marit Stud 16:8. https://doi.org/10.1186/s40152-017-0062-8

Blakely EJ, Snyder MG (1997) Fortress America: gated communities in the United States. Brookings Institution Press, Washington

Bowen GA (2009) Document analysis as a qualitative research method. Qual Res J 9:27–40. https://doi.org/10.3316/QRJ0902027

Chettiparamb A (2016) Articulating 'public interest' through complexity theory. Environ Plan C Govern Policy 34(7):1284–1305. https://doi.org/10.1177/0263774X15610580

Coccossis H (1985) Public policy for coastal areas. In: Hall DO, Myers N, Margaris NS (eds) Economics of ecosystems management. Tasks for vegetation science, vol 14. Springer, Dordrecht. https://doi.org/10.1007/978-94-009-5502-8_9

Congleton RD (2011) Perfecting parliament: constitutional reform, liberalism, and the rise of Western democracy. Cambridge University Press, Cambridge

Davoudi S, Evans E, Governa F, Santangelo M (2008) Territorial governance in the making. Approaches, methodologies, practices. Boletin de la AGE 46:33–52

Dinzey-Flores Z Z (2013) Locked in, locked out: Gated communities in a Puerto Rican city. 1st ed. Philadelphia: University of Pennsylvania Press (The city in the twenty-first century). https://doi.org/10.9783/9780812208207

Dostál P (2002) Territorial government and flexibility: a critical assessment. Belgeo. 3, 227-242.

Durington M (2009) Suburban fear, media and gated communities in Durban, South Africa. Home Cult 6(1):71–88

Faludi A (2012) Multi-level (territorial) governance: three criticisms. Plan Theory Pract 13(2):197–211. https://doi.org/10.1080/14649357.2012.677578

Fourie F (2018) Aligning South Africa's national development plan with the 2030 Agenda's sustainable development goals: guidelines from the policy coherence for development movement. Sustain Dev 2018(26):765–771. https://doi.org/10.1002/sd.1745

Garland D (2001) The culture of control: crime and social order in contemporary society. Oxford University Press, Oxford

Jeans DN (1990) Beach resort morphology in England and Australia: a review and extension. In: Fabbri P (ed) Recreational uses of coastal areas. Kluwer, Dordrecht

Jerzak K, Shrayer MD, Kośnicka KA, Lorens P, Zaucha J (2019) The essence of marine and coastal space—an interdisciplinary perspective. EUROPA XXI 36:15–33. https://doi.org/10.7163/Eu21.2019.36.2

Kranz E (2009) Sand for the people: The continuing controversy over public access to Florida's beaches. Florida Bar Journal, 83(6).

Kronenburg García A, van Dijk H (2020) Towards a theory of claim making: Bridging access and property theory. Society and Natural Resources 33(2):167–183. https://doi.org/10.1080/08941920.2018.1559381

Laher S, Fynn A, Kramer S (eds) (2019) Transforming research methods in the social sciences: case studies from South Africa. Wits University Press, Johannesburg. https://doi.org/10.18772/22019032750

Landman K, Schönteich M (2002) Urban fortresses: gated communities as a reaction to crime. Afr Secu Rev 11(4):71–85

Leedy PD, Ormrod JE (2014) Practical research: planning and design, 10th edn. Pearson Education, Edinburgh Gate

Lemanski C (2004) A new apartheid? The spatial implications of fear of crime in Cape Town, South Africa. Environ Urban 16(2):101–112

Lemanski C, Landman K, Durrington M (2008) Divergent and similar experiences of 'gating' in South Africa: Johannesburg, Durban and Cape Town. Urban Forum 19(2):133–158

Low S (2003) Behind the gates: life, security and the pursuit of happiness in fortress America. Routledge, London

Madanipour A (2010) Whose Public Space?: International Case Studies in Urban Design and Development (1st ed.). Routledge. https://doi.org/10.4324/9780203860946

Mafumbu L, Zhou L, Kalumba AM (2022) Exploring coastal access in South Africa: a historical perspective. Sustainability 14(7):3971

Makhale S, Landman K (2018) Gating and conflicting realities: challenges in practice and theoretical implications. Int Plan Stud 23(2):130–143

McKeon S (1970) Public access to beaches, Stanford Law Review. 22: 564–586

Monnakgotla HTT (2023) Negotiating privilege: Unpacking Public Coastal Access in the Kwadukuza, Overstrand, and Swartland coastal municipalities of South Africa. Doctoral thesis, University of Pretoria

Myers R, Hansen CP (2020) Revisiting a theory of access: a review. Soc Nat Resour 33(2):146–166. https://doi.org/10.1080/08941920.2018.1560522

Naar N, Mahenge J (2014) Finding a middle ground: conservation challenges among stakeholders in Coastal Tanzania. Anthropol Today 30(1):9–14. https://doi.org/10.1111/1467-8322.12084

Petr Dostál, "Territorial government and flexibility: a critical assessment", Belgeo [En ligne], 3 | 2002, mis en ligne le 01 octobre 2002, consulté le 21 août 1984. https://doi.org/10.4000/belgeo.15719

Phoenix C, Bell SL, Hollenbeck J (2021) Segregation and the sea: toward a critical understanding of race and coastal blue space in Greater Miami. J Sport Soc Iss 45(2):115–137. https://doi.org/10.1177/0193723520950536

Pow CP (2009) Gated Communities in China: Class, privilege and the moral politics of the good life (1st ed.). Routledge. https://doi.org/10.4324/9780203876206

Ribot JC, Peluso NL (2003) A theory of access. Rural Sociol 68(2):153–181

Sack RD (1986) Human Territoriality: Its theory and history. Cambridge: Cambridge University Press

Sowman M, Malan N (2018) Review of progress with integrated coastal management in South Africa since the advent of democracy. Afr J Mar Sci 40(2):121–136. https://doi.org/10.2989/1814232X.2018.1468278

Stead D (2013) Dimensions of territorial governance. Plan Theory Pract 14(1):142–147. https://doi.org/10.1080/14649357.2012.758494

Stead D (2014) The rise of territorial governance in European policy. Eur Plan Stud 22(7):1368–1383. https://doi.org/10.1080/09654313.2013.786684

Tesner-Smith D (2019) Towards a narratology of planning: stories of a South African gold mining town. Doctoral thesis, University of Pretoria

Yin RK (2009) Case study research: design and methods, 4th edn. Sage, Thousand Oaks

Zaucha J, Böhme K (2020) Measuring territorial cohesion is not a mission impossible. Eur Plan Stud 28(3):627–649. https://doi.org/10.1080/09654313.2019.1607827

Chapter 10
Provincial Public Transport Governance Systems and Urban Spatial Transformation Nexus in South Africa: Experiences and Lessons Learnt

Trynos Gumbo, Retsepile Kalaoane, and Kiara Lawrence

Abstract The 1994 democratic dispensation in South Africa ushered in a renewed impetus to redress previously skewed spatial, economic, and social systems. Previously, access to and movement within urban areas had been restricted for non-white people. Consequently, efforts were made to correct these inequalities and serve the entire population. Notably, legislative and policy instruments were enacted to regulate government interventions at all levels. These included policies, strategies and plans to guide and support public transport systems development. What remains to be documented adequately is the role of the governance systems and the public experiences of the structures at territorial levels. This chapter discusses the role, influences, and outcomes of urban public transport systems in South Africa. It focuses on the achievements that have been recorded in the Gauteng and KwaZulu-Natal provinces in formulating and implementing public transport policy and planning. Results reveal meaningful progress in formulating plans and strategies at these provincial levels which have significant potential for transforming urban spaces through large-scale urban public transport programme implementation. Spatial inequality remains a significant issue, with many communities still facing barriers to accessing economic opportunities and essential services. Long-standing structural and systemic barriers continue to impede the full realization of the goals set by these policies.

Keywords Public transport · Governance systems · Provincial spheres · South African cities · Spatial transformation

T. Gumbo (✉) · R. Kalaoane
Urban and Regional Planning Department, University of Johannesburg,
Johannesburg, South Africa
e-mail: tgumbo@uj.ac.za

K. Lawrence
Town and Regional Planning, Durban University of Technology, Durban, South Africa

D. du Plessis (ed.), *Territorial Governance and Spatial Transformation in Post-apartheid South Africa*, Local and Urban Governance,
https://doi.org/10.1007/978-3-031-85489-7_10

10.1 Introduction

In 1994, South Africa underwent a significant transformation from the apartheid era to democracy (Mosala et al. 2017). This crucial period saw changes in public transport instigated across different sectors of the country. Before 1994, the public transport system in South Africa served the primary purpose of transporting non-white people who were located on the fringes of urban areas to their places of work (Thomas 2016). Transport infrastructure and services were designed to separate non-white people economically and physically from white-only areas which were often characterised by improved infrastructure, economic opportunities, and services (Khosa 1998; Lemon 2021; Strauss 2023). These generated socio-economic exclusions, racial inequalities, spatial inefficiencies, and increased transportation costs. Spatial planning pre-1994 key events have been classified by Schoeman (2010) as follows:

- 1910 to 1930, spread of the British planning influence
  During the Union of South Africa's formation in 1910, the problem of misalignment between the country's several tiers of government were first identified (Schoeman 2015). Garden cities, town planning and zoning laws were highlighted in the introduction of British urban planning concepts (Howard et al. 2013; Coetzer 2016). These concepts primarily focusing on managing urban growth, the division of land uses and the establishment of suburban residential areas for the white population were put into practice in cities like Johannesburg and Cape Town (Howard et al. 2013; Coetzer 2016; Wood 2019). Three critical areas of development were the creation of garden suburbs which were influenced by Ebenezer Howards ideas resulting in the implementation of zoning laws and the introduction of formal town planning schemes. From 1910 to 1930 British planning influence expanded throughout South Africa solidifying racial segregation in transportation infrastructure and resulting in inadequately integrated urban transit systems and unequal access for non-white populations (Massey and Gunter 2020). Non-white South Africans experienced lengthy commutes and restricted mobility as a result which widened socioeconomic gaps.
- 1930s, Second World War and the post-war reconstruction efforts
  Urban planning has been widely regarded as a key instrument for societal reconstruction, as suggested by Mabin and Smit in 1997. South Africa's economy was significantly impacted by the Great Depression, resulting in heightened urban migration and strain on city infrastructure (Fredriksson and Moritz 2019). In response, the state escalated its involvement in urban planning to control growth and generate jobs via public works initiatives. The war redirected resources, diminishing focus on urban development, yet industrial expansion during this period spurred urbanization and city growth.
  After the war, the emphasis shifted to reconstructing and enhancing infrastructure to facilitate economic recuperation and expansion (Verney 2024). After taking office in 1948, the National Party implemented apartheid policies. Urban planning was impacted by the grand apartheid policies which enforced racial

segregation through the Group Areas Act of 1950 and other laws. Key developments were forced removals and resettlement of non-white populations and creation of racially segregated townships on the outskirts of cities.

- The period following the Soweto Uprising in 1976
  The 1976 Soweto Uprising was a significant anti-apartheid protest by black South African students marking increased resistance to apartheid policies (Brown 2016; Ndlovu 2017). Following the uprising some policies were reexamined considering the shortcomings and injustices of apartheid urban planning. Some apartheid policies were loosened because of growing international pressure and internal discontent.
- Late apartheid reforms after 1985
  To contain unrest, the apartheid regime implemented certain reforms in response to mounting internal resistance and international sanctions. Reforms included measures to improve urban conditions and integrate some aspects of urban life, but apartheid's core policies remained (Simon 2002; Strauss 2023). Important developments include the repeal of some apartheid laws like the Pass Laws and efforts to improve urban infrastructure and services in black townships. Therefore, urban planning contributed to the racial segregation of communities by marginalizing and relegating the South African black population to impoverished areas. Significant urbanization was brought about by industrialisation in the early twentieth century especially in mining regions like Johannesburg (Meier zu Selhausen 2022). The ability of planning authorities to provide sufficient housing and services was outpaced by this rapid growth. Zoning regulations with British influences frequently severed land uses disregarding mixed-use projects that could be advantageous to local communities. This occasionally resulted in ineffective urban planning and exacerbated segregation.
  Despite these disadvantages, the dismantling of apartheid policies in the post-apartheid era paved the way for remarkable strides in public transport governance systems. These facilitated the implementation of policies advocating for the development of an integrated transport system and the regulation of minibus taxis (Wilkinson 2010; Musakwa and Gumbo 2017; Diallo 2022).
  This shift towards an integrated and efficient transportation system reflected South Africa's commitment to rectifying previous historical inequalities. Nevertheless, South African public transport is still grappling with various challenges emanating from a lack of accessibility and higher transportation costs especially for the urban poor (Aivinhenyo and Zuidgeest 2019). For instance, conventional public transport is slowly losing market share to the minibus taxi industry. In South Africa, more than 50% of people are heavily reliant on minibus taxis as their main mode of mobility (Moyo et al. 2022; Walters and Pisa 2023). Urban spatial transformation has been widely documented during the apartheid era (Thomas 2016; Monama et al. 2022; Thumbran 2022). However, the dynamics of the interplay between provincial public transport governance systems and urban spatial transformation in South Africa since 1994 remains largely unexplored in the current literature. Using the cities of Johannesburg and Durban as case studies, this chapter documents the roles and outcomes of urban

public transport systems towards an integrated and sufficient public transport system as experienced by commuters. In the subsequent sections, this chapter discusses urban public transport systems in South Africa, followed by case study discussions. Section 10.4 focuses on materials and methods used while Sect. 10.5 present's findings and discussions.

10.2 Urban Public Transport Systems in South Africa

South African cities are characterized by low-density sprawl which developed over many years during the Apartheid era, and which created a culture of travelling long distances to work among black South Africans (Donaldson 2006). The challenges continued post-apartheid, with those living in developed areas of the city having access to private vehicles, and the poor living in underdeveloped areas on the outskirts of the city. These citizens were forced to use a combination of transport modes such as buses, minibus taxis, trains, bicycles, trucks, cars and even commuting on foot to get to work (Mwanyepedza 2017). These challenges should have been alleviated post-apartheid, as transport was listed as a basic need by the African National Congress (ANC) and the Department of Transport (DOT) in 1992. Changes were intended to be achieved through the creation of the National Transport Policy Forum (NTPF) and the introduction of the Reconstruction and Development Programme (RDP) of 1994. However, it was only after the 1994 election that transport policies were given due attention (Jennings 2015).

The National Transport Policy White Paper (1996) was implemented and the main goals under the Land Passenger Policy were to ensure:

> "the promotion of a safe, reliable, effective, efficient, co-ordinated, integrated, and environmentally friendly land passenger transport system in South African urban and rural areas, and the southern African region, managed in an accountable manner to ensure that people experience improving levels of mobility and accessibility".

The policy has remained in place but was updated in 2021 to include the provision of standard legislation to guide its previous goals (Lawrence 2022). In 2007, the Public Transport Strategy and Action Plan was released and policy around Integrated Public Transport Networks (IRPTN) was included within the plan. The National Land Transport Act, 2009 was enacted to ensure that the Minister of Transport provided sustainable and efficient transport. The result of this was that planners are required to provide Integrated Transport Plans to the Minister of Transport every 5 years.

Additionally, the Spatial Planning and Land Use Management Act is a significant piece of legislation in South Africa that governs and provides guidance for land use management, land development and spatial planning (SPLUMA). It was passed in 2013 and has significant implications for the way the transportation system operates and how cities evolve (Nel 2016). SPLUMA provides a comprehensive framework for managing land use and spatial planning. Its impact on urban transformation and transportation infrastructure is concentrated on creating more equitable, sustainable and efficient urban environments. Significant spatial and social inequality have resulted from the profound shaping of South Africa's urban landscapes by the legacies of colonialism and apartheid (Denoon-Stevens et al. 2022).

Consequently, the Official Guide to South Africa 2017/18 Transport contains the coveted Bus Rapid Transit (BRT) programme which was approved under the IRPTN policies released in 2007. The guide alluded to the introduction of a BRT system which would be able to create an integrated network in cities by 2020 (Lawrence 2022). Furthermore, the National Development Plan (NDP) 2030 stemmed from the newly appointed National Planning Commission which identified the main transport challenges since 1994. Those challenges influenced the core NDP 2030 goals identified by the National Development Plan 2030 (2017: 185–186) around urban transit. These were to:

- Increase investment in public transport and resolve existing public-transport policy issues,
- Devolve transport management to local government,
- Provide incentives for public-transport use,
- Improve road infrastructure, and
- Renew the commuter train fleet.

Presently, the multi-level governance theory rests on the premise that the national level of authority is not the primary or sole decision-maker. Instead, each level of governance such as the national, provincial, and municipal/district levels, has controlling power over certain policies and decision-making (Veeneman and Mulley 2018). Therefore, the term 'governance' usually does not refer to one entity but instead to a group of entities (Westman et al. 2019). Moreover, funding between different levels of government to implement policy is also a key factor (Veeneman and Mulley 2018). The South African government is comprised of the national, provincial and local spheres (Steytler 2017). This multi-level governance approach adopted in South Africa is thus also applicable to transport affairs.

However, most of the funding sources are generated by the national governanment which transfers funds to both provincial (nine provinces) levels and municipal levels of government. South Africa adopted a decentralised system of local government autonomy in 1996 which established metropolitan municipalities that could generate and control their own finances through mechanisms such as property taxes (Pieterse 2019). Rural municipalities however have limited capacity to generate income and the implementation of public transport is lacking in these areas compared to urban public transport within metropolitan municipalities. The national level of government (National Department of Transport) is responsible for policy-making and legislation concerning transport, while the implementation of transport projects is overseen by State Owned Enterprises, provincial and municipal government (Veeneman and Mulley 2018).

10.3 Case Studies: Johannesburg and Durban

Johannesburg and Durban are used as case studies for analysing provincial public transport governance systems in this chapter (Fig. 10.1). Durban (located in the Kwazulu-Natal province) and Johannesburg (located in the Gauteng Province) are two of the largest and most economically significant cities in South Africa (Fig. 10.1). Johannesburg is the largest and the economic hub of South Africa (Mashamaite 2021; Risimati et al. 2021) while Durban is a significant port and the third-largest city in the country after Cape Town.

In terms of public transportation, Johannesburg and Durban have taken different stances with the goal of enhancing public transportation efficiency and lowering congestion. The GO!Durban project is an example of a contemporary Bus Rapid Transit (BRT) system. In contrast Johannesburg boasts a more varied public transportation system that includes both formal and informal transport systems like buses, minibus taxis, BRT and the Gautrain rapid rail system. Comparing these two cities enables a thorough analysis of various approaches, strategies, and their outcome. Durban and Johannesburg are thus chosen as case studies for comparing public transport systems due to their prominence, contrasting approaches, significant infrastructure investments, and unique challenges.

10.4 Research Methods and Tools

This study employed the Preferred Reporting Items for Systematic Reviews and Meta-Analyses (PRISMA). This method provides a comprehensive and objective evaluation of the literature on the topic under study. It is further defined by Selçuk (2019) as a statistical method which can be used to scan available literature and find answers to precisely defined research questions. Therefore, this study employed PRISMA to analyse the state of knowledge of spatial transformation and public transport in South Africa (Fig. 10.2), aimed at understanding how public transport has shaped spatial transportation in South Africa. All studies between 1994 and 2023 were eligible for review. The systematic review was conducted from July 2023 to May 2024.

The search was conducted using Scopus, Web of Science (WoS) and Google Scholar as databases. While Scopus offers a multidisciplinary coverage of the literature, WoS provides high-quality scientific literature (Martín-Martín et al. 2021). The combination of both major databases provided a comprehensive overview within the area of focus. Additionally, Google Scholar was used to enhance comprehensiveness of the literature. The following keywords were used during the search:

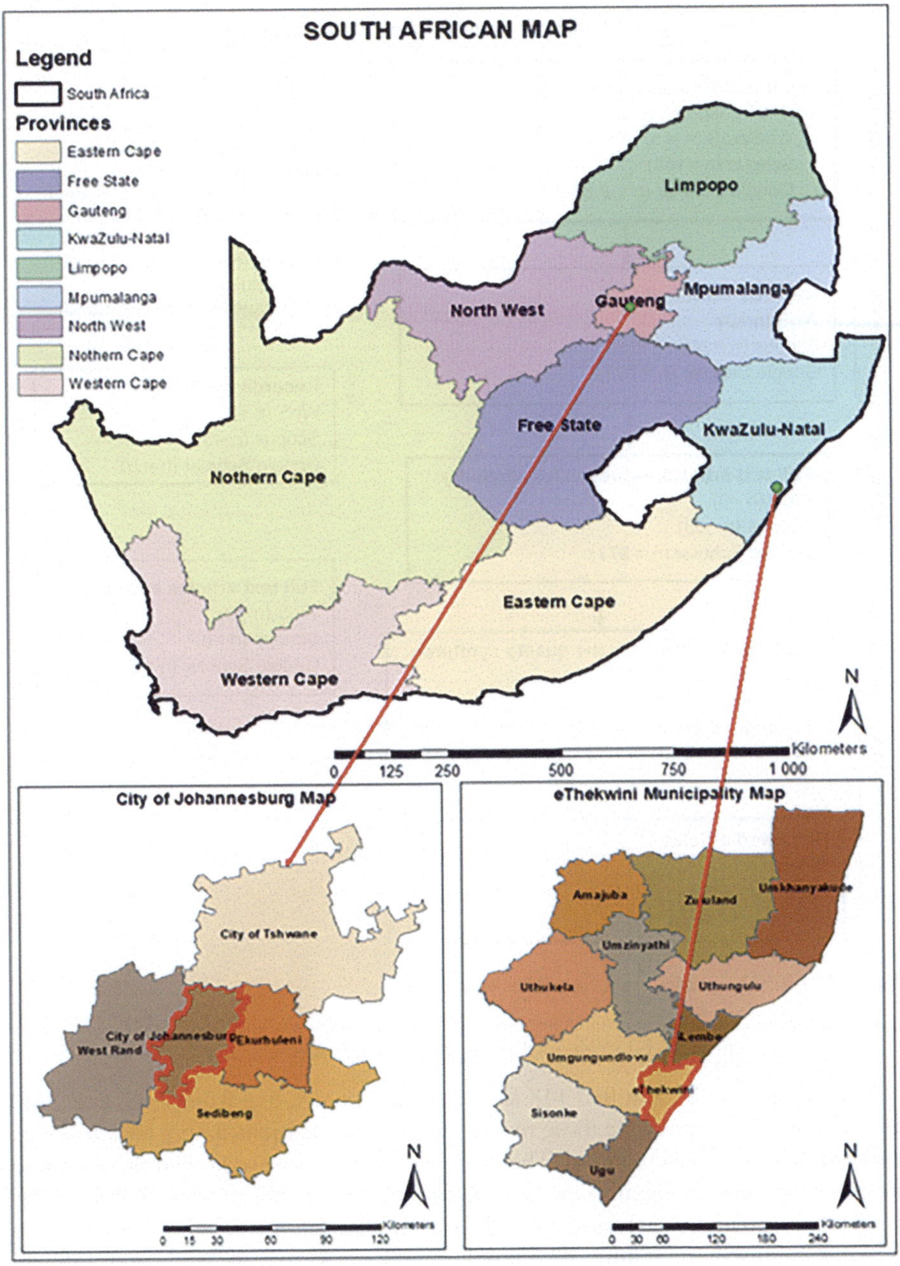

Fig. 10.1 Map showing the city of Johannesburg and Durban (original to the author)

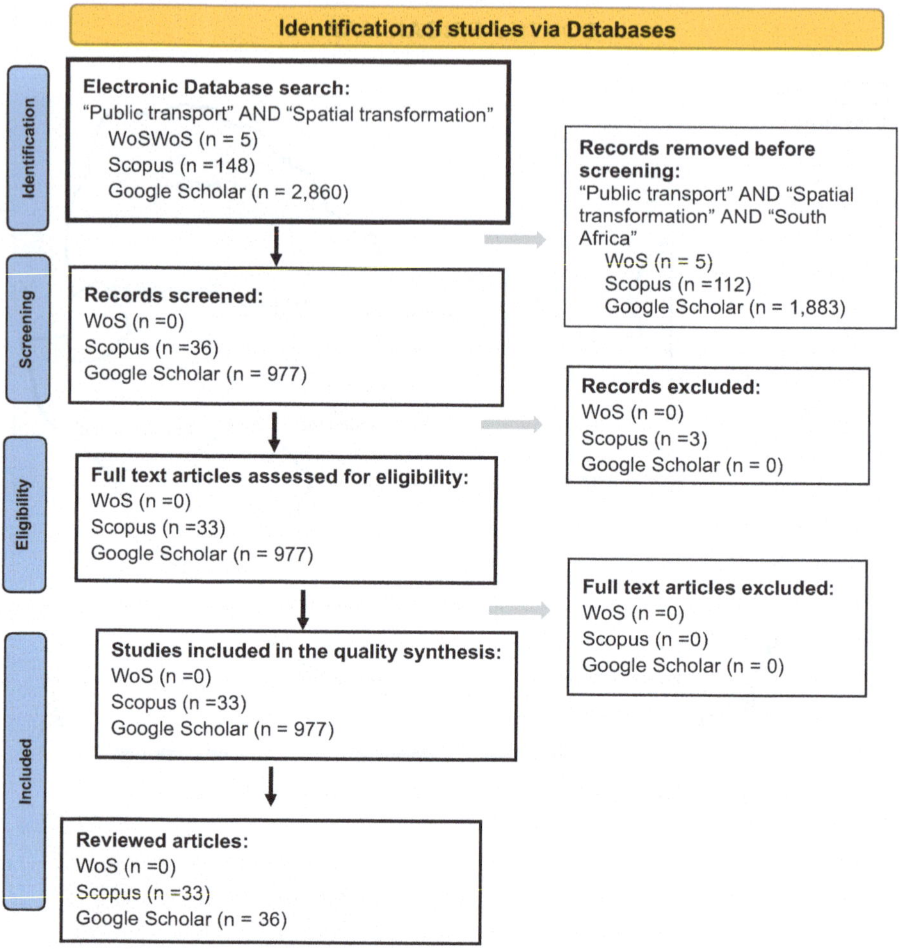

Fig. 10.2 PRISMA guidelines (original to the author)

- "Public transport" AND "Spatial transformation"

The initial search identified 148 articles from Scopus, five from WoS, and 2860 from Google Scholar. While these results covered publications dating back to 1998, further refinement was required. Therefore, the focus narrowed down the search to articles specifically related to the area of study, being South Africa ("Public transport" AND "Spatial transformation" AND "South Africa").

After extensive screening, 112 articles in Scopus, five in WoS, and 1883 in Google Scholar were excluded because the screening focused on refining the search to South Africa. As a result of this focus, a total of 2000 articles were excluded. This left 36 articles from Scopus, zero from WoS, and 977 from Google Scholar for further full-text screening. A further three articles were excluded from Scopus because they were outside the topic of focus. The remaining 33 articles from Scopus and 977 from Google Scholar are included in the study. Lastly, the final reviewed articles from the database search included 33 in Scopus and 36 from Google Scholar.

The publications that were included in this study were analysed using VosViewer which enables analysis of vast samples of academic records. It provides an efficient approach to visualizing the existing literature through map-based analysis (Arruda et al. 2022). VosViewer was used to identify keywords and identify emerging trends in research through co-occurrence network analysis. The results are presented in the following sections.

10.5 Findings and Discussions

10.5.1 Publication Years

The body of literature concerning spatial transformation and public transport in South Africa has developed over many decades, mirroring shifts in policy, urban planning, and societal values (Fig. 10.3). Significant works range from the late 1990s to more recently, showcasing an increasing focus in academia and policy-making on tackling spatial disparities through public transport initiatives.

There has been a steady increase in publications from 2018 to 2024, due to global trends in sustainable urban development which has prompted further research in South Africa. Increased urbanization has also prompted further exploration of the need for sustainable mobility options. These research publications highlight the persistent challenges and ongoing need for innovative solutions to tackle complexities in public transportation within South Africa.

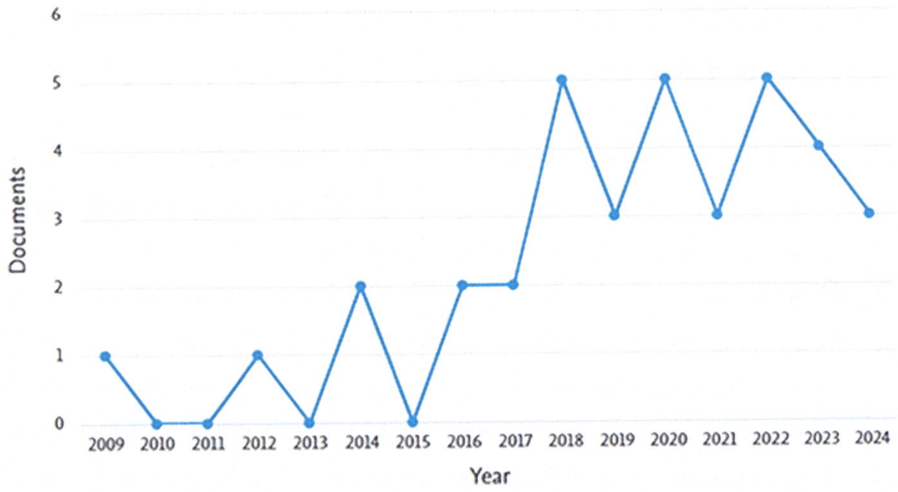

Fig. 10.3 Years of publication, based on Scopus data (2024) (original to the author)

10.5.2 Keywords Analysis

A co-occurrence keyword analysis was performed which resulted in two clusters (Fig. 10.4). The red cluster mainly focuses on urban transport and spatial transformation and explores these concepts within the framework of transit-oriented development (TOD) in South African cities. This underscores the obstacles encountered in the execution of TOD and attributes the primary problem to the challenges of attaining sustainable integration locally rather than to the incorrect application of TOD principles. The research stresses that the intricacies of the local urban landscape significantly impede the effective establishment of TOD initiatives. Nevertheless, cities are adopting strategies to address spatial development challenges and promote sustainable economies through improved policies. Urban public transportation is thus crucial in integrating communities and facilitating efficient mobility (Setati and Ogra 2018). Indeed, efficient public transport aims to offer an accessible, affordable, safe, and sustainable mobility option (Venter et al. 2021). Moreover, the cluster denoted by the colour green focuses on Johannesburg and urban development. The spatial landscape plays an important role in shaping mobility patterns in the city of Johannesburg. For example, the study by Luke (2024) conducted in Johannesburg, highlights spatial disparities in access to public transport services, with residents in different areas experiencing difficulties in accessing public transport, including varying levels of availability and safety concerns.

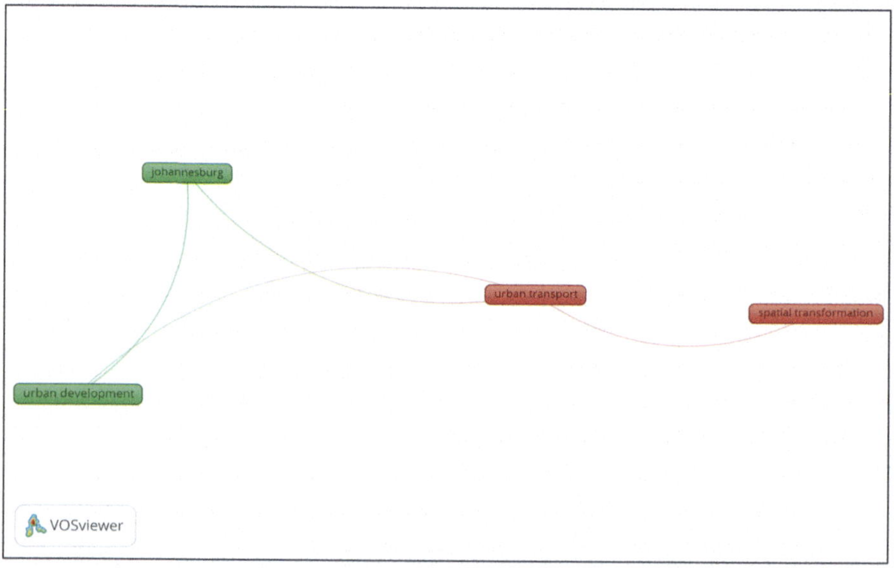

Fig. 10.4 Keyword analysis on public transport and spatial transformation in South Africa (original to the author)

The studies underscore the crucial role public transport can play in transforming the physical, social, and economic landscapes of South African cities. However, despite more than 25 years after South Africa's shift to democracy in 1994, advancements in transportation, spatial planning, and development have made limited progress in addressing inequalities and promoting transformative growth (Chakwizira et al. 2020).

10.5.3 Emerging Trends

The tremendous change of South African transport was witnessed by the turn of the twenty-first century. Most research between 2016 and 2017 focused primarily on urban transport (Fig. 10.5). In addition, 2018 to 2019 focused on urban development. However, between 2017 and 2019 this shifted more towards issues of spatial transformation. Currently, studies are increasingly concentrating on Johannesburg as the area of focus. Over the period between 2016 and 2024, studies examining public transport and spatial transformations in South Africa have revealed a dynamic landscape of emerging trends and evolving perspectives. For example, since 2016, discussions have been instrumental in recognizing the need to integrate transportation planning with urban development, emphasizing accessibility as a barrier for marginalized communities in ensuring equitable access to health care services (Mubaiwa and Mokonyama 2017). Subsequent years saw a deepening exploration

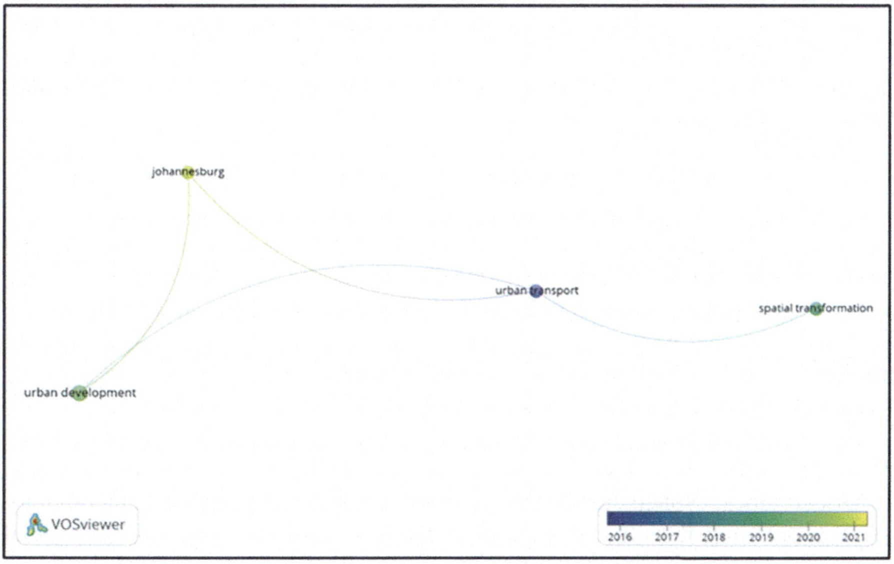

Fig. 10.5 Emerging trends on public transport and spatial transformation in South Africa (original to the author)

of urban development and spatial transformation with a nascent focus on transit-oriented development (TOD) principles (Harber 2020). Efforts have been made to provide better access to socio-economic opportunities through shared transportation modes like BRT systems, and metro railway services have been introduced to improve urban transport (van der Berg and du Plessis 2022).

In one study, Khumalo and Ogra (2018) investigated the effectiveness of the BRT system in the city of Johannesburg. The key finding from the study was that sustainable transportation continues to be a long-term objective for developing nations along with recommended changes to further transportation enhancements for sustainability. Although the establishment of the BRT system in Johannesburg has been effective in improving the city's mobility and development challenges, the city continues to be fragmented, limiting access to public transport services and ultimately, economic opportunities (Kotze and De Vries 2019; Chakwizira et al. 2020). Moreover, challenges exist due to fragmented regulatory frameworks and informal economy operations (van der Berg and du Plessis 2022). These studies thus show a need for coordinated planning efforts and innovative solutions to address the complex challenges of urban mobility and spatial transformations. Emerging technologies and promoting inclusive, sustainable development require specific attention. In addition, transport infrastructure and mobility patterns have the power to transform land potential, rendering it suitable for various uses depending on its accessibility (Harber 2020).

Current trends present opportunities to create sustainable, inclusive and resilient South African cities using public transportation as a tool. Thus, prominent public transport initiatives aimed at transforming urban spaces have been created through Integrated public transport networks (IRPTN). These networks in South Africa refer to Bus Rapid Transit (BRT) systems and the Gautrain. Cabinet approved the road-based BRT Public Transport Strategy in 2007 with the purpose of ensuring that the majority of residents in cities where the BRT was implemented in would be within 500 m of a BRT station.

The IRPTN is funded by national government, managed by provincial government, and planned and implemented by the responsible municipality (South African Cities Network 2016). The programme confirms that universal design aspects are incorporated into the projects to ensure that the BRT is accessible to all (Official Guide to South Africa 2017/18 2017; Department of Transport 2023). BRT systems, such as Rea Vaya and Gautrain in Johannesburg and the Go-Durban BRT are among the most prominent public transport initiatives. These systems provide affordable, efficient, and reliable transport options, thereby reducing travel times and costs for commuters from marginalized areas.

The critical challenges posed by insufficient transport services, have increased urban sprawl and exacerbated longer commuting hours and raised safety concerns in African cities. Nonetheless, there remains a notable scarcity of contemporary research detailing the effects of these factors on access to economic and social opportunities. Additionally, there have been limited studies in Africa that explore the intricacies of public transport and spatial transformation in detail. For example, the issues surrounding Durban's Bus Rapid Transit (BRT) system and the changing dynamics of spatial transformation and public transport receive insufficient coverage in the current literature on South Africa. Moreover, Johannesburg, although extensively researched, still contends with pressing issues on public and spatial transportation. Therefore, the current study highlights the state of knowledge, but also paves the way for further advancement on the dynamics of spatial transformation and public transport in South Africa.

10.5.4 Public Transport Governance Systems and Urban Spatial Transformation in Johannesburg

Johannesburg's public transport network is characterized by a mix of formal and informal modes of transportation (Figs. 10.6 and 10.7). Formal systems include Rea Vaya Bus Rapid Transit (BRT) system (Fig. 10.6a), Gautrain rapid rail network (Fig. 10.7a) and the Metro bus (Fig. 10.6b). In addition, minibus taxis (Fig. 10.7b) provide an informal mode of public transport.

10.5.4.1 Gautrain

The Gautrain is an 80 km mass rapid transit (MRT) system linking three areas of Gauteng (Pretoria, Johannesburg and OR Tambo International Airport (ORTIA)). There are two components to the Gautrain, an airport service (connects Sandton and Marlboro to the ORTIA) and a commuter service (Rhodesfield, Marlboro, and Sandton and Park, Rosebank, Sandton, Midrand, Centurion, Pretoria, and Hatfield).

Fig. 10.6 Rea Vaya (**a**) & Metro bus (**b**)

Fig. 10.7 Gautrain (**a**), and Minibus taxis (**b**) (original to the author)

The project provides commuters between Pretoria and Johannesburg with another mode of public transport to reduce traffic congestion between them (Official Guide to South Africa 2017/18 2017; Uys 2021). The Gautrain transports 1.2 million people a year (including weekends) with institutional support such as a new website and software applications, Gautrain buses and integrated parking facilities (excluding those at the ORTIA).

The Gautrain directly affects five aspects: "transport impact, job creation, rising property values and commercial activity, contribution to Gauteng's GDP and latent environmental impact" (Doyle 2020). It has created jobs for South Africans at various stages, including 35,000 jobs during construction, 10,900 jobs during operation and an expected 66,000 jobs indirectly through commercial nodes within 2 km of the Gautrain (Gautrain Management Agency 2019).

10.5.4.2 Rea Vaya

The Rea Vaya in Johannesburg was the first public BRT system of its kind launched in South Africa in 2009, ahead of the 2010 FIFA Soccer World Cup. The system became operational in October 2013 (Khumalo and Ogra 2018) and has been overseen by 4 Executive Mayors since its implementation.

The Rea Vaya BRT also required the services of existing public transport operators to transport Rea Vaya BRT users to and from the BRT (Gumbo and Moyo 2020). After lengthy negotiations, the existing operators became shareholders in the project through these feeder bus operating companies. This ensured that the livelihoods of the existing public transport operators were secured through job creation for bus drivers and compensation for loss of income due to the implementation of the BRT (Mjiyako 2021).

The Rea Vaya BRT experienced issues of accessibility for buses and users due to poor demarcation of bus lanes, consequently, the BRT implemented NMT lanes for users to be able to access buses. However, due to the poor demarcation of lanes, the demand for cyclists using the NMT lanes provided is low, but the demand for pedestrians using these NMT lanes remains high (Okoro and Lawani 2022). Transit Oriented Development (TOD) can also be seen along the routes of the Rea Vaya BRT NMT routes in the form of bicycle repairs, maintenance and refreshment stalls because of the implementation of the BRT and NMT (Lawrence et al. 2022).

10.5.5 Public Transport Governance Systems and Urban Spatial Transformation in Durban

Durban falls within Kwazulu-Natal guided by the eThekwini municipality's Integrated Development Plan (IDP) to guide all development projects. One of the main goals of the IDP is to create, "a compact, productive and liveable city that connects citizens with economic opportunities" (eThekwini Municipality 2023: np.). The minibus taxi industry is dubbed the pillar of the public transport sector by the province of KwaZulu-Natal, irrespective of its tumultuous unregulated history. The province has therefore taken the necessary steps to regulate the taxi industry and make it more profitable for owners and not only reliant on fares (Department of Transport 2023). The taxi association has since been organised into regional councils and has elected a Provincial Taxi Council. Operation Shanela was started in the 1990s by the KZN province to ensure the roadworthiness of vehicles and was mainly applied to taxis (Capital Newspapers 2019).

The BRT works towards connecting corridors and includes feeder routes which transport users from stations to residential areas and business districts (GO!Durban 2013). The aim of the GO!Durban project is to increase the accessibility and availability of transport in various parts of the municipality (eThekwini Municipality 2023: 14). The GO!Durban BRT system falls within the eThekwini municipality and was supposed to be launched in November 2023 after being delayed twice previously in 2022, though the IRPTN policy was approved by cabinet in 2008 (Magubane 2023). However, the eThekwini Transport Authority set a launch date of February 2024 which did not materialise, and no further launch date has since been set (Naidoo 2023). These delays were due to disputes between the municipality and taxi operators over ownership licences of the buses (Magubane 2023).

The project also aims to upgrade its rail network as part of Phase 1 of the development. The rail network runs from Bridge City and KwaMashu through to Berea Road and to Umlazi and Isipingo (eThekwini Municipality 2023). A bike-sharing pilot project was also introduced in the KwaMashu-Bridge City area to get more information on the implementation of the full programme. The pilot includes 50 bicycles with tracking units and involves training 50 participants from the GO!Durban sites. Although the GO!Durban BRT is not operational yet, Pillay

(2019) notes the five key successes of the construction phase that have been achieved by the BRT. The successes are:

- The implementation of 10 state of the art stations along the Bridge City—Pinetown corridor.
- Urban regeneration in the densely populated area of Pinetown, as the BRT made provision for NMT which promotes walkability, safety and urbanity. This makes the city more accessible to the public and businesses which in turn makes the precinct more liveable.
- The construction of two major bridges which link the northern areas to the rest of eThekwini. These bridges are M25/Bhejane Underpass Bridge and Cornubia Bridge which aim to redress the past spatial constraints which still poses many challenges in terms of linkages.
- Incorporating the latest technologies into the BRT system such as Kassel kerbs for bus docking at stations, tactile paving, and Universal Access design tools.
- The partnerships between different spheres of government as well as public-private partnerships were a key driver of the construction of the major bridges. These collaborative partnerships led to not only the linkages of different areas but promoted mixed use development, creating inclusive spaces and linking people to opportunities.

10.5.6 Key Lessons Learnt

The Johannesburg integrated transport system experience offers various insights. Learning from financial turmoil in Johannesburg, it will be imperative for Durban to successfully establish and run its BRT system by navigating financial obstacles. Johannesburg also encountered opposition to change as a challenge during implementing BRT and Durban might experience similar opposition from different stakeholders. Informal transport operators in Johannesburg opposed BRT because they saw it as a threat to their interests or current means of transportation. In order to establish confidence and garner support for the new system, Durban must anticipate and resolve these concerns through efficient stakeholder engagement and communication tactics (Fig. 10.8). Another issue that Durban and Johannesburg face in common is limited infrastructure. While establishing its BRT system Johannesburg had to deal with constrained road space and incompatible land use patterns. Similar difficulties might arise in Durban fitting BRT infrastructure to the city's current morphology. To secure the smooth integration of BRT into the city's transportation network it will be necessary to find creative planning and design solutions suited to Durban's particular context to overcome these infrastructure limitations. Furthermore, public opinion and participation played a significant role in Johannesburg's BRT rollout highlighting the significance of efficient communication and participation tactics in fostering public knowledge confidence and support for the new Durban system. Taking cognisance of these lessons Durban can thus

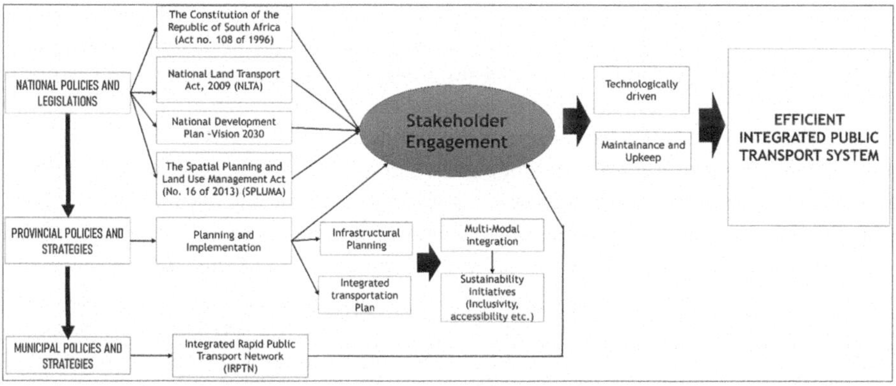

Fig. 10.8 Efficient integrated public transportation system framework (original to the author)

better prepare for the successful implementation of its own BRT system and taking proactive measures to address these potential obstacles.

10.6 Towards a Framework for Improved Provincial Public Transport Governance Structures

The governance of public transport in South Africa involves a multi-tiered approach, with responsibilities shared between national, provincial, and municipal authorities. At the national level, the Department of Transport (DoT) sets policies and provides funding, while provincial and municipal governments are tasked with implementation and management. This multi-layered governance system requires strong coordination and collaboration to ensure that public transport initiatives effectively contribute to spatial transformation. These initiatives to reduce socio-economic inequalities led to establishment of the BRT system and high-speed rail. Nevertheless, despite tremendous efforts, inadequate infrastructure, financial constraints, safety, availability and accessibility remain a barrier for public transport users. These highlight the need for an integrated and efficient public transport system for all. Figure 10.8 presents a proposed public transportation framework. This outlines an efficient approach to provide clear guidance to existing public transport operators for improved integrated transportation systems in South Africa.

Each element in this framework is interconnected (Fig. 10.8). The core element throughout this framework is stakeholder engagement which requires all stakeholders to take on their responsibilities and participate in all the activities. Government agencies, local authorities, private sector, transport operators, civil society and communities should be consulted about policy development to periodically address their concerns and keep them informed. The issue of stakeholder engagement in transportation planning is emphasised by different researchers (Le Pira et al. 2016;

Esztergár-Kiss and Tettamanti 2019; Litman 2022). This suggests that the successful execution of transportation initiatives involves an inclusive and transparent method, requiring the endorsement of all stakeholders, including grassroots communities. Such endorsement can be achieved through different governance frameworks like participatory decision-making methods, community forums, advisory panels with varied members, and cooperative alliances among government entities, local administrations, civil society groups, and community figureheads. These frameworks promote significant involvement, guarantee the reflection of varied viewpoints, and facilitate the integration of community input into the planning and execution of transportation strategies.

The National policy serves as a blueprint for transportation systems in Johannesburg and Durban. Regional priorities are addressed at the provincial level through governance mechanisms like intergovernmental committees, transportation authorities, and stakeholder forums. These entities promote collaboration among national, provincial, and local governments, and facilitate dialogue with key stakeholders such as community groups, businesses, and transport specialists. Alignment with national policies ensures that provincial transportation plans are clear, while also catering to unique regional challenges and opportunities. This is also clearly stipulated in SPLUMA which is a national policy guiding the provincial and local laws for land management in South Africa. It is based on the five principles of spatial sustainability, efficiency, spatial justice, spatial resilience, and good governance. It involves the development of an integrated master plan for promoting efficient infrastructural planning. The two provinces already have an existing integrated transportation master plan, however, failures of these plans are often associated with lack of implementation and not adhering to realities on the ground (Watson 2014). Therefore, this framework (Fig. 10.8) shows an effective approach for transport planning which involves multi-modal integration to ensure a more connected, inclusive, and accessible infrastructure. Multi-modal transport encourages seamless interaction between different modes, while encouraging sustainable use of mobility options.

Additionally, technological advancement and constant maintenance ensure efficient transportation operations. These involve real time use of scheduling and digital communication tools while sustaining existing public transport systems. Regular adjustments of strategies, considers feedback from all stakeholders and the constant technological improvements in the transportation sector. The fundamental aspect of the framework is to provide efficient and sustainable public transport systems in Gauteng and KwaZulu-Natal. This can be accomplished by integrating transparent and inclusive collaboration of stakeholders throughout the entire transport planning process while providing a platform for innovative solutions to the existing problems.

10.7 Conclusion

A key element of improved public transport operations is cooperation between provincial and municipal governance along with the participation of citizens. The bus and taxi industries play a significant role in South African public transportation. However, with only 17% of South Africans not utilizing minibus taxi's, it seems to be the most accessible and popular choice among both rural and urban households. Moreover, around 50% of South Africans report not having access to bus services, while 81% of households do not have access to train services (Kerr 2018). There are thus clear issues around accessibility of different public transport modes, and each province must remedy the current issues faced through incorporating the taxi and bus industries in new innovative plans. The City of Johannesburg looks to MRTs and BRTs to connect users from areas neighbouring the city to and from Johannesburg CBD. These new developments include participation from the taxi and bus industries as there are still major problems around spatial inequality and transportation provision. Durban is still in the initial stages of their IRPTN implementation, nevertheless, the city is working to create safer and more integrated systems along with citizen participation. In order to create efficient public transport systems, there needs to not only be participation from rail, bus, minibus taxis, metered taxis and NMTs but also between provincial and municipal governance and its citizens.

References

Aivinhenyo I, Zuidgeest M (2019) Transport equity in low-income societies: affordability impact on destination accessibility measures. In: Measuring transport equity. Elsevier, pp 111–128

Arruda H, Silva ER, Lessa M, Proença D Jr, Bartholo R (2022) VOSviewer and bibliometrix. J Med Library Assoc 110(3):392

Brown J (2016) The road to Soweto: resistance and the uprising of 16 June 1976. Boydell & Brewer

Capital Newspaper (2019) Department to strengthen taxi unit The KwaZulu-Natal Department has dismissed talk of a possible shutting down of one of its flagship programmes. https://www.citethisforme.com/cite/sources/websitemanualcite. Accessed 30 Sept 2023

Chakwizira J, Bikam P, Adeboyejo TA (2020) Transport realities and challenges for low income peripheral located settlements in Gauteng Province: are we witnessing the genesis of a new transport order or consolidation of the old transport order? In: Advances in human factors of transportation: proceedings of the AHFE 2019 international conference on human factors in transportation, 24–28 July 2019, Washington DC, USA, vol 10. Springer International Publishing, pp 618–630

Coetzer N (2016) Building apartheid: on architecture and order in imperial Cape Town. Routledge

Denoon-Stevens SP, Andres L, Nel V, Jones P (2022) Unpacking planners' views of the success and failure of planning in post-apartheid South Africa. Cities 130:1–9

Department of Transport (2023) Bus rapid transit system (BRT). https://www.transport.gov.za/bus-rapid-transport. Accessed 9 Sept 2023

Diallo FD (2022) Conflicted translations: an analysis of the bus rapid transit policy adoption process in Cape Town. Territ Politics Gov 12(8):1236–1254. https://doi.org/10.1080/21622671.2022.2099967

Donaldson R (2006) Mass rapid rail development in South Africa's metropolitan core: towards a new urban form? Land Use Policy 23(3):344–352

Doyle N (2020) South Africa's construction of a developmental state? The case of the Gautrain mega-project. Doctoral dissertation, University of the Witwatersrand. https://wiredspace.wits.ac.za/server/api/core/bitstreams/599e3097-e67d-406f-9bb0-9247d08210a3/content. Accessed 14 Sept 2023

Esztergár-Kiss D, Tettamanti T (2019) Stakeholder engagement in mobility planning. In: Autonomous vehicles and future mobility. Elsevier, pp 113–123

eThekwini Municipality (2023) Integrated development plan 5 year plan: 2023/24 to 2027/28. https://www.durban.gov.za/storage/Documents/Integrated%20Development%20Plans%20IDP%20-%20eThekwini%20Municipality/Final_IDP_2023_2024.pdf. Accessed 13 Sept 2023

Fredriksson P, Moritz L (2019) South Africa 1990: pressure for change. In: Apartheid in theory and practice. Routledge, pp 293–358

Gautrain Management Agency (2019) The economic and social impact of the Gautrain. Gauteng Provincial Government, Republic of South Africa. https://gma.gautrain.co.za/development/Shared%20Documents/Economic%20and%20Social%20Impact%20of%20Gautrain%20-%202019.pdf. Accessed 14 Sept 2023

GO!Durban (2013) About us. https://godurban.co.za/about-us/. Accessed 14 Sept 2023

Gumbo T, Moyo T (2020) Exploring the interoperability of public transport systems for sustainable mobility in developing cities: lessons from Johannesburg Metropolitan City, South Africa. Sustainability 12(5875):1–16

Harber J (2020) Building capacity for the governance of space: the case of Johannesburg's BRT. Urban Governance Paper Series, p 61

Howard E, Osborn FJ, Mumford L (2013) Garden cities of tomorrow. Routledge

Jennings G (2015) Public transport interventions and transport justice in South Africa: a literature and policy review. In: Southern African Transport Conference. http://hdl.handle.net/2263/57779. Accessed 28 July 2023

Kerr A (2018) Minibus taxis, public transport, and the poor. Working paper: World Bank

Khosa MM (1998) 'The travail of travelling': urban transport in South Africa, 1930–1996. Transp Rev 18(1):7–33. https://doi.org/10.1080/01441649808716998

Khumalo TN, Ogra A (2018) Effectiveness of Rea Vaya Bus Rapid Transit System (BRTS) in the City of Johannesburg. In: Proceedings of the international conference on industrial engineering and operations management. Pretoria, October, vol 29, pp 1306–1329

Kotze N, De Vries L (2019) Resuscitating the African giant: urban renewal and inner city redevelopment initiatives along the 'Corridors of Freedom' in downtown Johannesburg. Geogr Pol 92(1):57–70

Lawrence K (2022) Planning for the integration of non-motorised transport and public transport: a case study of the Harambee Bus Rapid Transit, Ekurhuleni. Doctoral dissertation, University of Johannesburg. https://ujcontent.uj.ac.za/esploro/outputs/graduate/Planning-for-the-integration-of-non-motorised/9928905407691. Accessed 6 Sept 2023

Lawrence K, Gumbo T, Jeeva Z (2022) The influence of non-motorised transport systems around the world: a case study of Singapore, Shanghai, Lagos, Jakarta, Johannesburg and Cape Town. In: Mobility, knowledge and innovation hubs in urban and regional development. Proceedings of REAL CORP 2022, 27th international conference on urban development, regional planning and information society. CORP–Competence Center of Urban and Regional Planning, pp 821–829. https://repository.corp.at/923/1/CORP2022_91.pdf. Accessed 14 Sept 2023

Le Pira M, Ignaccolo M, Inturri G, Pluchino A, Rapisarda A (2016) Modelling stakeholder participation in transport planning. Case Stud Transp Policy 4(3):230–238. https://doi.org/10.1016/j.cstp.2016.06.002

Lemon A (2021) The apartheid city. In: South African urban change three decades after apartheid: homes still apart? Springer International Publishing, Cham

Litman TM (2022) Evaluating transportation equity: guidance for incorporating distributional impacts in transport planning. Inst Transp Eng 92(4):43–49

Luke R (2024) Transport-related social exclusion and mobility in developing countries: the South African case. Front Sustain Cities 6:1–12. https://doi.org/10.3389/frsc.2024.1199055

Magubane T (2023) New date set for Go!Durban to go live. https://www.iol.co.za/mercury/news/new-date-set-for-godurban-to-go-live-a71ec6ea-475b-4fa2-9d33-fd78552c4e21. Accessed 13 Sept 2023

Martín-Martín A, Thelwall M, Orduna-Malea E, Delgado López-Cózar E (2021) Google Scholar, Microsoft Academic, Scopus, Dimensions, Web of Science, and OpenCitations' COCI: a multidisciplinary comparison of coverage via citations. Scientometrics 126(1):871–906

Mashamaite MM (2021) Managing the impacts of climate change on roads infrastructure in the City of Tshwane. In: Handbook of climate change management: research, leadership, transformation. Springer International Publishing, Cham, pp 1–21

Massey R, Gunter A (2020) Urban geography in South Africa: an introduction. In: Maharaj B (ed) The apartheid city. Urban geography in South Africa: perspectives and theory, pp 39–54

Meier zu Selhausen F (2022) Growing cities: urbanization in Africa

Mjiyako DN (2021) An assessment of the extent to which the city of Johannesburg's Rea Vaya bus rapid transit system is achieving its intended socio-economic and political objectives. Master's thesis, Faculty of Commerce. http://hdl.handle.net/11427/35924. Accessed 14 Sept 2023

Monama SA, Mokoele NJ, Mokgotho KD (2022) South African spatial planning fragmentation: repealing the apartheid planning imprint. Int J Entrepreneurship 26:1–11

Mosala SJ, Venter JCM, Bain EG (2017) South Africa's economic transformation since 1994: what influence has the National Democratic Revolution (NDR) had? Rev Black Political Econ 44(3–4):327–340. https://doi.org/10.1007/s12114-017-9260-2

Moyo T, Mbatha S, Aderibigbe OO, Gumbo T, Musonda I (2022) Assessing spatial variations of traffic congestion using traffic index data in a developing city: lessons from Johannesburg, South Africa. Sustainability 14(14):1–16. https://doi.org/10.3390/su14148809

Mubaiwa T, Mokonyama MT (2017) Implications of chronic disease patient travel to healthcare facilities on the design of national health insurance in South Africa—a preliminary review. In: Southern African Transport Conference, 10–13 July 2017, CSIR Convention Centre, Pretoria

Musakwa W, Gumbo T (2017) Impact of urban policy on public transportation in Gauteng, South Africa: smart or Dumb City systems is the question. In: Álvarez Fernández R, Zubelzu S, Martínez R (eds) Carbon footprint and the industrial life cycle: from urban planning to recycling. Springer, Cham, pp 339–356

Mwanyepedza R (2017) Solutions to public transport challenges. In: Proceedings of the 36th Southern African Transport Conference (SATC 2017), Pretoria, South Africa

Naidoo J (2023) R8 billion spent on Go!Durban bus project, but still no movement. https://www.iol.co.za/news/south-africa/kwazulu-natal/r8-billion-spent-on-godurban-bus-project-but-still-no-movement-39386cab-1750-4aab-893e-bf0e23d0c3ae#:~:text=According%20to%20Bethuel%20Manthoadi%2C%20senior,February%202024%2C%20IOL%20previously%20reported. Accessed 8 May 2024

National Development Plan 2030 (2017) National Development Plan 2030: our future—make it work. https://www.gov.za/sites/default/files/gcis_document/201409/ndp-2030-our-future-make-it-workr.pdf. Accessed 11 May 2023

Ndlovu S (2017) The Soweto uprisings: counter memories of June 1976. Pan Macmillan South Africa

Nel V (2016) SPLUMA, zoning and effective land use management in South Africa. Urban Forum 27:79–92

Official Guide to South Africa 2017/18 (2017) Transport. https://www.gcis.gov.za/sites/default/files/docs/resourcecentre/pocketguide/24-Transport-1718.pdf. Accessed 27 May 2023

Okoro C, Lawani K (2022) Optimising sustainable mobility: a performance assessment of non-motorised transport infrastructure in Johannesburg, South Africa. J South Afr Inst Civ Eng 64(2):67–76. https://doi.org/10.17159/2309-8775/2022/v64no2a6

Pieterse E (2019) Urban governance and spatial transformation ambitions in Johannesburg. J Urban Aff 41(1):20–38. https://doi.org/10.1080/07352166.2017.1305807

Pillay S (2019) IMESA conference 2019. https://www.imesa.org.za/wp-content/uploads/2019/10/paper-8-ethekwini-municipalitys-go-durban-brt-programme.pdf. Accessed 14 Sept 2023

Risimati B, Gumbo T, Chakwizira J (2021) Spatial integration of non-motorized transport and urban public transport infrastructure: a case of Johannesburg. Sustainability 13(20):1–17. https://doi.org/10.3390/su132011461

Schoeman C (2015) The alignment between spatial planning, transportation planning and environmental management within the new spatial systems in South Africa. Town Reg Plan 2015(67):42–57

Schoeman CB (2010) Urban and regional planning and the interface with environmental management and transportation planning. Noordwes-Universiteit, Potchefstroomkampus (Suid-Afrika), Potchefstroom. https://repository.nwu.ac.za/bitstream/handle/10394/8570/Schoeman_CB.pdf?sequence=1. Accessed 8 May 2024

Selçuk AA (2019) A guide for systematic reviews: PRISMA. Turkish Arch Otorhinolaryngol 57(1):57–58

Setati MA, Ogra A (2018) Performance of BRTS using Spatial Planning and Land Use Management Act (SPLUMA) principles: a case study of City of Johannesburg. Proc Int Conf Indust Eng Operat Manag 2018(November):860–876

Simon D (2002) Reform in South Africa and modernization of the apartheid city. In: Urban and regional change in southern Africa. Routledge

South African Cities Network (2016) State of South African Cities report 2016. http://www.socr.co.za/wp-content/uploads/2016/06/SoCR16-Main-Report-online.pdf. Accessed 14 Sept 2016

Steytler N (2017) The constitutional court of South Africa: reinforcing an hourglass system of multi-level government. University of Toronto Press. http://hdl.handle.net/10566/4742. Accessed 31 July 2023

Strauss PJ (2023) The reform of apartheid in the 1980's: the political approach of PW Botha and the ecclesiastical contribution by JA Heyns. Tydskrif vir Geesteswetenskappe 63(2):402–419

Thomas DP (2016) Public transportation in South Africa: challenges and opportunities. WORLD 3(3):1–15. https://core.ac.uk/download/pdf/268085252.pdf. Accessed 27 July 2023

Thumbran J (2022) Transporting the "Bus Stop Republic"–resilience and apartheid's transport infrastructure, 1979 to present times. Soc Dyn 48(1):16–28. https://doi.org/10.1080/02533952.2022.2054144

Uys T (2021) Challenges in the social impact assessment of the Gautrain Rapid Rail Link Project in South Africa. Environ Sustain Develop Perspect Iss. https://books.google.co.za/books?hl=en&lr=&id=mc5JEAAAQBAJ&oi=fnd&pg=PT188&dq=The+Gautrain+is+an+80km+mass+rapid+transit+system+linking+three+areas+of+Gauteng+(Pretoria,+Johannesburg+and+OR+Tambo+International+Airport).+There+are+two+components+to+the+Gautrain,+an+Airport+service+(connects+Sandton+and+Marlboro+to+the+ORTIA)+and+a+co&ots=frUFMhMPR&sig=J292vYov6_YJ9EeyFCl1BNjnPd8&redir_esc=y#v=onepage&q&f=false. Accessed 9 Sept 2023

van der Berg A, du Plessis A (2022) Shared mobility towards urban climate resilience in the city of Johannesburg. In: Urban climate resilience. Edward Elgar Publishing, pp 325–357

Veeneman W, Mulley C (2018) Multi-level governance in public transport: governmental layering and its influence on public transport service solutions. Res Transp Econ 69:430–437. https://doi.org/10.1016/j.retrec.2018.07.005

Venter C, Mahendra A, Lionjanga N (2021) Urban expansion and mobility on the periphery in the global South. In: Urban form and accessibility. Elsevier, pp 243–264

Verney K (2024) The great depression and the Second World War, 1930–1945. In: The debate on black civil rights in America, 2nd edn. Manchester University Press, pp 84–113

Walters J, Pisa N (2023) Review of South Africa's public transport system. Res Transp Econ 100:101322. https://doi.org/10.1016/j.retrec.2023.101322

Watson V (2014) African urban fantasies: dreams or nightmares? Environ Urban 26(1):215–231

Westman LK, Broto VC, Huang P (2019) Revisiting multi-level governance theory: politics and innovation in the urban climate transition in Rizhao, China. Polit Geogr 70:14–23. https://doi.org/10.1016/j.polgeo.2019.01.002

Wilkinson P (2010) The regulatory cycle stalled? An assessment of current institutional obstacles to regulatory reform in the provision of road-based public transport in Cape Town. Res Transp Econ 29(1):387–394. https://doi.org/10.1016/j.retrec.2010.07.049

Wood A (2019) Circulating planning ideas from the metropole to the colonies: understanding South Africa's segregated cities through policy mobilities. Singap J Trop Geogr 40(2):257–271

Chapter 11
Institutional Arrangements that Influence the Planning of Airport-Led Development: Lessons from the Ekurhuleni Aerotropolis, South Africa

Masilonyane Mokhele

Abstract Encapsulated in the normative spatial planning models of airport-led development, the concentration of urban development near airports is a prominent subject in academia, policy and practice. Following global trends, South Africa is not spared in the hype of airport-led development. However, there is a paucity of literature that analyses the institutional arrangements that influence airport-centric development planning. Using the case study of the Ekurhuleni Aerotropolis, which is centred on OR Tambo International Airport, South Africa, the study analysed the institutional arrangements that affect the planning of development that hinges on airports. Framed under institutional economic geography and collaborative planning, the study aimed to analyse the stakeholders' roles, interdependencies, synergies, and contentions in the Ekurhuleni Aerotropolis planning. The study was based on a qualitative design, which entailed interviews with four key informants selected through snowball sampling technique. The study found the project's reliance on a triad of primary stakeholders: Ekurhuleni municipality, Gauteng provincial government and the Airports Company South Africa. The intra- and inter-organisational disagreements between the stakeholders centred on the project's geographical scope, ownership, land-use allocation, and relevance of the project relative to pressing socio-economic issues in the municipality. To transcend the narrow municipal boundaries, it is recommended that provincial governments facilitate, through regional spatial development frameworks, the planning of aerotropolis in a functional region context.

Keywords OR Tambo International Airport · Ekurhuleni municipality · Aerotropolis · Institutions · Airport-led development · Consensus-building

M. Mokhele (✉)
Faculty of Informatics and Design, Cape Peninsula University of Technology, Cape Town, South Africa
e-mail: mokhelem@cput.ac.za

D. du Plessis (ed.), *Territorial Governance and Spatial Transformation in Post-apartheid South Africa*, Local and Urban Governance,
https://doi.org/10.1007/978-3-031-85489-7_11

11.1 Introduction

Evocative of Markusen's (1999) notion of fuzzy concepts, airport-led development has gained global prominence in development policy and practice. Epitomised by 'airport city' and 'aerotropolis', numerous normative models are used to promote the idealised economic activity mix and spatial form of airports and their environs (Freestone and Baker 2011; Pereira et al. 2023). The proponents use the airport city model to conceptualise urban development in the immediate surroundings of airports (Conway 1993; Kasarda 2009). As more economic activity gravitates towards airport cities, the urban form of aerotropolis emerges, consisting of development that could stretch for miles from airports (Kasarda 2009; Kasarda and Lindsay 2011). The aerotropolis, therefore, conceptualises the airport and the surrounding airport city as the centre of development in a metropolis or region (Banai 2017). Discussions about airport-led development are closely intertwined with the debates on the airfreight catchment of airports, which can span different geographical scales: airports and their immediate surroundings, metropolitan areas, functional regions, countrywide locations, areas across national borders, and even locations across continents (Mokhele 2022).

Following global trends, South Africa is not spared in the hype of airport-led development, and several airport environs, metropolitan areas and regions in the country have been branded airport cities or aerotropoli. As noted by Mokhele (2016), these initiatives largely hinge on the following airports, with their respective municipalities or regions indicated in brackets: OR Tambo (Ekurhuleni, Gauteng City-Region), King Shaka (eThekwini), Cape Town (City of Cape Town), Bram Fisher (Mangaung) and Lanseria (City of Johannesburg, Gauteng City-Region).

Despite a range of insightful literature that analyses the spatial economic attributes of airports and surrounds in South Africa (for instance, see Coetzee and Swanepoel 2017; Mokhele 2018; Mokhele and Mokhele 2022), there is a paucity of studies analysing the institutional arrangements that influence the planning of airport-led development. Using the study area of the Ekurhuleni Aerotropolis, which centres on OR Tambo International Airport, the study analysed the institutional arrangements that affect the planning of airport-led development. The study's specific objective was to examine the stakeholders' roles, interdependencies, synergies, and contentions in the planning of the Ekurhuleni Aerotropolis.

Addressing this research objective would provide insights into the institutional challenges and prospects of integrating land use and transport infrastructure (i.e., airports in the context of this chapter). As argued by Giuliano (1989), relationships between land use and transportation have important policy and practical implications. In the same vein, the National Planning Commission (2012) argues that improved integration is required between spatial and transport planning for South Africa's development vision to be realised. As potential catalysts of spatial development (Kamruzzaman et al. 2021), airports are crucial transport infrastructure in metropolitan areas, taking on the features of central business districts (CBDs) and acting as regional development poles (Boquet 2018; Peneda et al. 2011) à la the normative models of airport-led development highlighted earlier.

This chapter is structured as follows: Section 11.2 presents the literature review, including the conceptual and theoretical frame of reference that informed the study. The third section introduces the study area, followed by the research design and the data collection and analysis methods in the fourth section. The fifth section presents the findings of the study. Section 11.6 concludes the chapter.

11.2 Literature Review

11.2.1 Institutional Arrangements and Airport-Centric Developments

Several studies describe the institutional arrangements that impact development on and around airports. According to Galvin (2010), France observed a significant transformation in the Orly Airport area's development starting in 2007, leading to its classification as a national project. National projects are situated in regions with substantial development potential and are under the purview of the French Ministry of Land Planning. Consequently, in the areas characterised by a high concentration of businesses, local planning entities were replaced by a state agency known as the *Establissement Public d'Amenagement* (EPA). This shift in responsibility means that policymaking and land-use planning are primarily managed by the state, with prefectures representing the state at the provincial level being the sole authorities issuing construction permits and granting project approvals within national projects. Local stakeholders play a role in the development implementation phase through an economic development agency, which acts as the central coordinating body bringing together institutional and market actors and aiding companies in identifying suitable premises (Galvin 2010).

In the context of Frankfurt Airport and its environs in Germany, Van Wijk (2007) notes that the state of Hesse holds exclusive responsibility for overseeing the aviation-related expansion processes. Conversely, the City of Frankfurt takes charge of real estate and infrastructure development, acting as a facilitator for developers and commercial stakeholders. Another significant player is the Rhein Main Verkehrsverbund (RMV), a publicly funded, independently operating body that comprises various local and regional governments as shareholders. RMV's involvement in the airport's development revolves around planning regional transport initiatives that connect the airport with its surroundings (Van Wijk 2008).

Regarding Amsterdam Schiphol region in the Netherlands, Van Boxtel and Huys (2005) observe that decision-making regarding airport development has been a source of controversy since the 1960s. Nonetheless, at certain junctures, consensus was achieved, culminating in the approval of a development plan for the airport surroundings in 1991. As pointed out by Galvin (2010), this development plan led to significant regional growth, accommodating a wide range of activities, from logistics to mixed-use projects. According to Van Wijk (2008), the province of North Holland plays a pivotal mediating and coordinating role, bridging the gap between local and national spheres of government and navigating the interests of private and public stakeholders. Furthermore, the province exercises oversight over local land-use plans to ensure their alignment with the regional plan. Within the Schiphol regional plan, the province establishes land-use and zoning regulations, including location approval criteria based on the relevance of the proposed land use to the airport. As a guiding principle, the regulations stipulate that the closer the land is to the airport, the greater the degree of airport-relatedness that the proposed development should possess. This zoning framework is designed to, among others, safeguard the airport and its surroundings from traffic congestion (Van Wijk 2008).

As presented by Galvin (2010), the province of North Holland is actively involved in two forums, where it assumes a coordinating role. Firstly, there is an administrative forum known as the Bestuursforum Schiphol (BFS), which serves as a platform for urban planning stakeholders in the airport region. The province chairs the BFS, with the municipalities of Amsterdam and Haarlemmermeer as its active members. Secondly, on a broader scale, there is an administrative coordination group, Bestuurlijke Schiphol, which encompasses neighbouring regions and municipalities that express interest in the development of the airport.

Furthermore, Galvin (2010) indicates that the public-private actor system is managed by two organisations during the development implementation phase. The first is the Schiphol Area Development Company (SADC), which is responsible for coordinating public guidelines and private interests to ensure cohesive development processes. SADC also has the authority to develop land. However, it is important to note, as mentioned by Van Wijk (2008), that SADC does not oversee real estate development; instead, this function is conducted by private real estate developers. The second coordinating body is the Amsterdam Airport Area, a public-private association comprising members from municipalities and port authorities. This association takes on the marketing responsibilities for the airport area (Galvin 2010).

Towards uncovering factors that resulted in the evolution of institutional arrangements in the environs of Schiphol Airport, the subsequent work of Van Wijk et al. (2014) identified six governance episodes in 5 years. The inaugural episode was marked by establishing two governance structures: the governance forum on the one hand and a separate governance agreement on the other. Although the two structures appeared to have overlapping roles, their complementarity was identified in the second episode. In the third period, there were tensions between the municipalities in the governance forum. The disagreements (including those between public and private entities) continued in the fourth era, subsequently introducing additional governance structures. The fifth phase was marked by the efforts to recover from the disputes of the preceding era, while in the last episode, the focus was on bringing down the curtain on the governance forum. This happened because, with the competing responsibilities in the previous episodes addressed, limited agenda items were the subject of the forum's engagements. Furthermore, changes in the key personnel played a part in the demise and disassembling of the governance forum (Van Wijk et al. 2014).

Van Wijk et al. (2014) note that one of the factors that had initially contributed to the success of the governance forum was the importance of connections with political heavyweights, including mayors, councillors, governors, ministers, and members of parliament. However, the political connections deteriorated over time, leading to a decline in the forum's success and relevance (Van Wijk et al. 2014).

In an African context, Arthur (2018) analysed the Accra Airport City, Ghana, through unpacking, among others, the partnership between the state and private sector in developing the airport city and analysed the nature of the environment that enabled the private sector to participate in the development of the airport city. It was noted that the development of the Accra Airport City was a combined effort between the two sets of key stakeholders, the state and private investors. The Ghana Airports Company Ltd. (GACL) acquired the project site, serviced it, and made it available to private investors on a leasehold basis. It was found that the provision of serviced land was instrumental in attracting investors. To ensure the airport city's upkeep, the investors formed the Airport City Association, which the GACL later joined to reinforce the partnership and perform a facilitative role (Arthur 2018).

Against the backdrop of the studies overviewed above, Freestone and Baker (2011) advocate for a regional perspective in airport development, emphasising the alignment between the planning of airport and surrounds and the broader regional plans, and the necessity for collaborative processes among public and private stakeholders. Freestone and Baker (2011) acknowledge that regional governance structures often struggle to effectively address the complexities of airport regions, which are often fragmented into multiple authorities and conflicting interests.

11.2.2 Conceptual and Theoretical Frame of Reference

11.2.2.1 Institutional Economic Geography

In its endeavour to analyse the institutional arrangements that influence airport-led development, this chapter is befitting the institutional turn in human geography or what could be referred to as institutional economic geography. The institutional focus is based upon the premise that the economic landscape cannot be fully understood without considering the institutions on which economic activity depends. The argument is that although institutions cannot be the sole cause of particular geographical patterns of development, they can enable or constrain development. This implies that economic activity is institutionally situated and, therefore, must be interpreted as intertwined with the broader structures of social, economic and political rules, procedures and conventions, which can be formal or informal (Martin 2000; Glückler et al. 2018).

Institutional economic geography delves into several core themes. Firstly, it endeavours to uncover the impact of institutions on the spatial economy, distinguishing between the institutional environment and institutional arrangements. The institutional environment encompasses informal customs, social practices, and formal structures of rules and regulations (often legally enforced), which shape socio-economic behaviours. The concept of institutional arrangements pertains to the organisational structures that emerge because of, and are influenced by, the institutional environment. Institutional economic geography is concerned with both these facets of the economic institutional framework, particularly their interplay. The institutional environment fundamentally influences economic organisations' formation, functioning, and evolution. Similarly, as institutional arrangements operate, they perpetuate and transform the institutional environment. The way this plays out across different regions and how it moulds economic outcomes are central focal points in institutional economic geography (Martin 2000).

Secondly, the institutional perspective within economic geography emphasises the economic landscape's evolution. Institutions, by their inherent nature, are characterised by persistence and endurance. These qualities provide structured frameworks for coordinating economic activities and enable exchanges across different periods and locations. Institutions exhibit a phenomenon known as path dependence, meaning that they tend to develop gradually and self-perpetuate over time (Martin 2000).

Thirdly, institutional economic geography concentrates on economies' social regulation and governance. National and local institutional arrangements are considered pivotal components of the spatial economy's institutional regulatory framework. While the primary scope of regulation is the national economic realm, economic geographers also pay close attention to local and regional aspects of regulation. This is because national mechanisms for socio-economic regulation directly impact the local level, and regulation possesses a micro-level dimension specific to each locality, reflecting the unique diversity and characteristics of local institutional structures (Martin 2000).

Economic geographers have endeavoured to develop distinct spatial concepts in examining institutions. Two notable concepts are institutional space and institutional thickness. Institutional space pertains to the specific geographic area within which a particular institution is established and exercises its influence. A hierarchy of institutional spaces can be identified, ranging from supra-national to national, regional, and local spaces. The arrangement and interaction of these institutional spaces, known as nestedness, varies from one location to another. Consequently, when transitioning from one area to another within a national economic domain, institutional nestedness changes, affecting the dynamics among the institutions. This leads to recognising diverse local institutional arrangements (Martin 2000).

Significant disparities in institutional arrangements also exist among regions and localities within national economies. Economic geographers analyse these distinctions through the concept of institutional thickness. Amin and Thrift (1995), cited in Martin (2000), define institutional thickness based on several components. The first component is the presence of robust institutional structures. The second element involves a high level of interaction among the institutions to foster networking, cooperation, and exchange, leading to mutual conformity among local institutional arrangements. Thirdly, institutional thickness relies on well-defined power structures, coalition-building, and collective representation to mitigate conflicts between institutions. This, in turn, gives rise to inclusiveness and collective mobilisation, where a shared sense of purpose emerges around commonly held agendas or projects related to regional or local development. The specific combinations of these elements within a locality determine the extent and character of local institutional thickness (Martin 2000).

The concept of territory suggests that it can, among others, be a social and political construct that underscores collective action. This collective action arises from the mobilisation of groups, organised interests, and territorial institutions, a process in which interactions among actors can result in various outcomes, including confrontation, cooperation, and conflict (Davoudi et al. 2008). Therefore, territorial governance refers to the method of organising the multitude of relationships that characterise interactions among stakeholders and diverse interests in a given area. The organisational aspects relate to the development of a common territorial vision, aimed at establishing territorial cohesion at multiple levels. The primary challenge for territorial governance is to foster both horizontal and vertical cooperation and coordination among various levels of government, sectoral policies that have a territorial impact, and governmental and non-governmental organisations and citizens. Achieving such coordination, both horizontally and vertically, leads to the integration and coherence of diverse responsibilities, competencies, and territorial visions (Davoudi et al. 2008).

11.2.2.2 Collaborative Planning

According to communicative theory, also known as collaborative planning, an urban and regional planner's role is to facilitate consensus among diverse perspectives (Fainstein 2000). Collaborative approaches underscore the significance of promoting cooperation among various stakeholders in policy formulation and implementation, acknowledging diverse forms of knowledge, and establishing social networks as a source of institutional capital. These approaches shift the role of planning from merely constructing places to nurturing the institutional capacity. They recognise that planning processes, much like governance and management processes in general, are dynamic social processes in which meanings are shaped through discourse (Healey 1998).

Furthermore, collaborative approaches to place-making aid in creating platforms that serve as educational arenas where stakeholders acquire fresh approaches. By cultivating an institutional capacity capable of forging consensus within a world characterised by diversity, there is the promise of nurturing a fertile foundation in which projects and partnerships can flourish. Such a capacity serves as a resource through which individuals can access knowledge and connections to support their initiatives, and it provides a means of addressing conflicts (Healey 1998). However, Watson (2009) argues that caution is warranted when striving for consensus in planning processes. While planning endeavours typically aim to avoid intentionally generating conflict (though, at times, conflict may be unavoidable), there can be situations in which consensus-driven approaches inadvertently lead to exclusion rather than inclusion. A central challenge, therefore, revolves around how to position planning within conflicting rationalities (Watson 2009).

11.3 Study Area

The study focused on the Ekurhuleni Aerotropolis project, centred on the OR Tambo International Airport, Gauteng province, South Africa. Owned and operated by the Airports Company South Africa (ACSA), OR Tambo Airport was opened in 1953, known then as Jan Smuts Airport. The airport is positioned in the jurisdiction of Ekurhuleni metropolitan municipality, approximately 23 km northeast of the Johannesburg central business district (CBD), 23 km from Sandton, and 40 km from Tshwane (Fig. 11.1). This relative proximity of the major economic and administrative centres shows that while the airport is positioned in Ekurhuleni and located in a multi-airport context, it serves the Gauteng City-Region and surrounding areas, including the country's wealthiest and most populous city, Johannesburg, and the administrative capital city, Tshwane (Mokhele 2017).

OR Tambo International Airport is surrounded by a diverse concentration of economic activities, which include firms involved in the following standard industrial classification (SIC) categories: manufacturing; transport, storage and related activities; wholesale and retail; accommodation and catering; and car repair, tyre dealer, testing and repair of machinery (Mokhele 2017). The varied economic activities are positioned on the airport landholding and in the industrial areas of Isando, Jet Park, Spartan, Meadowdale, Elandsfontein, Pomona and Bartlett, which are contiguous with the airport (Fig. 11.2). The industrial areas around the airport were, among other zones, promoted for industrial development from the 1940s to drive the economy when gold mining declined (Fair 1956).

OR Tambo Airport was considered an appropriate study area for analysing the institutional arrangements that influence airport-led development initiatives for two reasons. First, it is the busiest airport in South Africa regarding the number of passengers and volume of airfreight processed, an attribute that increases its potential impact on the spatial economy of the Ekurhuleni municipality and the Gauteng City-Region. Second, OR Tambo Airport had been the subject of planning efforts in the manner of the normative models of airport-led development, wherein the proponents claim that the industrial areas near the airport are the epitome of Africa's first aerotropolis (see, for instance, Ekurhuleni Municipality 2013; Mail and Guardian 2013a, b). Although it is difficult to pinpoint the dates with certainty, the Ekurhuleni Aerotropolis planning project commenced around 2011 and concluded in 2015 with the approval of the aerotropolis master plan (also refer to the introductory note of Sect. 11.5).

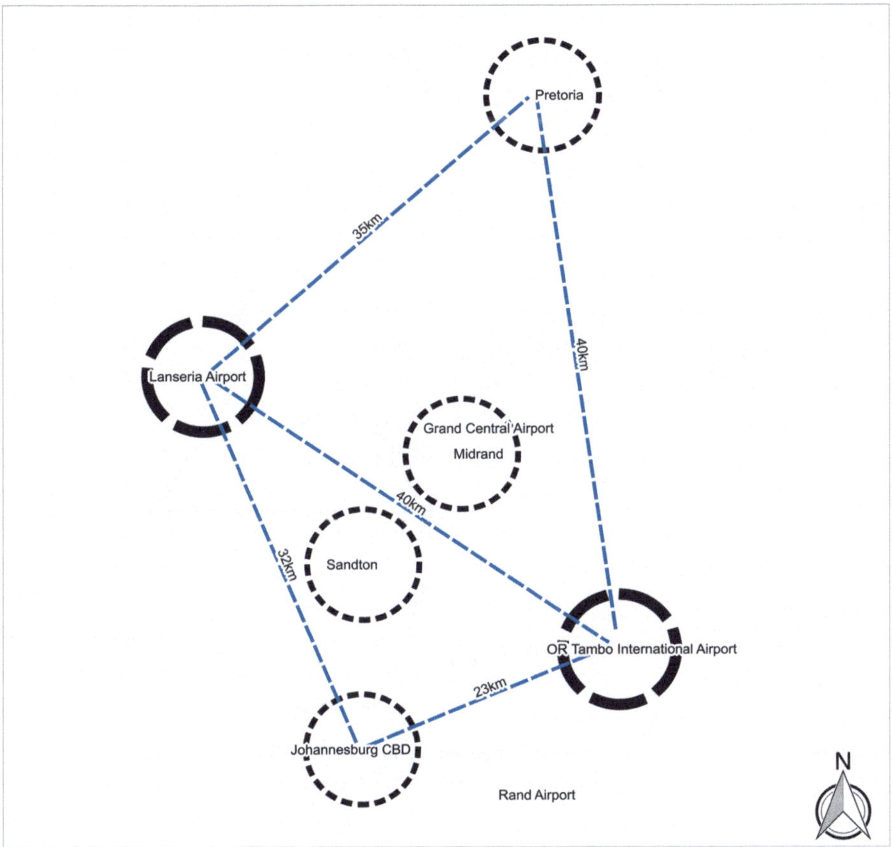

Fig. 11.1 OR Tambo International Airport within the broader Gauteng City-Region. (Adapted from Mokhele 2016)

11.4 Research Design and Methods

The study was based on a qualitative design, which entailed interviews with four key informants/respondents knowledgeable about the Ekurhuleni Aerotropolis planning process. The informants were selected because they had been part of the institutional structures of the aerotropolis project. With crucial institutional memory, the respondents were, therefore, in a position to provide first-hand insights into the institutional dynamics of the project. The identification and selection of the key informants hinged on the non-probability sampling technique of snowballing. Based on the snowball analogy, the sampling process began with identifying two key informants and spread out based on the links, recommendations and referrals made by the individuals contacted (Gobo 2004; Neuman 2014). The snowball sampling technique implied that the respondents were not randomly selected and would have been professionally connected in one way or another (Neuman 2014).

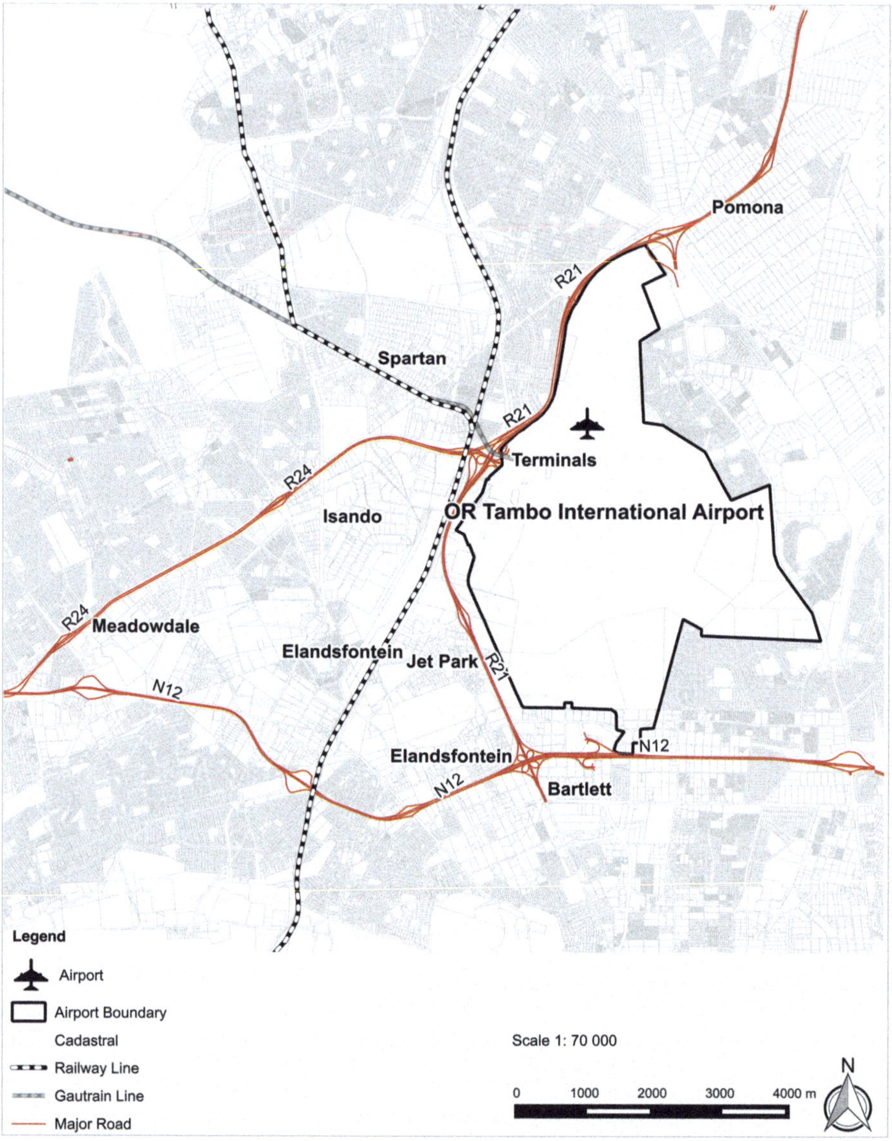

Fig. 11.2 OR Tambo International Airport and its environs. (Adapted from Mokhele 2016)

11.4.1 Data Collection

In preparation for the interviews, the researcher sent email requests to the key informants, introducing the study's aim and enquiring about their willingness and availability to participate in the research. Informed by the availability of the informants, the semi-structured interviews were conducted virtually in July and August 2023 via the Microsoft Teams platform. Virtual interviews are akin to telephonic interviews, whose advantages are largely similar to face-to-face interviews and more efficient in terms of time and cost savings (Neuman 2014).

Conducted by the author, each interview lasted between 30 and 40 min. The discussions were guided by a semi-structured questionnaire, which primarily entailed open-ended questions. Permission was obtained from the respondents for the interviews to be recorded to aid the analysis later in the research process. The respondents were assured that confidentiality and anonymity would be maintained. Therefore, identifying information is omitted from this chapter, such as the names of the respondents' organisations and the years they were involved in the Ekurhuleni Aerotropolis project. Instead, the key informants are assigned codenames Respondent A, B, C and D. Widely used in qualitative research, codenames are index terms allocated to the participants (Heaton 2022).

11.4.2 Data Cleaning and Analysis

The data collected from the qualitative interviews with the key informants were transcribed through Microsoft Teams and cleaned in Microsoft Word. To improve the quality and accuracy of the data, the cleaning process entailed listening to the audio recordings, removing duplicated words or stutters in the transcript, amending the words captured incorrectly during the transcription process and removing parts of the discussion that were general conversation unrelated to the topic at hand. The responses or remarks of the respondents were left verbatim instead of being rephrased or paraphrased. The researcher intentionally used the laborious manual cleaning of the transcripts to get used to the data collected.

The data-cleaning process was followed by thematic analysis, a qualitative method used to identify, analyse, and report patterns or themes within data (Braun and Clarke 2006). The analysis entailed reading through the data and looking for patterns in the stakeholders' roles, interdependencies, synergies and contentions during the Ekurhuleni Aerotropolis planning process.

11.5 Findings and Discussion

This section presents the study's findings, organised around the following aspects: composition of the institutional structure, coordination of the stakeholders, contentions among the stakeholders, consensus building, and the respondents' perception of the success of the aerotropolis planning process. As noted in the study area overview, the origins of the Ekurhuleni Aerotropolis planning process can be traced to around 2011 when, during the preparation of the Ekurhuleni metropolitan spatial development framework (MSDF), the OR Tambo International Airport was identified as an essential node towards restructuring the municipality (Respondent B).

11.5.1 Composition of the Institutional Structure

The Ekurhuleni Aerotropolis planning process comprised a triad of main stakeholders: Ekurhuleni municipality, Gauteng provincial government, and ACSA (Fig. 11.3). The former represented two spheres of government, and the latter a state-owned company mandated with the efficient operation and profitability of the airport. These main parties, therefore, had different organisational mandates, priorities, interests, and territorial spheres of interest, which are discussed later with the contentions among the stakeholders.

Ekurhuleni municipality was regarded as the primary stakeholder because it conceptualised and initiated the aerotropolis project, and the airport was positioned within its jurisdiction. The units or departments of the municipality involved in the aerotropolis planning process included City Planning, Economic Development, Roads and Transport, Integrated Development Planning, and Information Technology. The municipality appointed an internal aerotropolis project manager and set up an office for the project (Respondent B).

The second main stakeholder, the Gauteng provincial government, got involved between 70% and 80% through the planning process after the municipality had laid the groundwork for the project. The involved departments at the provincial government level included the Premier's Office and the Planning Department. The inclusion of the provincial government in the project was strategically meant to achieve two interrelated objectives: to, on the one hand, leverage political support at a level higher than Ekurhuleni municipality and, on the other, to augment the project's significance relative to other aerotropolis projects in South Africa (Respondent B), particularly at King Shaka and Lanseria airports mentioned in the introduction.

The third stakeholder, ACSA, was involved as the owners and managers of OR Tambo International Airport. Given that the aerotropolis concept intrinsically hinges on the airport, it was essential for ACSA to be included in the initial stages of the planning process. It was acknowledged that development could not occur on the airport without ACSA; hence their participation and buy-in were required to enable

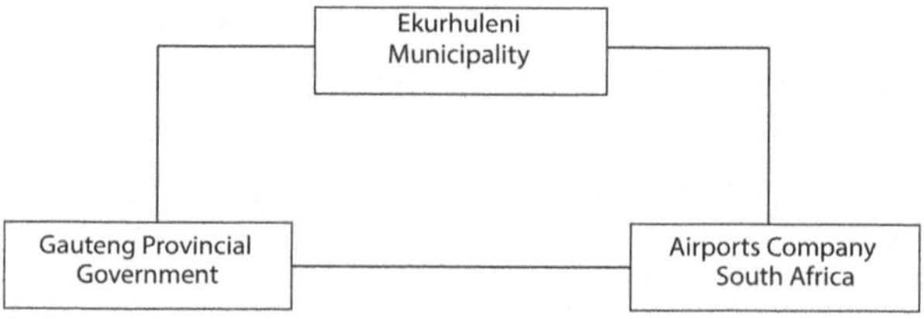

Fig. 11.3 Main stakeholders in the Ekurhuleni aerotropolis project (original to the author)

implementation (Respondent B) and ensure success of the project. The relationship between ACSA, the municipality and the provincial government was formalised by signing a memorandum of understanding (Respondent C).

Other stakeholders were identified and incorporated into the project through the interactions of the three primary stakeholders. They included the National Department of Transport, the business chamber of Ekurhuleni, various associations related to OR Tambo Airport (like the airfreight association and the pilots' association) and the residents' associations and groups. Some of these stakeholders were involved mainly for supporting roles and did not contribute directly to setting the aerotropolis strategy (Respondent B). For instance, the National Department of Transport was involved to some extent because several roads had to be upgraded to unlock development within the study area (Respondent D). It is noted that despite the literature's acknowledgement of the crucial role of the private sector in airport-led development (e.g., Arthur 2018; Galvin 2010), relative to the three primary stakeholders, the private sector did not play a leading role in the planning of the Ekurhuleni Aerotropolis.

Although the respondents had divergent views on the inclusiveness of the project, it could be ascertained that the necessary stakeholders were included in the project in one way or another at particular periods of the planning process. The respondents acknowledged that including all stakeholders in the process was crucial. As Respondent A put it, "The aerotropolis thing is so big...", hence the participation of all stakeholders was critical.

11.5.2 Coordination of the Stakeholders

The structures for coordinating the stakeholders evolved during the planning process and primarily occurred through two committees: the joint planning committee and the steering committee. These committees differed in the centre of governance, being either at the municipal or provincial government level (Fig. 11.4).

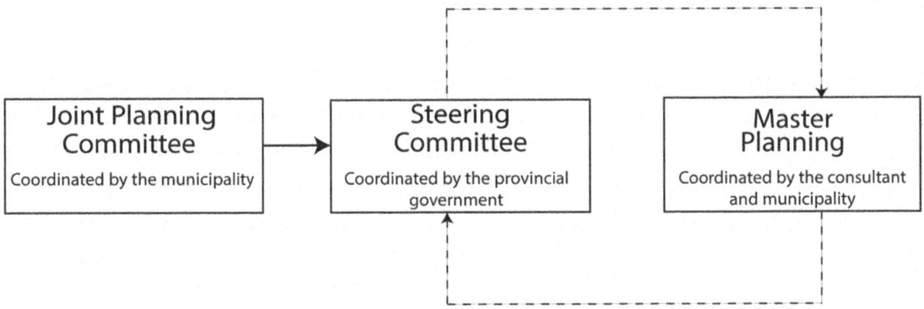

Fig. 11.4 Structures that facilitated stakeholder engagements (original to the author)

The first structure, the joint planning committee, was established by the Ekurhuleni municipality to coordinate the aerotropolis planning. From the early days of the planning process, a joint planning committee comprising the municipality's and ACSA's representatives convened monthly and sometimes more often (Respondent B). It should be noted that this committee was established before the provincial government joined the project.

The second structure, the technical steering committee, was set up later in the process as a joint decision of the Gauteng province premier and the executive mayor of Ekurhuleni municipality. The early joint planning committee that comprised representatives from ACSA and the municipality was incorporated into the new larger steering committee coordinated by the provincial government. This was after Ekurhuleni municipality formally adopted the aerotropolis through a council resolution, hence the focus of the steering committee was more on the implementation and funding mechanisms (Respondent B). Integrating the joint planning committee into the new structure was appropriate as it negated a situation similar to the case of Schiphol International Airport, where there were, at some point, parallel structures with overlapping roles (Van Wijk et al. 2014). The technical steering committee comprised representatives of the following stakeholders: Ekurhuleni municipality, Gauteng provincial government, Gauteng growth and development agency, ACSA, Denel, Transnet, and the Passenger Rail Agency of South Africa (PRASA).

In 2013, the Ekurhuleni municipality appointed a consulting firm to coordinate the processes of formulating the Ekurhuleni Aerotropolis master plan. The consultants were tasked with the responsibility of developing the 25-year master plan and the 5-year implementation plan (Respondent D). The firm, in collaboration with the municipality and the technical steering committee, was crucial in, among others, conducting the stakeholder engagements to develop the master plan (Respondent C) and accordingly had engagements with communities and business representatives (Respondent D). The consultants also engaged the political steering committee, which considered the submissions from the technical steering committee.

The consultants completed their assignment around 2015 when the aerotropolis master plan was submitted to Ekurhuleni municipality for approval.

11.5.3 Contentions Among the Stakeholders

Similar to the findings of Van Wijk et al. (2014) and the argument of Freestone and Baker (2011) on the challenges of regional governance structures, the Ekurhuleni aerotropolis project was not immune to stakeholder disagreements and conflicts. As Respondents A and C ably put it, "… there was a tug of war…" among the stakeholders in the project. The disagreements pointed to the lack of institutional thickness expounded by Martin (2000). It can be argued that the main stakeholders' distinct organisational mandates, responsibilities, priorities and territorial spheres of interest partly influenced the contentions. However, other respondents were

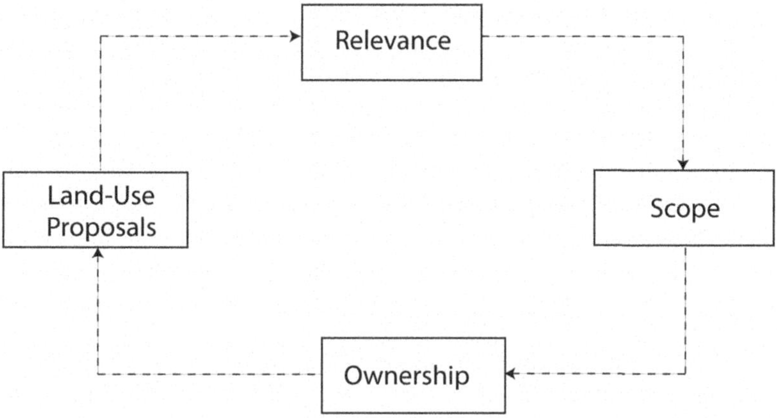

Fig. 11.5 Areas of contention (original to the author)

optimistic and felt some disagreements were merely divergent perspectives that possibly strengthened the planning process (Respondent D).

The contentions during the planning process revolved around the following intertwined aspects: scope and relevance of the project, ownership and driving of the project, and land-use allocation (Fig. 11.5).

11.5.3.1 Relevance, Importance and Urgency of the Aerotropolis Project

There were disagreements between some of the stakeholders about the relevance and urgency of the aerotropolis project. Reflecting conflicting rationalities (Watson 2009), some stakeholders argued there were more pressing socio-economic issues or problems in the municipality, such as providing housing and infrastructure, rather than the preoccupation with the aerotropolis. As narrated by Respondent A, at some point, one leader remarked, "… the aerotropolis is really nice but it is not building houses. We need houses, we need services, we do not need this …". Similarly, the communities were not concerned with the technicalities of the aerotropolis. For them, what was important was the creation of employment opportunities (Respondent D). The business representatives also wanted initiatives and proposals to provide opportunities for local businesses instead of prioritising international anchor tenants for the aerotropolis (Respondent D).

11.5.3.2 Geographical Scope of the Project

There were disagreements on the appropriate geographical scope of the aerotropolis project. On the one hand, as the project's initiator, Ekurhuleni municipality held a strong view that the project should be restricted within the municipal boundaries.

This showed that Ekurhuleni wanted to maximise the benefit of the project for the municipality (Respondent D). On the other hand, the provincial government believed the project scope should extend to the broader Gauteng City-Region. The province even wanted the project to be named the Gauteng City-Region Aerotropolis (Respondent C). Relatedly, a form of aerotropolis was proposed around Lanseria Airport, in the neighbouring City of Johannesburg municipality, which resulted in tensions between Ekurhuleni and the City of Johannesburg as some actors within Ekurhuleni believed there could only be one aerotropolis in the city-region (Respondent C). These tensions between the provincial and municipal spheres of government reflected what could be referred to as the scramble for the aerotropolis.

As noted in the introduction, the airport catchment or sphere of influence could span multiple geographical scales or territories, namely airports and their immediate surroundings, metropolitan areas, functional regions, countrywide locations, areas across national borders, and even across continents. It is also acknowledged that airport-related economic activities, such as logistics firms, are not necessarily positioned near airports (Mokhele 2022). However, metropolitan and local municipalities would, because of specific mandates, be narrowly interested in the territories of the airport environs and metropolitan areas.

11.5.3.3 Roles, Responsibilities and Ownership of the Project

Related to the contention on the geographical scope of the Ekurhuleni Aerotropolis, there were tensions on who the project owner or driver should be, intra- and inter-organisationally. This shows that there were no well-defined power structures to mitigate conflicts between the stakeholders (see Martin 2000).

Even though a dedicated office was set up for the project within Ekurhuleni municipality, there was a sign of internal squabble on whether the aerotropolis project was the purview of Spatial (City) Planning or Economic Planning departments. Each department wanted to oversee the project (Respondent A), once again reflecting the scramble for the aerotropolis.

Inter-organisationally, the provincial and municipal interests were at loggerheads. For instance, there was an idea of setting up an entity that would run the aerotropolis project in the future, and the two spheres were in an arm wrestle on whether that entity would be set up at the provincial government or the municipality (Respondent D).

Some key informants believed the decision on who the driver should be among the primary stakeholders boiled down to who contributed the most financial resources to the project. In this regard, Respondent C argued that the municipality utilised substantial funds to develop the aerotropolis master plan. However, ACSA and the provincial government did not contribute much financially; hence the municipality felt entitled to own the project and set the agenda. Relatedly, there were disagreements on who was responsible for funding particular infrastructure upgrades. Therefore, many disagreements and conflicts related to funding (Respondent B).

The above shows plenty of differences in opinion, which often boiled down to the different stakeholders' roles, responsibilities and priorities (Respondent B). It should also be noted that the timing of the province's involvement could have partly led to disagreements on the project details. The provincial government joined the process towards the conclusion of the planning process; when the aerotropolis had already been endorsed by the Ekurhuleni municipal council, hence it would have been difficult for them to make fundamental changes to the project.

11.5.3.4 Land-Use Allocation

There were plenty of disagreements on the land-use proposals and the allocation of particular economic activities in specific landholdings, zones or precincts. Because of the diverse organisational imperatives and interests, some stakeholders believed certain key land-use categories were appropriate on particular landholdings, while others held a different view. One big issue at some point was why the project was planned on privately-owned land and whether the aerotropolis was used to force the private sector into a particular development direction (Respondent C). Relatedly, some stakeholders wanted particular development on their land (Respondent D). For instance, it required significant effort to convince one key stakeholder not to develop a retail mall on their land, as it would have had a better value for non-retail uses (Respondent B), and retail activities were more appropriate elsewhere.

The examples above show many disagreements around the land use proposed on particular land, the project timing, and the subsequent implementation framework (Respondent B).

11.5.4 Relationship Between the Stakeholders

Relationships between the stakeholders evolved during the planning process, and although the key informants had different views on the nature of the relationships, the institutional arrangements' inertia and durability (see Martin 2000) were questionable. This characteristic showed the lack of path dependence.

Respondent B believed that the relationship between the stakeholders fluctuated during the planning process. Reflecting on the importance of political heavyweights in the manner of the findings of Van Wijk et al. (2014), Respondent B noted, "I think what helped a lot was that there was a good political relationship between the premier and the executive mayor". As a result, the relationships "... improved over time ... because we started finding one another and seeing the commonality" (Respondent B). Respondent B emphasised the importance of personal relationships among the actors:

> "… It came to a point where you started feeling like colleagues in the same organisation, so you came to a point where you know the other people so well that you were actually glad to see them when the meetings took place. So those personal relationships were quite important to understand. Who's got what perspective? What value do they bring to the project? What is their personal view about it? So, knowing and understanding those people on a more personal level actually assisted quite a lot" (Respondent B).

Reflecting a different view, Respondent A noted that "… some stakeholders had more of a voice than others. Specific departments were stronger than others." Respondent D, therefore, felt the relationship between the stakeholders deteriorated over time. "… I suspect it worsened. There was a big power struggle between the provincial and municipal interests… So, I think things actually got worse between those parties rather than better."

11.5.5 Consensus Building Among the Stakeholders

Despite the questionable inertia and durability, divergent perspectives and disagreements between the stakeholders, there were signs of collective mobilisation and the emergence of a shared sense of purpose around the project (see Martin 2000). As Respondent A put it: "Despite the differences … everybody supported the idea…" hence there was a consensus among the stakeholders.

> "There were (also) degrees of compromise involved, but … it was to the benefit of the project. So, in some cases … you had to give a bit to get a bit… we started supporting some of the initiatives to get them to support some of our initiatives" (Respondent B).

Relatedly,

> "The master plan was also adjusted to provide benefits to all of the stakeholders. So, that was probably the foundation for us reaching an agreement. So, probably the most significant thing was adjusting the plan so there was something for everyone" (Respondent D).

As the sign and culmination of consensus among the stakeholders, the aerotropolis master plan was adopted as the base document for driving the aerotropolis implementation plan (Respondent B). However, the claims for consensus should be interpreted with caution (Watson 2009), particularly given the assertion by the stakeholders in the previous discussion that some actors had more voice and power than others.

11.5.6 Perception of the Success of the Aerotropolis Planning Process

On the background of the contentions among the stakeholders and the subsequent perceived consensus, it was essential to ascertain the perception of the key respondents on the success of the Ekurhuleni Aerotropolis planning process. In reflecting on the success, the respondents referred to the approval and adoption of the master plan and the subsequent ability of the plan to guide development as measures of success.

Referring to the master planning, Respondent A remarked, "... That process was very robust. They went through all the committees. It was very well managed. The document was well done..." Similarly, Respondent D remarked, "I think the planning process was absolutely a success, and the measurement for that is... we had a plan that got sign-off from all stakeholders."

Respondent C concurred and argued that "... the plan truly has unlocked a lot of ... development ... at least guided it in the correct direction from a planning point of view..." However, it should be acknowledged that development near OR Tambo Airport emerged long before the introduction of the models of airport-led development in South Africa. From the 1940s, industrial development was promoted around the airport, among other areas, to drive the economy when the gold mining declined (Fair 1956). The concentration of development near the airport intensified since then, hence it is difficult to attribute the contemporary development to the aerotropolis planning. As Respondent C commented "I personally think the market would still go there, even without the [aerotropolis] master plan... but raising the profile of the city through the aerotropolis... definitely did help".

11.5.7 Possible Ways of Improving the Planning Process

Although the planning process was arguably successful regardless of the stakeholders' contentions and fluctuating relationships, certain aspects of the project could have been approached differently. These included the driving of the project and the decentralisation of the power base.

Firstly, some key informants believed the Gauteng provincial government should have been given more authority to drive the project, as they had more political influence and interest that transcended the Ekurhuleni municipal boundaries.

> "So, if you had the province at the time leading, ... [the project] could have possibly progressed much further than ... it did. So, I would say from the planning point of view, everything was done right but give a stronger lead to the province to drive it" (Respondent C).

This approach would have acknowledged the regional perspective of airport-related development planning and improved governance arrangements (see Freestone and Baker 2011). The recognition of the aerotropolis as a provincial project would be akin to the environs of Orly Airport in France, which were identified as a national project, driven by the national government (Galvin 2010) or the case of Amsterdam Airport, where at some point, the province of the North Holland played a mediating and coordinating role between local and national levels of governance (Van Wijk 2008).

Secondly, some respondents felt the power base inside the municipality was too centralised. It would have, therefore, been better for the municipality if a broader set of actors participated in driving the project. Similar to Van Wijk et al. (2014) findings on the impact on the governance arrangements of changes in the key personnel, Respondent D noted that political changes were part of the problem because the departure of certain leaders set the project back quite significantly. As argued by Forester (1995), planners cannot ignore those in power lest they become powerless and frustrated. There should have, therefore, been continuous education of the wide range of political actors on the nature of the aerotropolis planning.

11.6 Conclusion

To draw lessons for territorial governance in airport-led development, the chapter analysed the stakeholders' roles, interdependencies, synergies, and contentions in the planning of the Ekurhuleni Aerotropolis, South Africa. A triad of primary stakeholders was identified: Ekurhuleni municipality, Gauteng provincial government and the Airports Company South Africa. Although the respondents argued that, besides the disagreements, there was ultimately consensus among the stakeholders, it can be discerned that there was no territorial cohesion and a shared vision for the Ekurhuleni Aerotropolis. There were strong disagreements on the extent of the territory, with some stakeholders arguing the territory should be the municipality as defined by the administrative boundaries. In contrast, others contended it should be the Gauteng City-Region defined by the functional linkages across municipalities. It is, therefore, recommended that, depending on the nature and extent of the airport's catchment, the sphere of governance beyond a municipality facilitates the planning of airport-related development initiatives. In South Africa, the focus could be on linking airport-centric development planning processes to regional spatial development frameworks instead of limiting them to municipal spatial development frameworks.

To address the chapter's limitation of focusing only on the planning process, it is proposed that future research focus on the institutional prospects and challenges of implementing airport-led development.

References

Arthur IK (2018) Exploring the development prospects of Airport City, Ghana. Area Dev Policy 3(2):258–273.

Banai R (2017) The aerotropolis: urban sustainability perspectives from the regional city. J Transp Land Use 10(1):357–373

Boquet Y (2018) From airports to airport territories: expansions, potentials, conflicts. Hum Geogr J Stud Res Hum Geogr 12(20):137–156

Braun V, Clarke V (2006) Using thematic analysis in psychology. Qual Res Psychol 3(2):77–101

Coetzee PJVV, Swanepoel PA (2017) Spatial relationship and movement patterns of the air cargo industry in airport regions. J Transp Supply Chain Manage 11:1–10. https://doi.org/10.4102/jtscm.v11i0.297
Conway M (1993) Airport cities 21: the new global transport centers of the 21st century. Conway Data, Atlanta
Davoudi S, Evans N, Governa F, Santangelo M (2008) Territorial governance in the making. Approaches, methodologies, practices. Boletin de la AGEN 46:33–52
Ekurhuleni Municipality (2013) Ekurhuleni aerotropolis planning and land use guidance document. Ekurhuleni Metropolitan Council
Fainstein SS (2000) New directions in planning theory. Urban Aff Rev 35(4):451–478
Fair TJD (1956) A planning survey of the Southern Transvaal: the Pretoria-Witwatersrand-Vereeniging area. The Government Printer, Union of South Africa
Forester J (1995) Planning in the face of power. In: Stein JM (ed) Classic readings in urban planning: an introduction. McGraw-Hill, pp 437–456
Freestone R, Baker D (2011) Spatial planning models of airport-driven urban development. J Plan Lit 26(3):263–279. https://doi.org/10.1177/0885412211401341
Galvin V (2010) Coordinating spatial development in airport regions. Embeddedness and experimentation at Paris Orly and Amsterdam Schiphol. Aerlines 48:1–5
Glückler J, Suddaby R, Lenz R (2018) On the spatiality of institutions and knowledge. In: Glückler J, Suddaby R, Lenz R (eds) Knowledge and institutions. Springer, pp 1–19
Gobo G (2004) Sampling, representativeness and generalizability. In: Seale C, Gobo G, Gubrium JF, Silverman D (eds) Qualitative research practice. Sage, pp 405–426
Giuliano G (1989) New directions for understanding transportation and land use. Environ Plann A 21(2):145–159.
Healey P (1998) Building institutional capacity through collaborative approaches to urban planning. Environ Plan A 30(9):1531–1546
Heaton J (2022) "*Pseudonyms are used throughout:" a footnote, unpacked. Qual Inq 28(1):123–132
Kamruzzaman M, Aston L, Baker D, Braun B, Shatu F (2021) Changes in land use typology of global airports: an empirical investigation with implications for the aerotropolis concept. J Transp Geogr 97:103217
Kasarda JD (2009) Airport cities. Urban Land 68:56–60
Kasarda JD, Lindsay G (2011) Aetrotropolis: the way we'll live next. Penguin Group
Mail & Guardian (2013a) Ekurhuleni ready to roll out the red carpet for the ACE 2013 conference, 19–25 April
Mail & Guardian (2013b) Ekurhuleni: Moving towards a world class city, 8–14 November
Markusen A (1999) Fuzzy concepts, scanty evidence, policy distance: the case for rigor and policy relevance in critical regional studies. Reg Stud 33(9):869–884
Martin R (2000) Institutional approaches in economic geography. In: Sheppard E, Barnes TJ (eds) A companion to economic geography. Blackwell Publishers, pp 77–94
Mokhele M (2016) Spatial economic attributes of airport-centric developments in Cape Town and Johannesburg. PhD thesis, Stellenbosch University
Mokhele M (2017) Spatial economic evolution of the airport-centric developments of Cape Town and OR Tambo international airports in South Africa. Town Reg Plan 70:26–36
Mokhele M (2018) Spatial economic attributes of O.R. Tambo and Cape Town airport-centric developments in South Africa. J Transp Supply Chain Manag 12:a344. https://doi.org/10.4102/jtscm.v12i0.344
Mokhele M (2022) The geography of logistics facilities relative to airports: taxonomy of literature and research agenda. J Transp Supply Chain Manag 16:a770. https://doi.org/10.4102/jtscm.v16i0.770
Mokhele M, Mokhele T (2022) Spatial configuration of logistics firms relative to Cape Town International Airport, South Africa. Logistics 6(3):49. https://doi.org/10.3390/logistics6030049
National Planning Commission (2012) National development plan 2030. Our future-make it work. National Planning Commission

Neuman WL (2014) Social research methods: qualitative and quantitative approaches, 7th edn. Pearson

Peneda MJA, Reis VD, Macário MDRMR (2011) Critical factors for development of airport cities. Transp Res Rec 2214(1):1–9

Pereira ACC, Milne D, Timms P (2023) Investigation of the aerotropolis concept and its transferability around the world. J Air Transp Manag 106:102271. https://doi.org/10.1016/j.jairtraman.2022.102271

Van Boxtel M, Huys M (2005) Unravelling decision making about the future developments of Amsterdam Airport Schiphol. Paper presented at the 45th European Regional Science Association, Amsterdam, 23–27 August 2005

Van Wijk M (2007) Airports as cityports in the city-region. Doctoral thesis, Utrecht University

Van Wijk M (2008) Development of airport regions: varieties of institutions in Schiphol and Frankfurt. Aerlines Magazine, Airports: e-zine edition 40. www.aerlines.nl

Van Wijk M, Van Bueren E, Te Brommelstroet M (2014) Governing structures for airport regions: learning from the rise and fall of the 'Bestuursforum' in the Schiphol airport region. Transp Policy 36:139–150

Watson V (2009) Seeing from the South: refocusing urban planning on the globe's central urban issues. Urban Stud 46(11):2259–2279

Chapter 12
Grey Governance: Micro-Regional Territorial Governance Through Customary Land Management Systems in Informal Peri-Urban Settlements in Cape Town

Herman Geyer

Abstract The research analyses the territorial governance of informal development in developing regions within a deregulated policy framework and a hybridisation of regulatory authorities. Grey governance is a collaborative and transcalar mode of governance between citizenry and the state, combining multi-level government jurisdictions with heterogeneous and specialised micro-regional community modes of governance. The research focuses on the co-federal collaboration between informal horizontal territorial regimes operating through customary land management systems (CLMS) in informal settlements in Cape Town, South Africa, and the formal vertical governance structure of municipal and provincial planning and policies. In this mode of territorial governance, the CLMS provides residents in micro-regions with accessible and transparent mechanisms to participate in the region's governance through SANCO. Formal regulatory authorities, in various municipal and provincial agencies and specialised units take on a new role in selectively enforcing policy regulations based on the normative principles of good governance by negotiating compliance and compromises with community representatives to effectively manage certain challenging aspects of informal development. The CLMS becomes an extension of the state by ensuring partial compliance and self-regulation of policy and practices. This reduces the regulatory burden of local authorities and endears the state to the community as a primary provider, higher authority, and final arbitrator.

Keywords Territorial governance · Grey governance · Informal · Peri-urban · Cape Town · Customary land management

H. Geyer (✉)
Centre for Regional and Urban Innovation and Statistical Exploration, Department of Geography and Environmental Studies, Stellenbosch University, Stellenbosch, South Africa
e-mail: hsgeyerjr@sun.ac.za

D. du Plessis (ed.), *Territorial Governance and Spatial Transformation in Post-apartheid South Africa*, Local and Urban Governance,
https://doi.org/10.1007/978-3-031-85489-7_12

12.1 Introduction

The major governance challenge in the Global South is the governance of 'ostensibly ungovernable territories'. These territories are characterised by unregulated peri-urban settlements in which the majority of urban development consists of informal or semi-formalised structures. Typical formal governance mechanisms are often ineffective when much of the land is undertitled. Undertitled land has limited occupation rights but lacks formal title deeds and secure tenure, especially when the ownership and occupation rights of land is uncertain or disputed. These settlements generally comprise of extremely socially diverse populations with a variety of ethnicities, cultures, citizenships and identities resulting in complex social structures that are often inadequately represented in formal governance structures. The adequate territorial governance of such regions requires a micro-regional approach in which numerous small local communities collaborate as a regional entity to redefine the appropriate scale of representation. Furthermore, these settlements often sprawl outside the administrative boundaries of cities, or are located in customarily entitled land in traditional authorities. These transboundary regulatory challenges require supra-regional territorial governance initiatives to maintain essential basic services, health and safety, and to protect the tenure of the vulnerable and marginalised residents.

The aim of the study is to evaluate grey governance as an informal transcalar mode of territorial governance operating between multiple political levels (micro-regional to supra-regional scales) and combining vertical and horizontal modes of governance. These modes of governance combine Type I formal multi-level governance structures with defined jurisdictions with Type II flexible and heterogeneous transjurisdictional modes of governance. The purpose is to mobilise collective action and to develop consensus between special interests through a highly informal and relational process of concerts building. The study evaluates the context of inefficient and neopatrimonial governments in Sub-Saharan Africa and pluralistic and politically contested cross-border subregions and micro-regional enclaves. Grey governance operates through territorial regimes utilising and extra-legal quasi-statal structures, particularly in micro-regions. However, as the literature will demonstrate there remains a degree of vertical integration with existing formal multi-level governance structures, particularly at the supra-regional level. In particular, the study evaluates how informal community-led governance operates through customary land use management systems (CLMS) and the interaction between CLMS and the state. The analysis section of the study evaluates the implementation of grey governance in terms of a combined informal horizontal governance regime operating through CLMS at a macro-regional level within peri-urban settlements and its collaboration with the formal vertical governance structure of the state operating at a supra-regional level.

12.2 The Transitions from Local Government to Territorial Governance

It has long been recognised that governments utilising legal-rational structures such as policy formulation and strategic planning does not automatically translate into efficient regional governance. Many local governments are characterised by bureaucratic fragmentation, regional introversion and declining executive capacities due to inescapable asymmetries in knowledge, sources of power, capital, and local resources (Wassenhoven 2008). Successful regions require mechanisms of public participation and conflict resolution that mobilise collective action through the development of long-standing coalitions between different stakeholders with parallel goals and aspirations for the region (Alaoui and Mamoun 2021). Thus, successful governance requires the active collaboration of local citizens, institutions, and external capital investment within a complex network of socio-political relationships (Robinson and Parsons 2006). Territorial governance is therefore an important factor in regional development, requiring the dynamic coordination of urban development processes between public and private stakeholders and institutions.

This concept of territorial governance is linked to the transition from hierarchical top-down government towards more participative and relational forms of governance between the state, local communities, and the market. It involves defining territorial capital assets that serve the common good of the region, the socio-political valorisation of territorial capital to generate localised socio-economic spillovers, and defining the territory in which these actions occur (Storper 2004). It also refers to the hollowing out of state authority structures from the post-war model of a managerial state, towards its present role as representative and facilitator of supra-local socio-economic networks within and between regions (Davoudi et al. 2008). However, this also results in the radical reconfiguration of state practices through which local government relates to its constituents. This includes public/private partnerships in regional development, as well as the broader consultation of stakeholders and special interests in the design and implementation of policy initiatives. Territorial governance therefore becomes enmeshed in utilising alternative mechanisms beyond typical state channels to integrate different groups of citizens with different territorial interests (Le Galès 2002).

The first transition is a shift towards the re-articulation of scale in terms of nodes and flows. Territorial governance is not limited to the historic administrative boundaries of the regional government bodies, but rather harnessing transboundary interactions to delineate socio-political space in terms of clusters and enclaves to anchor economic growth (Scholte 2017). Territorial governance thus requires a transcalar approach to development. Transcalar modes of governance operates within and between multiple political levels, reflecting a different scope of activity at different scales. This includes local community activism that influences regional and supra-regional policies and vice versa (Pallas and Bloodgood 2022). Each location, albeit the neighbourhood, the city, or the region, cooperates and competes with its peers to attract mobile capital and state resources within polycentric urban regions (Parr

2004). In this process, local governance is re-scaled. It is deterritorialised, upscaling its decision-making powers beyond its limited administrative boundaries into new regionalist coalitions. It is also re-territorialised, downscaling decision-making to micro-level territorial special interests through increasingly powerful homeowners associations, business chambers, and community forums (Mayer 2013). This results in complex collective action challenges, since local governments are required to generate new modes of coalition from increasingly fragmented special interest groups that have either regional transboundary objectives or local micro-regional priorities based on complex socio-political constructions of territory (Davoudi et al. 2008).

A second transition is a shift from top-down policy directives to the mobilisation of collective actions between various special interests to build consensus amongst stakeholders and to define the territory within a common vision (Davoudi et al. 2008). While the different special interests share certain territorial objectives, these invariably also have certain conflicting goals and aspirations. The socio-political nature of territorial governance requires a participative process in which various stakeholders cooperate to achieve convergent objectives and negotiate compromises between the conflicting spatial objectives. Therefore, the role of the state is to construct and steer a polity of regional interests to create a city as a collective (Le Galès 2002). The overall objective is to develop the… social, physical, and economic capital in the region by harnessing the geopolitical and economic location of the territory, the size of its territory, and its natural, socio-cultural, and productive assets. Together, this generates a milieu that has endogenous potential in terms of enhancing the quality of life of its residents and has exogenous potential in terms of valorising the knowledge, production outputs, and assets of the region to actors beyond the region.

A third transition is the combination of vertical and horizontal coordination processes. Horizontal coordination entails less-institutionalised forms of cooperation, either formally in terms of interregional coalitions, or informally in terms of intra-regional cooperatives between stakeholders within the region. Horizontal coordination also involves the devolution of executive powers by integrating diverse citizenry in the collective management of the region (Wassenhoven 2008). While horizontal territorial governance results in the hollowing out of the local government, it does not represent a shift to complete regional autarky. There is still a vertical measure of hierarchical authority in which the state has defined policing powers in terms of enforcing top-down policy, and which requires some measure of subsidiarity from stakeholders (Infranca 2019). Moreover, there is a bottom-up process of clientelism, in which local government increasingly represents the interests of citizens in inter-regional coalitions. Although there is a measure of vertical devolution of executive powers and control, the execution of territorial governance still requires the technical competency of officials within a vertical framework of multilevel governance. Accountability is ensured through the vertical coordination of policy-making through public participation, and horizontal community stakeholder identification and participation in regional governance (Davoudi et al. 2008). These two modes of coordination are termed Type I and Type II governance, in which Type I governance represents multi-level forms of government with clearly delineated jurisdictions, exclusive memberships, and highly regulated policy structures. Type II governance primarily refers to specialised and heterogeneous jurisdictions with lean and flexible administrations, intersecting memberships, and cross-jurisdictional territories (Faludi 2012).

One recurring challenge in territorial governance is the decentralisation of governance away from mere placatory public-participation checklist procedures towards real stakeholder identification processes, inclusionary participation, delegated power, collaborative planning, and active citizenship (Wassenhoven 2008). Yet, increasing citizen control risks diluting the accountability and competency of decision-making. Another challenge is defining the appropriate spatial scale for the devolution of power, with lower-scale decentralisation risking the hijacking of public policy by special interest home voters, while higher-scale decentralisation risks marginalising the interests of micro-regions and small communities (Fishell 2002). Furthermore, there is a challenge of selecting the appropriate mode of participation, whether through formal codified bureaucratic procedures or through informal workshops and community consultation sessions. Regardless of the degree, scale, or mode of stakeholder participation, there will be a certain measure of public exclusion and marginalisation between joiners, who are effectively able to participate within the governance processes, and non-joiners, who are effectively unable to provide a meaningful contribution due to cognitive, language and mobility barriers (Davoudi and Petts 2000). It is acknowledged that most stylised territorial governance models are scaled to supra-local levels, with participatory devolution mostly relegated to public-private partnerships and involving highly codified bureaucratic procedures. This challenges the coherence and legitimacy of such processes since most territorial governance is consequently nothing more than rhetoric for bureaucratic top-down governance. Developing truly devolved modes of governance is time-consuming since it requires the nurture of an active culture of citizenship and strong intra-local coalitions and involves the development of extensive communication networks. Moreover, the decentralisation of powers and competencies is often constrained by the existing policy and legislation. Territorial governance is often merely a product of internal governmental rationalities, legitimising only certain regional governance issues through its internal policy-making processes (and thereby marginalising other regional governance issues). Instead, pertinent regional governance issues should be defined through a broad-based consultation with local stakeholders themselves (Flyvbjerg 1998).

12.3 Territorial Regimes in Sub-Saharan Africa

Territorial governance is emerging as an important research theme in Sub-Saharan Africa because the distinction between local government and territorial governance often stands in stark contrast to each other. While local governments outwardly conform to international legal-rational criteria in terms of government elections and policy mechanisms, there is often little incentive for maintaining coherent territorial governance. The ability of local governments to govern is often relatively poor, with many of these operating as mere clientelist neopatrimonial regimes modifying and manipulating existing policy regulations to accommodate the windfalls of official development assistance initiatives and foreign capital investment for political patronage (Robinson and Parsons 2006). The participatory relationships between state actors and local communities are primarily defined in terms of the exchange of marginal public benefits and development spillovers for political patronage, and thus there is a lack of real vertical multi-level territorial governance (Ọmọbọwale and Olutayọ 2010). Most of the touted decentralisation measures in Sub-Saharan Africa are relatively ineffectual because the central government is unwilling to relinquish powers resulting in a model of decentralisation that neither delegates decision-making nor decentralises participation (Ofoulhast-Othamot 2018). Due to the neopatrimonial nature of many local governments in Sub-Saharan Africa, vertical integration rarely generates social and economic spillovers, nor does it create more flexible policies or greater territorial cohesion (Holanda Maschietto 2016). Vertical clientelist relationships serve to initiate territorial development projects, but most of the participatory governance exercises are anything more than symbolic measures to reinforce neopatrimonial regimes and invariably result in uncompleted projects with no real territorial development. Although these agendas have arbitrary outcomes it still serves to reduce the unpredictability of local government, strengthen vertical interpersonal relations and limit capitalist accumulation (Sabourin 2022).

Critical failures in local government often necessitate separate modes of territorial governance operating parallel the traditional vertical governance structures (Sabourin 2022). With incoherent formal governance structures, territorial governance is often accomplished through informal networks of individual local government officials, community associations, and business interests, collectively cooperating to govern the region. This generates decentralised forms of territorial governance structures known as territorial regimes (Burroni and Crouch 2008). This is often legitimised by the fact that much of the land is under communal tenure with most residents still recognising the authority of traditional community leaders over that of non-elected local government officials (Chitonge 2019). Territorial regimes are co-federalist, defined as the substitution of ineffectual vertical governance structures and policy mechanisms, with strong horizontally-embedded modes of informal territorial governance operating semi-independently of the state (Erk 2019). Whilst policy mechanisms may be valid on paper, often local customary institutions have more effective de facto authority than local government itself (Lund and Boone 2013). Hooghe et al. (2010) describes co-federalism as new-medievalism, a complex network of public and private actors operating in continually shifting coalitions within fuzzy boundaries. Since the various local institutions individually have low levels of power and utilise different sources of power, the territorial regime is highly fragmented along socio-political lines. However, through strategic alliances and negotiations, territorial regimes create a polycentric governance network to provide informal but effective regulatory control within the local community (Rodríguez-Pose and Storper 2006).

Since territorial regimes are specific to the socio-political context of the region, one territorial regime does not easily carry over to other areas as a model of territorial governance, but similar regimes are replicated in other areas based on their socio-political contexts (Burroni and Crouch 2008). The lack of territorial cohesion is often because most of the borders of territories were inherited from colonial administrations with little concern for the rich tapestry of ethnicities and cultures transcending these boundaries. Regional administrative boundary delineations are often internally fragmented into pluralistic and politically contested cross-border subregions and micro-regional enclaves which invariably do not correspond to the historic administrative boundaries of cities and regions (Söderbaum 2004). Consequently, these regions share a metageography, in which the real territory conceptualised in terms of a macro social space is detached from delineated administrative districts (Scholte 2017). Territorial jurisdictions are often vague, situated in a complex layered network of traditional, community, and statal governance structures creating various alternate boundaries and social groupings, often competing over functional territorial jurisdictions (Lund and Boone 2013).

Common property regimes in communal land and traditional authority areas in Sub-Saharan Africa, often transcend the ad hoc administrative boundaries established in the colonial era. The prevalence of common property regimes requires informal governance structures to ensure the effective functioning of territorial regimes between territories. Despite inhibiting property commodification, the costless access to resources in common property enables the growth of robust local economies that are highly resistant to economic recessions, and which generate significant local synergies and economic spillovers (Balakrishnan and De Schutter 2014). By controlling the rights of entrance and use, territorial regimes prevent the overutilisation of resources in common property and bind various beneficiaries into forms of voluntary cooperation and self-regulation needed for sustainable economic growth. By internalising externalities, territorial regimes regulate both the distribution of benefits accruing from common property and the negative externalities of certain activities without the need for vertical state intervention. Through negotiations within the polycentric authority structures in territorial regimes democratise the governance of the territory. Moreover, by utilising structured negotiations and verbal interactions territorial regimes de-escalate local tensions in common property (Dagan and Dagan 2020).

An example of a territorial regime is the dual tenure system in Nigeria, combining customary tenure and statutory tenure. The Land Use Act of 1978 (LUA-1978) integrated the diverse land tenures under custodianship of provincial (regional) governors who administer the land in trust through a five-member Land Use Allocation Committee. This Act grants customary right of occupancy (CRO) to rural areas and statutory right of occupancy (SRO) in urban areas in terms of land titles which is administered by the Provincial Development Authorities and the Ministry of Lands and Urban Development. Under the LUA-1978, CRO provides a delegation of regional governance to traditional tribal authority and provides the statutory protection of communal land ownership. However, the Land Use Allocation Committee also has powers to revoke SRO and CRO for overriding public interest in terms of the Urban and Regional Planning Law of 1992 (Tagliarino et al. 2018; Onyebueke and Ikejiofor 2017; Onyebueke et al. 2020). This generates a territorial regime in which the vertical governance of regions is implemented through the provincial governor, the Land Use Allocation Committee, the Provincial Development Authorities and the Ministry of Lands and Urban Development at a supra-regional level. However, the numerous urbanised communal lands are also horizontally governed through traditional authorities operating in a micro-regional context. Since SRO entitlements are poorly enumerated and often lacking in urban areas and CRO only provides usufruct rights, but judicial law prioritises occupancy rights through legal precedent, the territorial regime operates tenuously through the negotiations between vertical and horizontal governance. This combines Type I formal multi-level governance structures with defined jurisdictions with Type II flexible and heterogeneous transjurisdictional modes of governance and combining formal legal-rational modes of governance with customary modes of governance.

Since many of the procedures of territorial regimes operate outside the legal-rational framework of formal policy structures, they utilise various extra-legal procedures that have no legal authority. This is referred to as territorial innovations defined as creative quasi-statal modes of governance utilising informal rules, tools, and mechanisms (Torre 2018). This often leads to the reinvention of established territorial governance structures (Dansou and Carrier 2023). The territorial regime operates on the principle of collective efficiency, whereby joint action is a product of evolutionary adaptations in an adverse political environment, limiting the role of the state to mediating conflicts (Schmitz 1999). These regimes are deeply embedded in local communities, based on long-term interpersonal relationships, customs, values, and practices, and thus the governance is strongly horizontal with a high level of passive internal enforcement and compliance. Despite the limited degree of vertical integration, state representatives perform symbolic roles within both the legal-rational territorial governance system and informal territorial regimes, both legitimising the territorial regime and legitimising the role of the state as a supra-regional representative of the territorial regime (Bauder and Mueller 2023). Although the structures within these regimes have club characteristics, limiting vertical integration into supra-regional networks outside the territory, it does serve to strengthen local communities and prevent external competition and capital accumulation.

A useful way to conceptualise these modes of territorial governance is metagovernance (Jessop 1998). Metagovernance is the self-organised steering of multiple institutions, systems, and actors that are autonomous from each other but coupled due to mutual interdependence. The horizontal density of governance networks serves to support the deficient vertical governance structures. This mode of governance operates semi-independently of vertical managerial control, and in turn local government needs the support of the territorial regimes to effectively implement important government policies due to the lack of state authority and capacity. Metagovernance results in greater participation and collective action, and interactions that are at the same time complex and conflictual (Sørensen and Torfing 2009). Often vertical integration is actively opposed at the local level, since the centralisation of decision-making will inevitably result in power asymmetries, resulting in the decline of community institutions, networks, and the autonomies necessary for effective micro-regional governance. Since territorial regimes generate Type II modes of governance, the replication of Type I governance is superfluous. The main prerogative of governance frameworks is not to create more efficient multilevel government structures, rather it is focused on strengthening horizontal networks of interaction in the absence of overall state authority (Hooghe et al. 2010).

12.4 Customary Land Management Systems in Informal Peri-Urban Settlements

A prevalent mode of horizontal governance in micro regions in Sub-Saharan Africa is customary land management systems (CLMS) (Geyer 2023). CLMS are informal land use regulatory structures that utilise a combination of municipal bylaw regulations, tribal law, and neo-customary tenure to regulate the complex socio-political contexts in informal peri-urban settlements. With 80% of land in Sub-Saharan Africa under various forms of informal entitlements or indigenous customary tenure, and with 62% of urban development in African cities consisting of unregulated informal development, coherent territorial governance requires unique modes of self-regulation (Wily 2018; Okyere et al. 2023). Although little is known of these regulatory systems beyond the Global South, it is recognised by the African Union as a prominent mode of urban governance (African Union 2009). CLMS is a form of territorial governance ubiquitous in informal peri-urban settlements in the Global South, particularly in Sub-Saharan Africa. These settlements generally comprise expansive piecemeal informal developments on state or communal land informally sold for urban development. With no real distinction between land uses, districts, or territories, the complexity of multi-ethnic social relationships requires a territorial regime that is flexible and contextually relevant.

CLMS utilise vernacular face-to-face interactions and non-statutory community regulations to govern land use management and informal development in areas categorised by neo-customary tenure, i.e., areas where traditional customary tenure is adapted for the urban areas with a heterogeneous multi-ethnic population (Onyebueke and Ikejiofor 2017). CLMS regulations combine customary regulations, social relations, and planning bylaws to create a governance regime mediated through a plurality of community representatives who strategically negotiate governance issues on behalf of their constituency through transparent open-ended negotiations. This creates a territorial regime that is adaptable to a rapidly changing socio-political environment and provides a basic regulatory framework for unregulated and undertitled land uses (Geyer 2023). Instead of the use of policy regulations and bureaucratic procedures, CLMS utilises modes of deregulation, in which territorial governance occurs through active and informal channels in the absence of effective state governance (Roy 2009). These procedures are often more democratic, organised, and transparent than formal regulations, enabling easy access for the poor and marginalised.

CLMS arises in the context of multi-ethnic peri-urban settlements developed on communal land surrounding Sub-Saharan African cities. Since most of the urban population consists of in-migrants who have no customary tenure on tribal land but acquire land informally through vernacular sales and usufruct agreements, these residents have no recourse to indigenous governance systems. However, since much of the land is legally entitled as communal land, which is legally protected from disenfranchisement, they also have no recourse to planning bylaws (Geyer 2023). Territorial governance is therefore exceedingly complex because of the great social heterogeneity of the population, the layers of overlapping property rights claims, and the lack of formal tenure and entitlements. Regularisation and servicing costs often exceed the value of the land itself. Amid often inefficient and corrupt local government, CLMS provides a cost-effective and accessible alternative platform to govern territories, and these systems often have more social legitimacy than state policies because of the community recognition and the authority of community leaders (Chitonge et al. 2017). It utilises various forms of legal codes and policies, neo-customary tenure, and dense networks of social relationships and patronages. The social heterogeneity spawns a polycentric authority structure in which the various social groupings rally behind various representatives who operate from different power bases. These community authorities negotiate the externalities of territorial issues and cooperate to direct the collective actions of the community, especially where informal regulations require enforcement through community sanction (Altrock 2016). In contrast to the often obscure and inaccessible mechanisms of formal governance, CLMS operates based on mutual commitments and obligations, reciprocity, and trust. With low entry barriers, low costs, and a balance of power bases in a plurality of authorities, CLMS provides a relatively equitable and accessible mode of territorial governance, ensuring the active collaboration of citizenry in governing regional development.

CLMS is, however, not conducive to the sustained economic growth and development of a region, and thus is not comparable to territorial governance in the Global North. CLMS undermines efforts to commodify land in terms of formal entitlements and capital investment. The plurality of authority structures can undermine a cohesive policy structure and external direct investment since rights and legitimacies have to be acquired from both the state and community representatives. The application of policy is difficult due to a lack of clarification regarding the correct regulatory procedures and processes. The multiplication of procedures and vague legal principles applied in the regulative processes results in a situation where rights and legitimacies are given to reinterpretation and reversals. Given the low level of technical skill and the poor understanding of supra-regional processes, there is often a lack of competency in decision-making in the CLMS, and decisions are often biased in terms of gender, ethnicity, status, or religion (Geyer 2023).

CLMS as a territorial regime can be described as grey governance, since it operates beyond defined administrative boundaries, within different levels of governance and often apart from these structures as an alternate mode of governance. It does not fully adhere to the logic and rationality of a defined legal-rational governance system and is often in abeyance of its policies and regulations. However, CLMS also better actualises the principles of good territorial governance, by mobilising the active collaboration of citizenry, ensuring broad-based public participation, decentralising governance within the community, strengthening vertical and horizontal linkages within the region, and applying transcalar approaches to governance. Consequently, CLMS is analogous to the Foucauldian notion of governmentality, which defines governance as an ensemble of institutions, procedures, analysis, and reflections, not reducible to the state, but resulting in complex realisations of different sets of powers in the urban space (Foucault 1991). While the complex social structures and fragmented urban development patterns in the Global South make formal multilevel governance a cumbersome and contested process of cajoling actors into a cohesive urban vision, CLMS uses collective action to refocus territorial governance into new distinctive rationalities more reflective of the territory (Otter 2004). This is because the region is not a unitary statal entity, but a deeply fragmented assemblage of special interests with multiple identities and meanings of border, territory, and citizenship (Ranganathan 2014).

12.5 Grey Governance Through Customary Land Management Systems

Since CLMS form territorial regimes operating semi-autonomously to formal multilevel governance structures it creates a complex, counterhegemonic mode of governance which is neither a regional autarky separate from multilevel governance nor a function of state governance. Instead, CLMS operates in a 'tenurial shell' outside the full sovereignty of either state authorities or the community because CLMS lies in a grey area in terms of scale and legal status: the governance of territory often transcends administrative boundaries to conform to social conceptions of territory, and both the CLMS and multilevel governance structures require tacit recognition from both the community and the state to maintain some semblance of public legitimacy (Berrisford and MacAuslan 2017). For this reason, there is often a tacit cooperation between the legitimate multilevel governance structures of the state and the informal community authorities in the CLMS. Moreover, because of the internationally recognised legitimacy of customary tenure, postcolonial states adopt a facilitative approach to incorporating community authorities in territorial governance processes, albeit informally (Durand-Lasserve and Royston 2012). Consequently, there is a high degree of tolerance of customary modes of territorial governance utilising a polycentric network of authorities, institutions, and procedures that do not intrinsically threaten the patrimonial status of local government authority (Chimhowu 2019).

Peri-urbanism presents a significant governance challenge due to the unstructured nature of these settlements, the absence of effective policy enforcement, and the strong statutory protections over customary entitlements and informal development. Since these settlements often fall outside the administrative boundaries of historic colonial cities, these settlements intrinsically are averse to Type I modes of territorial governance. Consequently, many local governments implement grey governance principles in a dualistic regulatory structure in peri-urban settlements. At the micro-level, territorial regimes operate Type II modes of governance, often facilitated by the local authorities. In turn, facilitated by territorial regimes, local governments implement Type I modes of governance at the supra-regional level (Onyebueke et al. 2020). While the state is not obligated to recognise the legal authority of territorial regimes, the state is obligated to include community authorities as legitimate public representatives within multilevel governance initiatives and mobilises these representatives to organise collective action within the community to effectively implement supra-regional policies. Moreover, while the CLMS is not obligated to incorporate the state in informal governance of the territory at the micro level, it invariably seeks the affirmation of the state to strengthen the legitimacy of its actions (Geyer 2023). This creates a symbiotic relationship between vertical and horizontal governance measures, enabling both the CLMS and local authorities to effectively govern territories.

Since territorial regimes are extra-legal, having no legal authority in territorial governance, the state is in an ambiguous position to CLMS. Multi-level governance structures are at liberty regarding how they address territorial regimes. They may normalise CLMS as an accepted mode of governance, tacitly assenting to their authority since it provides a measure of territorial governance in otherwise ungovernable regions without committing the resources of the state. Local government may often mobilise and exploit these processes for its own political goals. However, it is also at liberty to actively repress extra-legal CLMS authorities and coerce the local communities into compliance when it serves the supra-regional objectives of government, especially when foreign capital investment and official development assistance require the compulsory acquisition of land for development. The manner of state intervention depends on the largesse of the state, and this is only granted through political patronage and subordination in a clientelist system (Azuela and Meneses-Reyes 2014).

The distinction and relationship between multilevel governance and CLMS is based on a difference between rules-in-use and rules-in-form (Ostrom 2014). Multilevel governance utilises formal policies as rules-in-form to provide a comprehensive system of territorial governance according to legal-rationalist criteria. However, these processes are often inflexible and inappropriate at a local level, which limits their effectiveness. CLMS utilises horizontal networks and collective action as rules-in-use to generate territorial efficient governance innovations, but it does not have the legitimacy of public policy, nor does it have formal systems of accountability and transparency. However, since the CLMS operates through a network of social relations with governance capabilities within an inefficient larger social matrix, it can generate efficient modes of governance in micro-regions and serves to effectively mobilise the local community within a supra-regional polity.

12.6 Grey Governance in the Informal Settlements of Cape Town Horizontal Governance Through Street Committees

The study evaluates territorial regimes and informal modes of territorial governance in the townships of Zola, Nomzamo, and Lwandle in the Helderberg district of Cape Town Metropolitan Municipality. The study area was selected since it is the major township within the municipal subdistrict (See Fig. 12.1). This township consists of formal low-income housing development, colloquially referred to as RDP housing from the Reconstruction and Development Programme public housing policy. The formal public housing is informalised into peri-urban modes of development through significant infill development of informal structures, which is sprawling into extensive informal settlements along surrounding road servitudes. The research consisted of an ethnomethodological analysis of over 50 in-depth interviews conducted with participants randomly selected between different ethnic and social groupings and selected evenly between the different sections of the study area. The random selection of residents included a random selection of homeowners, tenants, business owners, South African and foreigners. The municipal officials participating in the study included local councillors, police officers and local officials in the informal trade, building control, law enforcement and planning units all operative within the study area. Sampling was conducted on the basis of multistage sampling, selecting random samples from randomly selected subgroups in various locations

Fig. 12.1 The study area

within the study area. The selection of officials involved in land use management involved snowball sampling, because other role players were revealed by respondents.

The first impression that there is no clear territorial governance mechanism operative in township. While the settlement consists of government-subsidised RDP housing on small urban plots, all open space surrounding the properties is developed into shacks for rental income. The predominant land uses are mixed, with certain residents operating small-scale enterprises serving the local community, and along the major thoroughfare roads the housing was developed into informal enterprises, many of them located within repurposed shipping containers. Every available space is utilised, and when possible, the activities spill over onto neighbouring properties. While most shacks are constructed from timber, plastic, and galvanized sheets, the houses are constructed from brick and cement. Certain properties are redeveloped into multilevel apartment blocks with up to 10 units on a 150 m^2 plot. Because of these characteristics, the study area can be considered as peri-urban development, including a range of informal and semi-informal developments in and around formal RDP housing. The settlement has extremely high densities, with an average population of 1550 persons/ha in a settlement designed for ±200 persons/ha.

This results in significant governance challenges regarding the regulation of space. One major challenge is the inability to secure public open spaces. Although there are many public parks and significant undeveloped open land outside the township, these are rarely utilised, since there are no micro-regional community organisations governing the space. The only accessible, safe public area within view of the local community is the local streets. The narrow 8–10-metre-wide road reserves are utilised as local meeting places and play areas for children. The lack of open space also limits the development of green infrastructure within the area. A particular challenge is the lack of trees. Formerly the area was a eucalyptus plantation that provided sufficient shade and rapidly drained flood water from the streets, but these invasive species were removed during construction process and could not be replaced because of the narrow road reserve and the tendency of endemic trees to uproot buildings and infrastructure. The lack of greening and prevalent asphalt and concrete structures creates a local heat island effect especially during the summer with ambient temperatures of over 40 °C, and flash flooding during the winter due to the number of paved surfaces and poor drainage.

Another major governance challenge is the handling of waste and sewage. The informal structure of the settlements is not suitable for conventional waste management, and thus much of household waste is dumped on street corners, open spaces, and along watercourses. This clogs stormwater drainage channels and local streams impedes the effective drainage of upstream rainwater regularly causing 100-200 mm flooding after major rains (Geyer and van Lille 2024). Another governance challenge is frequent fires because of the lack of open fire breaks between adjacent properties and the use of flammable temporary construction materials. Finally, there are numerous regulatory challenges related to land use and boundary contraventions in the dense settlements. Since RDP units are not supplied with fencing, there are numerous encroachments over invisible property boundaries and infrastructure servitudes, particularly where a couple of centimetres appropriated on any boundary could enable the development of an additional room or structure. There are also issues regarding the appropriate use of building materials since temporary building materials have poor conduction, resulting in high ambient temperatures in summer, low ambient temperatures in winter, and fire risks.

These issues illustrate how complex territorial governance challenges in informal peri-urbanist settlements are. Typical solutions such as building line enforcement and building control do not work because these measures will reduce the supply of housing, resulting in homelessness, a loss of rental income, and the growth of informal settlements elsewhere. To overcome these challenges, the township, like most other townships in South Africa, has developed a complex CLMS composed of street committees, local business forums, community policing forums, and taxi associations. The primary organisation is the street committees, informally known as "Streets". These are voluntary civic organisations with no legal powers, representing between 40 and 100 residents sharing a common street frontage. These micro-regional organisations are based on personal associations with neighbours and regulate civic issues within the Street. However, all the Streets are amalgamated into SANCO (South African National Civic Organisation), an organisation representing the interests of thousands of Street committees within local and national government. SANCO is a democratic organisation with each Street providing one committee member representing one vote in decision-making processes. The Street members make comments or lodge complaints to a Street committee representative, which either handles micro-regional issues internally or forwards these decisions or complaints to municipal councillors, the police, or the broader municipality through SANCO. This structure enables the local community to directly access complex multi-level governance structures through intermediaries.

Street members are property owners who debate and vote on local policy and regulations affecting the Streets in weekly community meetings. The regulations and rules differ from Street to Street and operate within different languages and customs based on the composition of the micro-regional community. Decisions are only made on a unanimous vote following lengthy negotiations and all decisions are carefully notarised in ledgers. Infractions are punitively dealt with through community-lodged fines or offenders handed over to the police or bylaw enforcement authorities and a repeat offender can be evicted from the community, although their property rights remain in place. The Street committee is elected from Street members who voluntarily provide magisterial functions on behalf of residents including negotiating with municipal representatives, acquiring information from authorities, or instituting legal proceedings over disputes. Typical issues are boundary disputes, tenancy arbitration, land use regulation, parking congestion, business operating hours, business competition arbitration, and public nuisance and noise abatement orders. They also collect the rent on behalf of absentee landowners, facilitate property sales and ownership disputes, conduct background checks on prospective tenants, and arrange funeral support for members who need to organise a funeral. Street committee members are strictly prohibited from receiving gifts or benefit financially from their position to ensure impartiality.

12.7 Vertical Governance Through Councillors and State Officials

Outwardly, the settlement operates within the vertical multilevel governance structure of electoral wards, regional districts, municipalities and provinces. These vertical governance structures operate in the region within Provincial and Municipal bylaws and planning structures with the study area falling within the Lwandle and Nomzamo Wards represented by councillors and the Helderberg district represented by the Helderberg district Council. Furthermore, at the supra-regional level these council members also report to the city of Cape Town municipal council, which in turn coordinates its decisions with the Executive Council of the Western Cape. Since Municipalities and Provincial Government are constitutionally mutually autonomous, they have separate delegated Provincial and Municipal governance functions. Governance follows a bottom-up structure where elected ward councillors discuss community issues with their constituents, which are included into the sub-regional Precinct Plans and the regional District Plans. These plans are integrated into the supra-regional level in the annually revised 5-year Municipal Integrated Development Plans, and territorialised within Spatial Development Plans and Integrated Urban Development Frameworks (Thompson 2014). The results from these plans and frameworks are aligned with the Western Cape Provincial Spatial Development Framework, the Western Cape Provincial Strategic Plan and other specific plans as designated by the provincial authorities and the collective governance decisions are enforced in a top-down manner through the alignment of the lower-level plans and policies. These decentralised governance frameworks are formalised under the Municipal Systems Act of 2000, prescribing an orderly governance process of aligning the different development priorities following extensive processes of public consultation (Thompson et al. 2018). However, this process of negotiation is difficult due to the inability of the residents, and even the councillors to understand the formal policy process. This results in an ineffective public consultation process, with improperly facilitated public feedback on policy initiatives and public feedback often consisting of unrealistic wish lists. Many of the formal processes are designed to limit participation to endorse decisions already made by higher tiers of government and do not provide channels for the effective devolution of public decision-making (Barolsky and Borges 2019).

While the original RDP housing is formal, and fully integrated into the vertical territorial governance structure, most development consists of informal infill development, including unapproved renovations, additions, rebuilds, and additional shack structures built in the open spaces adjacent to the buildings. The settlement was granted a flexible zoning category: Single Residential Zoning 2 (SR2), developed around the Less Formal Township Establishment Act (LFTEA) 113 of 1991, to permit the incremental development and regularisation of informal development. These concessions were designed to enable the flexibility of development, protecting tenure and providing RDP housing a degree of sovereignty from land use policies as a measure to facilitate the incremental formalisation and eventual integration of the peri-urban periphery into the rest of the city. Originally, the LFTEA provided another layer of sovereignty between the settlement and the Local Municipality enabling micro-regional governance through the CLMS since LFTEA was governed by the autonomous Provincial Authorities. However, these provisions fell away with the implementation of the Spatial Planning and Land Use Management Act (SPLUMA) 16 of 2013, through which informal development is solely delegated within the purview of the City of Cape Town's Development Management Department, but certain other land use activities are governed by various Provincial and Municipal departments.

In this hybridised urban structure, territorial governance consists of a combination of strong informal horizontal governance occurring through the Street Committees/SANCO CLMS, and a formal vertical multilevel governance structure consisting of councillors, municipal officials, and provincial government. Despite the apparent logical inconsistencies of such a structure, territorial governance actually operates relatively efficiently because the different roles and responsibilities are clearly defined and delineated between the different stakeholders. Moreover, the different parties have mutual recognition and respect for each other's interests and are generally willing to make concessions where it is practical and reasonable. This complex governmental arrangement serves as a territorial innovation to substitute for ineffective public interventions by the state, resulting in a shift from government delivery to territorialised governance arrangements (Amin and Cirolia 2018).

The primary interface between the vertical multilevel governance structures of the Municipal and Provincial authorities and the horizontal informal community governance structures is the local councillors. These are resident members of the community who represent the community in the various councils. Councillors are politically selected by the local ruling political party and SANCO to stand unopposed for election during local government elections to facilitate governance within the township. Even though one section of the township technically falls under the purview of another councillor in an adjacent electoral district, that councillor has conceded his effective control over that section of the township to the councillors in the study area to facilitate the governance in that section, even though they represent different political parties. The local councillors have a top-down role in coordinating all strategic development initiatives at the supra-regional level with the local community through SANCO. However, the councillors also have a bottom-up role in petitioning for projects, services, and employment from the municipality and provincial government, funnelling these toward the community. The councillors also serve as a coordinator between the different municipal and provincial departments operating within the township. The counsellor thus has an autonomous role partially separate from both the vertical and horizontal governance structures, but fully integrated within both.

Due to the challenging nature of governance within this territory, several specialised units operate within the township. This includes municipal law enforcement, the anti-invasion unit, the informal trade unit, units regulating taxes and licensing; the metal theft unit, building control, the roads and stormwater departments, the parks department, the waste department, the water services department, and the electricity unit. An unusual governance instrument is the local police. While the police do not actively work as a function of territorial governance, they perform an important role in monitoring and pacifying the often-violent interactions within the governance regime. This serves as a last resort to prevent any infringement on rights, particularly in terms of enforcing tenure agreements. Together, these different agencies and departments provide security of tenure and a basic level of services within the township. Although many of the activities infringe upon the policies and regulations of the governing authorities, the state has retreated to only implementing those policies and regulations that best serve the community and do not onerously impede access to housing, services, and income-generating activities.

Certain provisions are strictly enforced, for instance, the regularisation of water and electricity utilities, since this may cause bodily and environmental harm if not properly regulated. Building lines and infrastructure servitudes are strictly enforced since these often result in conflicts between residents or impede the municipality in providing services to the community. Tenure is also strictly enforced, even if the tenure is informal since the municipality is mandated to protect the most vulnerable members of the community from exploitation. The regularisation of permanent structures is also strictly enforced since the collapse of an informal housing block recently caused the death of one of the neighbours. Furthermore, land invasions and theft are highly regulated since they destroy the infrastructure needed to provide basic services and significantly increase the cost of housing development. However, the authorities often turn a blind eye to certain activities to facilitate governance. For instance, informal non-permanent structures transgressing building lines are tolerated to enable the maximum provision of housing; small informal enterprises evading tax are permitted to operate to provide informal income-generating activities; and the dumping of household waste is permitted in certain locations accessible to waste removal vehicles since these settlements are often unable to utilise removable waste bins due to the absence of fencing to secure the bins.

This can be regarded as grey governance, a mixed formal and informal mode of governance in which state officials and policymakers provide some degree of concession and leniency regarding policy to accommodate the community and utilises the structure of the territorial regime as an informal executive arm of the state (See Fig. 12.2). The grey governance structure provides sufficient protections to secure tenure but prohibits the state from assuming novel policing powers when it will prohibit the provision of housing or employment. It permits informal activities necessary to provide adequate housing and income for the urban poor, whilst retaining enough regulatory provisions to regulate the most egregious activities that reduce service provision and living quality. The grey governance structure is not a statutory or legal partnership, but a product of an un-codified informal consensus between the CLMS and the state.

Grey governance structures provide the state with a wide selection of responses to different challenges in peri-urban settlements. Since the activities are unregulated and technically illegal, the state has the right to punitively enforce its policies where needed but also enables them to ignore the enforcement of policies when it would best enable the continued well-being of the community. The integration of the

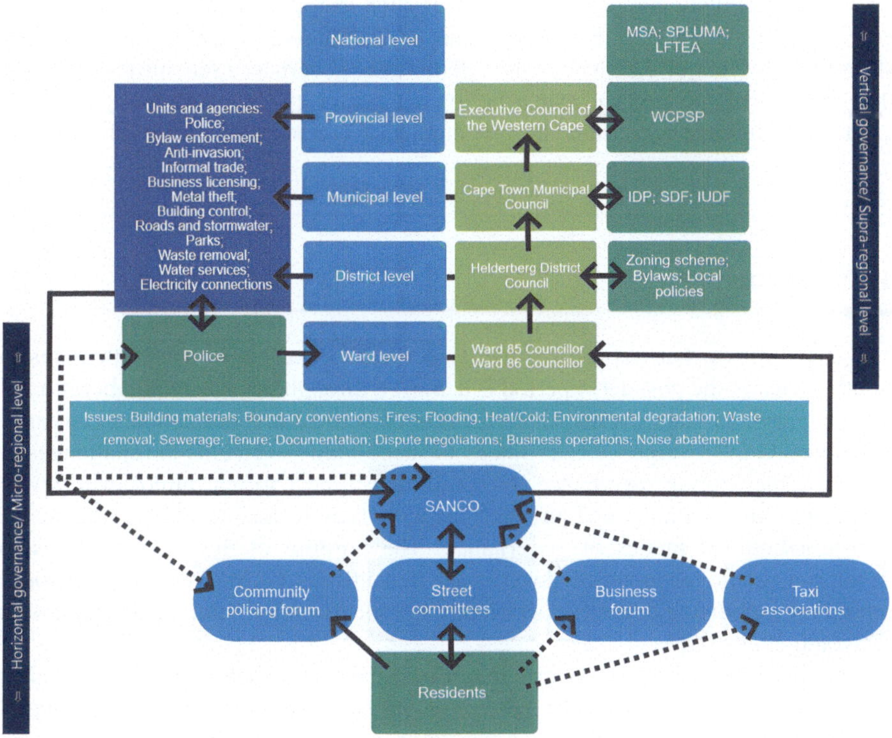

Fig. 12.2 An organogram of the grey governance system (original to the author)

CLMS in the form of Street committees and SANCO provides buy-in from the community, and these representatives provide an easy mechanism to access communities to negotiate consent in policy initiatives. Furthermore, the CLMS often enforces policy on behalf of the authorities to regularise the communities, lowering the burden of responsibility of the state. In many cases, it enforces public compliance to many regulations, since non-compliance could result in punitive measures that could negatively affect some of the rights and liberties enjoyed by others in the township. This serves to incrementally regularise the township and generates a greater level of public cooperation in public policy initiatives.

In turn, the state serves as a final arbitrator and tribunal for particularly challenging decisions, such as property ownership or tenure disputes, when the community cannot come to a consensus, and which could create tensions that would destabilise the community. Generally, state authorities are held in high regard and their decisions are respected because the community realises that the state only intervenes when the infractions will lead to health and safety risks or impede the provision of basic services. When the state decides the intervene or to implement policy, it follows a comprehensive public participation process in which the councillors are consulted, who convey the information to the rest of the community. Furthermore, the state provides many compensatory services such as toilet cleaning services, child-care support, health care distribution and basic building materials for rebuilding structures after major disasters. Although this grey governance scheme results in the deregulation of governance and a hollowing out of the state, it strengthens the capacity of the state to facilitate development based on normative principles of good governance.

12.8 Conclusions

The study shows that territorial governance in informal peri-urban settlements in South Africa in the post-1994 period is achieved through the active collaboration of the citizenry and state within a combination of vertical and horizontal governance structures. Grey governance is a strategy to successfully overcome the challenges of successful territorial governance by combining Type I formal multi-level governance structures with defined jurisdictions with Type II flexible and heterogeneous transjurisdictional modes of governance. The purpose of these structures is to develop a relational mode of governance that utilises CLMS within a territorial regime by mobilising collective action through the reconfiguration of state practices (Davoudi et al. 2008).

One important observation was that a transcalar approach is employed (Pallas and Bloodgood 2022), with the vertical governance structure of the state implementing supra-regional policies in a top-down manner, whilst the horizontal governance structure of the CLMS practices governance in a bottom-up manner from micro-community Street committees that are integrated into a decentralised supra-regional collective of SANCO (Mayer 2013). In this manner, territorial governance policies are both deterritorialised and re-territorialised into a metageography that integrates the different social conceptions of territory within a common vision for the area. By combining vertical and horizontal coordination processes, the territorial governance structure in this case study generates a highly participative process, in which ordinary citizen concerns are translated into actions through the mediation of the Street committees and councillors (Wassenhoven 2008).

While the CLMS has a less-institutionalised mode of operation, based on strong intraregional interaction and collective management through SANCO, the vertical structures of the state are reorientated to provide a mode of state clientelism, with regular and informal modes of interaction to ensure that all governance initiatives are transparent and inclusionary (Robinson and Parsons 2006). This generates a high level of competency in decision-making because of the overt formal structures of multilevel governance, and it also includes necessary safeguards to prevent the hijacking of territorial governance by special interests in the CLMS. With the use of a decentralised set of representatives representing micro-communities, the challenge of non-joiners is effectively eliminated since there are strong interactions between actors in the Street Committees (Davoudi and Petts 2000).

The case study has a highly accountable and responsible vertical state government without the neopatrimonial economic dependencies and political exclusion evident in various cities in the Global South. There is also a horizontal CLMS governance structure with decentralised governance networks, creating a democratic and transparent territorial regime. This highlights that co-federalism and meta governance are important aspects of successful territorial governance in the Global South (Erk 2019). It cannot be minimised that there are serious administrative difficulties within the study area, with regular occurrences of fires, security issues, climactic issues, waste, and flooding. However, despite these challenges, the territorial innovations employed in the study area, i.e., the tacit cooperation between informal quasi-statal modes of governance with strong vertical state representation, resulted in greater participation and collective action. The CLMS framework incorporating the Street committees, SANCO, councillors, business forums, community policing forms, the police, and the various municipal and provincial agencies and departments creates a decentralised and polycentric administrative structure that is both relatively coherent and in terms of policy but is also adaptable and flexible to provide a responsive mode of governance with low entry barriers and participation costs for the urban poor. This has minimal impact in terms of producing new policies or plans in the post-1994 South African experience, but creates a parallel mode of governance that compliments the implementation and framing of existing policies and plans.

Despite the peri-urban characteristics of the township, in terms of infill informality within a formal township substructure, presenting unique governance challenges, the combination of a vertical Type I formal multilevel governance and horizontal Type II informal territorial regimes into a mode of grey governance presents state authorities with a range of possible actions in terms of permitting certain practical policy contraventions whilst enforcing other policies when it is necessary to maintain the continued provision of basic services. This reduces conflicts between residents and maintains services and security of tenure for the most vulnerable and marginalised residents. In turn, the CLMS also becomes an extension of the state, pressuring the citizenry to comply with the most necessary policies and programmes of the state, thereby reducing the regulatory burden of multilevel governance structures. This generates a new role for the state authorities as a higher tribunal and final authority in an effectively community-run system of CLMS governance, increasing the relevancy and validity of the state in informal settlements.

References

African Union (2009) Framework and guidance on land policy. African Union, Addis Ababa

Alaoui Y, Mamoun SM (2021) The pivotal position of the citizen in the concept of territorial governance. J North Afr Res Bus 20:1–8

Altrock U (2016) Conceptualising informality: some thoughts on the way towards generalisation. In: McFarlane C, Waibel M (eds) Urban informalities: reflections on the formal and informal. Routledge, London, pp 171–194

Amin A, Cirolia LR (2018) Politics/matter: governing Cape Town's informal settlements. Urban Stud 55(2):274–295

Azuela A, Meneses-Reyes R (2014) The everyday formation of the urban space: law and poverty in Mexico City. In: Braverman I, Blomley N, Delaney D (eds) The expanding spaces of the law: a timely legal geography. Stanford University Press, Redwood, CA, pp 167–189

Balakrishnan R, De Schutter O (2014) Property rights from below: an introduction to the debate. In: De Schutter O, Rajagopal P (eds) Property from below: rethinking use of land and related natural resource. Routledge, pp 1–19

Barolsky V, Borges D (2019) Is social cohesion the missing link in preventing violence?: case studies from South Africa and Brazil. In: Salahub JE, Gottsbacher M, De Boer J, Zaaroura MD (eds) Reducing urban violence in the global South. Routledge, pp 104–132

Bauder H, Mueller R (2023) Westphalian vs. Indigenous sovereignty: challenging colonial territorial governance. Geopolitics 28(1):156–173

Berrisford S, MacAuslan P (2017) Reforming urban laws in Africa: a practical guide. African Centre for Cities, Cape Town

Burroni L, Crouch C (2008) The territorial governance of the shadow economy. Environ Plan C Govern Policy 26(2):455–470

Chimhowu A (2019) The 'new' African customary land tenure. Characteristic, features and policy implications of a new paradigm. Land Use Policy 81(6):897–903

Chitonge, H. (2019). Land governance in Africa: The state, traditional authorities and the control of customary land (Doctoral dissertation, Tokyo University of Foreign Studies)

Chitonge H, Mfune O, Kafwamba D, Kajoba G (2017) Hybrid land markets: monetarised customary land transactions in Zambia. Canadian J Afr Stud/Revue Canadienne des Études Africaines 51(1):123–143

Dagan H, Dagan T (2020) Facilitating the commons inside out. In: De Schutter O, Rajagopal P (eds) Property rights from below. Routledge, pp 153–169

Dansou HD, Carrier M (2023) Decentralization, institutional innovation and governance of interterritorial relations: a view from Benin. Cities 133(2023):104115

Davoudi S, Petts J (2000) How to include the public in waste planning. In: Proceedings of the fifth national conference in strategic and local planning for waste, MEL research, 5–6 July, Birmingham

Davoudi S, Evans N, Governa F, Santangelo M (2008) Territorial governance in the making. Approaches, methodologies, practices. Boletín de la Asociación de Geógrafos Españoles 46:33–52

Durand-Lasserve A, Royston L (2012) International trends and country contexts–from tenure regularization to tenure security. In: Durand-Lasserve A, Royston L (eds) Holding their ground. Routledge, London, pp 17–50

Erk J (2019) Comparative territorial politics in Sub-Saharan Africa. In: Detterbeck K, Hepburn E (eds) Handbook of territorial politics. Edward Elgar Publishing, pp 354–376

Faludi A (2012) Multi-level (territorial) governance: three criticisms. Plan Theory Pract 13(2):197–211

Fishell W (2002) The home voter hypothesis: how home values influence local government taxation, school finance, and land-use policies. Harvard University Press, Cambridge, MA

Flyvbjerg B (1998) Rationality and power. The University of Chicago Press, Chicago/London

Foucault M (1991) Governmentality. In: Burchell G, Gordon C, Miller P (eds) The Foucault effect: studies in governmentality. Harvester Wheatsheaf, Hemel Hempstead, pp 87–104

Geyer HS (2023) Conflicts and synergies between customary land use management and urban planning in informal settlements. Land Use Policy 125:106459

Geyer HS, van Lille G (2024) The medicalisation of solid waste: coordination challenges of domestic biomedical risk waste in illegal dumping sites in informal settlements in Cape Town, Working Paper. Stellenbosch University

Holanda Maschietto R (2016) Decentralisation and local governance in Mozambique: the challenges of promoting bottom-up dynamics from the top down. Conflict Secur Dev 16(2):103–123
Hooghe L, Marks G, Schakel A (2010) The rise of regional authority: a comparative study of 42 democracies. Routledge
Infranca J (2019) The new state zoning: land use pre-emption amid housing crisis. Boston Coll Law Rev 60(3):823–888
Jessop, B. (1998). The Rise of Governance and the Risks of Failure: The Case of Economic Development. International Social Science Journal 50 (155): 29–45
Le Galès P (2002) European cities: social conflicts and governance. OUP Oxford
Lund C, Boone C (2013) Introduction: land politics in Africa—constituting authority over territory, property and persons. Africa 83(1):1–13
Mayer M (2013) First walled urban activism: beyond austerity urbanism and creative city politics. City 17(1):5–19
Ofoulhast-Othamot G (2018) The conundrum between political and sectoral decentralizations: the case of Cameroon. Dev Policy Rev 36(3):347–367
Okyere SA, Frimpong LK, Diko SK, Mensah SL, Pedrosa EL (2023) Informal settlements and pro-poor in urban planning in African cities. In: Cobbinah PB, Gaise E (eds) Reimagining urban planning in Africa. Cambridge University Press, Cambridge, MA pp 43–59
Ọmọbọwale AO, Olutayọ AỌ (2010) Political clientelism and rural development in South-Western Nigeria. Africa 80(3):453–472
Onyebueke VU, Ikejiofor C (2017) Neo-customary land delivery systems and the rise of community-mediated settlements in peri-urban Enugu, Nigeria. Int Dev Plan Rev 39(3):319–340
Onyebueke V, Walker J, Lipietz B, Ujah O, Ibezim-Ohaeri V (2020) Evicting the poor in the 'overriding public interest': crisis of rights and interests, and anti-planning in Nigerian cities. Cities 101:1–14
Ostrom E (2014) Do institutions for collective action evolve? J Bioecon 16(1):3–30
Otter C (2004) Cleansing and clarifying: technology and perception in nineteenth-century London. J Br Stud 43(1):40–64
Parr J (2004) The polycentric urban region: a closer inspection. Reg Stud 38(3):231–240
Pallas CL and Bloodgood EA (2022) New Transcalar Advocacy and Evolving Patterns of Advocacy beyond the Boomerang. In Pallas, C. L. and Bloodgood, E. A. (eds.). Beyond the Boomerang: From Transnational Advocacy Networks to Transcalar Advocacy in International Politics. University of Alabama Press, pp. 1–23.
Ranganathan M (2014) 'Mafias' in the waterscape: urban informality and everyday public authority in Bangalore. Water Alternat 7(1):89–105
Robinson JA, Parsons QN (2006) State formation and governance in Botswana. J Afr Econ 15(1):100–140
Rodríguez-Pose A, Storper M (2006) Better rules or stronger communities? On the social foundations of institutional change and its economic effects. Econ Geogr 82(1):1–25
Roy A (2009) Why India cannot plan its cities: informality, insurgence, and the idiom of urbanization. Plan Theory 8(1):76–81
Sabourin E (2022) Public policies for sustainable territorial development in Brazil: between clientelism and participation. Sustain For 14:3058
Schmitz H (1999) Global competition and local cooperation: success and failure in the Sinos Valley, Brazil. World Dev 27(9):1627–1650
Scholte JA (2017) Globalization: a critical introduction. Bloomsbury Publishing
Söderbaum F (2004) Modes of regional governance in Africa: neoliberalism, sovereignty boosting, and shadow networks. Glob Gov 10(4):419–436
Sørensen E, Torfing J (2009) Making governance networks effective and democratic through metagovernance. Public Adm 87(2):234–258

Storper M (2004) Territories, flows, and hierarchies in the global economy. Reading Econ Geogr 3:271
Tagliarino NK, Bununu YA, Micheal MO, De Maria M, Olusanmi A (2018) Compensation for expropriated community farmland in Nigeria: an in-depth analysis of the laws and practices related to land expropriation for the lekki free trade zone in Lagos. Land 7(1):1–38
Thompson L (2014) Agency and action: perceptions of governance and service delivery among the urban poor in Cape Town. Politikon 41(1):39–58
Thompson L, Tapscott C, Tsolekile-De Wet P (2018) An exploration of the concept of community and its impact on participatory governance policy and service delivery in poor areas of Cape Town, South Africa. Politikon 45(2):276–290
Torre A (2018) Les moteurs du développement territorial. Rev Econ Reg Urbaine 4:711–736
Wassenhoven L (2008) Territorial governance, participation, cooperation and partnership: a matter of national culture? Boletín de la Asociación de Geógrafos Españoles 46:53–76
Wily LA (2018) Collective land ownership in the 21st century: overview of global trends. Land 7(2):68–94

Chapter 13
Transformation of Territorial Governance Under Traditional Authority in Peri-Urban Areas of Gauteng and the Associated Spatial Implications for Urban Change

Mfaniseni Fana Sihlongonyane and Avhatakali Sithagu

Abstract It is widely acknowledged that urban areas in the African continent have witnessed a resurgence of traditional authorities. No matter how contentious their role under the colonial and apartheid systems of government was, the Constitution of the new South Africa (1996) recognised traditional authorities and afforded them opportunities for a political resurgence. This is occurring in a period of rapid urban growth especially in peri-urban areas within the sphere of extended urbanisation. While municipal governance attempts to control urban growth using legal and planning instruments, traditional authorities are shaping the location, direction, and expansion of urban growth through customary systems of tenure rights and responsibilities. This chapter uses the case of Thembisile Hani Local Municipality (THLM), a peri-urban area in the Mpumalanga Region, South Africa, to discuss the factors shaping the transformation of territorial governance under traditional authority in peri-urban areas and the associated spatial implications for urban change. The central argument of the chapter is that the changing status of traditional authorities and their integration into global/economic dynamics of urban change have made them counterparts in shaping urban transformation, especially in peri-urban areas.

Keywords Traditional authority · Gauteng · Territorial governance · Peri-urban · Urban change

---

M. F. Sihlongonyane (✉) · A. Sithagu
School of Architecture and Planning, University of the Witwatersrand, Johannesburg, South Africa
e-mail: mfaniseni.sihlongonyane@wits.ac.za

D. du Plessis (ed.), *Territorial Governance and Spatial Transformation in Post-apartheid South Africa*, Local and Urban Governance,
https://doi.org/10.1007/978-3-031-85489-7_13

13.1 Introduction

It is widely acknowledged that urban areas in the African continent have witnessed a resurgence of traditional authorities (Van Kessel and Oomen 1997; Pierre 2002; Ubink 2008b; Ainslie and Kepe 2016). No matter how contentious their role under the colonial and apartheid systems of government was, the Constitution of the new South Africa (1996) recognised traditional authorities and afforded them opportunities for a political resurgence. This is occurring in a period of rapid urban growth especially in peri-urban areas within the sphere of extended urbanisation. While 284 municipal governments in South Africa control urban growth using formal statutory planning instruments, traditional authorities estimated to be 14 kings and several traditional councils who compose of 844 senior traditional leaders (BusinessTech 2018) located in former homelands making up 33% of the total population are shaping the location, direction, and expansion of urban growth through customary systems of tenure rights and responsibilities.

Using a case study of the former KwaNdebele homeland (locally known as KwaMhlanga) within the THLM, Mpumalanga Region, South Africa, this chapter draws on qualitative methods to gather evidence including document review, focus group discussions, interviews of residents, institutional consultation and field observation. The chapter contributes to the understanding of existing urban governance processes and the competing interests of urban governance actors in peri-urban areas as influenced by "actually existing neoliberalism" (Brenner and Theodore 2002: 351). It is also providing a window through which to understand peri-urbanization, in its territorial, functional, and transitional vectors of urban governance. Furthermore, it offers some insights into the rural-to-urban transformation, interaction and their associated spatiotemporal dynamics in the global south. Unlike much literature that has emphasized approaches which theorize the urban and rural as areas that are in competition over resources and services, this chapter looks at the cooperative factors and the influences that are reshaping the extension of urban change in peri-urban areas. In addition, much of the urban studies literature has emphasised the tension between traditional authorities and municipal institutions e.g. ward councillors, municipal laws, urban projects (Rugege 2003; Ntsebeza 2004; Knoetze 2014) and there has been very little research on the adaptation of traditional authorities to urban environments in Africa.

The first section of the chapter based on literature reviews looks at the transformation of chieftaincy governance, initially during the pre-colonial, colonial and in particular during the post/colonial period against the urban growth and capitalist forces. The second section explores the encounters between the resurgence of chieftaincy, rapid growth and neo-liberal forces. The third section deals with the transformation of territorial governance and their spatial implications in THLM; and then, concluding remarks.

13.2 Traditional Authorities in Rural Environments During the Precolonial Era

According to Rugege (2009: 172), pre-colonial Africa was governed "by kings supported by a hierarchy of chiefs and councillors or advisors" ... and "the traditional authorities served as political, military, spiritual and cultural authorities and were regarded as custodians of the values of society." Ntsebeza (2003: 172) points out that the "traditional societies were composed of groups that were under the authority of independent chiefs (traditional authorities); with common acceptance that all land is held in trust by the chief for the people. The chief was obliged to consult with his group of councillors and there were clearly stipulated conditions determining where and when the chief could appropriate land." This form of organic governance enabled chiefs to allocate land use rights as the land use management system was vested with the traditional authority (Southern African Regional Poverty Network 2002). Thus, "traditional authorities provided for some organic form of governance that is closer to the people" (Claassens 2014 in Sithole and Mbele 2008: 26).

When the white settlers initially came, there were notable differences in terms of how leaders assume their positions, how much power they commanded, and how accountable they were to their communities (Mengisteab 2019). On the one hand, the institution of traditional leadership represented an early form of societal organisation during the pre-colonial era. The rural chiefs were traditional authorities determined through birth lineage in the different chiefdoms. They were not elected, but the son would inherit the father or uncle's leadership position. Chieftaincy constituted a political and administrative centre of governance for traditional rural communities. It embodied the protection of culture, traditions, customs and values. The chieftaincy system was based on the principle of governance of the people, where a traditional leader was accountable to his people. On the other hand, white settler areas were located on farms. While they were widely dispersed far away from each other, they were the focus of the trade and exchange. Many of these later formed the nuclei of urban areas. Hence, South Africa was characterised by the development of islands with a well-developed infrastructure where white settlers were located e.g. Pretoria, Lydenburg, Potchefstroom, etc., surrounded by large rural zones of territory with almost no infrastructure. Both urban and rural areas not only reflected types of government functions and activities in those areas, as narrowly defined by functionalist approaches, but also represented the perceptions of people who live in those areas and how these were governed (Sadiki and Ramutsindela 2002).

13.3 Shifts in Governance of Traditional Authorities During the Colonial Period

The introduction of colonialism in Africa distorted the powers of the traditional authorities and transformed them into their agents responsible for mobilising the subjects for labour and being accountable to the colonial administration as opposed to the subjects (Mamdani 1996; Ntsebeza 2004). In many cases in the continent, chieftaincy was no longer solely based on hereditary, but the colonial state could appoint some people into the office. The colonial state, for example, invented chiefs where there were no centralized authority systems and imposed them on the decentralized traditional systems, as among the Ibo of Eastern Nigeria, the Tonga in Zambia, various communities in Kenya, and the communities in Somalia. In the centralized systems also, traditional leaders of various titles were reduced to "chiefs" and the colonial state modified notably the relations between the chiefs and their communities by making the chiefs accountable to the colonial state rather than to their communities (Coplan and Quinlan 1997). In that way, indirect rule was established to administer Africans under the colonial administration rather than give them the right to vote.

In South Africa, the Black/Native Administration Act (1927), gave the British colonial government the authority to select and appoint traditional authorities but also relocate the traditional authority's areas of jurisdiction according to Section 2(7) and (8) of the Black Administration Act 38 of 1927. Therefore, traditional authorities reported to the colonial government and not their respective communities, as the chiefs had been incorporated into the colonial government administration system (ibid). According to Butler (2002), the Act (1927: 14) "formally imposed the white governor general as the appointed supreme chief with powers to appoint chiefs and *izinduna* and regulate their roles and privileges; to regulate land ownership in the reserves through the Department of Native Affairs; intervene in local governance and declare new tribal boundaries and force tribes to move between different areas." The chiefs were paid salaries for maintaining law and order (Keulder 1998). This enabled "autocratic traditional authorities to do as they wished in their own administrative areas" (Knoetze 2014: 167). Therefore, the promulgation of the Native Administration Act in 1927 constitutionally amplified the process of incorporating 'traditional authorities' into the apartheid body politic.

In 1948, the National Party under the leadership of DF Malan came to power. The traditional leadership system fitted its plan of separate development. In 1951, the Bantu Authorities Act was passed, making traditional leaders the administrative agents of the Apartheid state in the designated reserve areas, starting a process of setting up new separate political institutions for the African population. This was reinforced by the passing of the Promotion of Bantu Self-Government Act in 1959, which provided for the establishment of 10 self-governing homelands or Bantustans. The Act cemented the establishment of African pseudo-states (Sadiki and Ramutsindela 2002). Various exclusionary measures were deployed to separate the 'white race' from non-white. These included the appropriation of land for white ownership; influx control and counter-urbanization strategies; separation of residential areas, places of work; and so on (Smith 1992; Christopher 1994). Fair et al. (1969) observed that these constituted modern core areas and peripheral areas.

The Apartheid government also took away certain rights such as control over the distribution and administration of land in Bantustans such as Transkei, Bophuthatswana, Venda and Ciskei. Many of the traditional authorities colluded in the segregationist policies of the South African government and were instrumental in coercing their subjects to comply with the colonialist's instructions (Mamdani 1996; Amanor and Moyo 2008). They served as agencies of the alien states and enjoyed more powers in controlling the African population whilst losing accountability to the subjects during the apartheid period (Ntsebeza 2004). According to Butler (2002: 16) governance of the Bantustans "continued to be heavily dependent on the indirect rule model and so the cooperation of traditional authorities was fundamental." For this reason, by and large, the traditional authorities lost legitimacy as they were viewed as collaborators of the alien systems.

13.4 Chieftaincy in Post Apartheid Era

In the run-up to the first democratic election, South Africa attempted to ignore the traditional institution (see Bank and Southall 2014). In response, the Congress of Traditional Authorities (CONTRALESA) was formed in 1987 to push for the recognition of traditional authorities and their institutions as the primary level of government in rural areas (Ntsebeza 2004: 270). With 'the promise of delivering their "block vote", chiefs began to assume a new importance to the ANC: no longer [were they seen as] relics of a feudal past, but [as] strategic allies in the conquest of state power' (Van Kessel and Oomen 1997: 571). It was envisaged that they would oversee traditional matters, such as local succession debates, cultural heritage, ritual ceremonies and calendrical events, and the entrenchment of indigenous languages. Consequently, the final South African Constitution of 1996 recognised the traditional authorities. Section 212(2)(a) of the Constitution provides that "national or provincial legislation may provide for the establishment of Houses of Traditional Leaders to deal with matters relating to traditional leadership."

Several laws and white papers around traditional institutions and land ownership were passed, bolstering the traditional authorities but bewildering both scholars and rural supporters of the ANC. These laws included: the National House of Traditional Leaders Act 10 of 1997; White paper on Local Government 1998; Municipal Systems Act of 1998; the Traditional Leadership and Governance Framework (TLGF) Act of 2003, the 2005 and the 2008 Traditional Courts Bill (Ainslie and Kepe 2016). The National House of Traditional Leaders was established in terms of the then National House of Traditional Leaders Act, 1997 (Act 10 of 1997). Its objectives and functions are to promote the role of traditional leadership within a democratic constitutional dispensation, enhance unity and understanding among traditional communities and advise national government. Provincial houses of traditional leaders were established in all six provinces that have traditional leaders. The national and provincial houses of traditional leaders facilitate the cooperative relationships within national and provincial government, while the establishment of local houses of traditional leaders deepens and cements the relationship between municipalities and traditional leaders on customary law and development initiatives.

Under the neo-traditionalist presidency of Jacob Zuma (2009–2018), traditional authorities have managed to consolidate key gains made in the early years of their political resurgence. Under Zuma, the national Department of Co-operative Governance and Traditional Affairs (COGTA) was established with its own minister in 2010. COGTA has since been renamed the Department for Traditional Affairs (DTA) and had its administrative brief extended (Ainslie and Kepe 2016). Provincial Houses of Traditional Leaders were finally set up in all six provinces where traditional authorities were formally recognised (Claassens 2014; Walker 2003). In Gauteng, a Traditional Leadership and Government Act 4 of 2010 was promulgated in the province. This followed a recognition of two AmaNdebele traditional communities in 2007 now appointed into the NCOP in 2017 namely: Kgosi Kekana of the Lebelo and Chief J.J. Mahlangu of the Sokhulumi. More requests for recognition are pending (Martins 2019). This has signalled institutional recognition of a kind that would have been impossible to orchestrate two decades earlier, when the hostility of 'civics' represented a serious threat to the existence of traditional authorities.

Broadly, the recognition of chiefs has been justified on the basis of decentralisation and in terms of the developmental state imperative. Decentralisation allows "the transfer of political power and authority from central to sub-national governments, and therefore involves the balancing of the exercise of power between various levels of government" (Siddle and Koelble 2012: 5). Developmental government is "a state that puts economic development as the top priority of governmental policy and can design effective instruments to promote such a goal" (Bagchi 2000: 398). Other justifications were that "when the traditional institutions are robust, they allow chiefs to mobilise communities behind particular projects and causes" (Baldwin 2015: 105). As the United Nations Economic Commission for Africa (UNECA) study (2007) notes, traditional leaders often operate as custodians of customary law and communal assets, especially land (United Nations Economic Commission for Africa 2007). Furthermore, Mengisteab (2019) observed that traditional leaders dispense justice, resolve conflicts, and enforce contracts, even though such services are conducted in different ways in different authority systems. They also serve as guardians and symbols of cultural values and practices.

13.5 The Transformation of Territorial Governance of Chiefs in Peri-Urban Areas

It should be noted, that chieftaincy governance has not been static since the 1990s. Governance under chieftaincy has been influenced by the adoption of the Growth, Employment and Redistribution (GEAR) a macroeconomic policy of South Africa in 1996. GEAR is a neo-liberal market-orientated policy that focuses on facilitating privatisation, competition, less government intervention, deregulating capital markets, and lowering trade barriers (Le Roux 1997). The adoption of GEAR led generally to the de-agrarianisation of rural areas instigating rapid rural-urban migration and income diversification (Bryceson 1996, 1997). Over the last 50 years, the urban population of South Africa grew substantially from 48% to 68.8% rising at an increasing annual rate that reached a maximum of 1.08% in 1986 and then decreased to 0.71% in 2023. As much as 67.35% of the total population was living in urban areas and cities in 2020 and this figure is expected to increase to 71.3% by 2030 (O'Neil 2024). By 2050, eight in 10 people will be living in urban areas and this will increase demand for basic infrastructure requirements (Parliamentary Monitoring Group n.d.).

Since the 1990s, many of the migrants were/are unable to settle in the city centres due to low income to afford land, rent and cost of living. As Macagnano (2002: 158) observed, peri-urban areas within the outskirts of South African cities have received the brunt of the 'invasion' with a rapid rise of large informal settlements, accommodating a multi-cultural variety of people living in sub-standard conditions (Macagnano 2002). Hence literature on extended urbanisation in the Gauteng City Region of South Africa reflects areas far beyond the city as the urban periphery. Indeed, many African cities are growing through peri-urbanisation. Chiefs as custodians of land allocation have had to contend with the rapid demand for land. In many cases, local governance of chiefdoms has embraced entrepreneurialism to position themselves. This entrepreneurialism has generated neo-customary practices whereby land is commodified through land sales, leases, contracts and rents. Through neo-customary practices, communal land rights, commonly practised in rural areas, are gradually changing to individualized land rights in peri-urban areas (Chauveau 2007; Alden and Hammond 2001; Amanor 2010). Indeed, Wehrmann (2008) observed that the emergence of peri-urban areas has come with new forms of land tenure systems, which may not be compatible with the customary land tenure system and, as such, contradict customary rules.

The commodification of customary land through various mechanisms of sale has become a normative function of peri-urban growth since the 1990s due to increased demand for land. These opportunities have included the emergence of new land markets attracting both local and urban-based individual investors- and transnational corporate investors (Cotula 2013; Scoones et al. 2016). The literature on extended urbanisation claim "such processes of urban expansion are linked to the ubiquitous reordering of the global urban periphery through complex processes of displacement of central populations to the margins and the creation of new functional centralities (jobs, infrastructures, densities) away from the traditional core in South Africa" (Connolly et al. 2021: 248). Peri-urban areas have become sites of accumulation, economic restructuring, globalization, territorialization, and innovation, revealing decidedly urban ambitions. Legal pluralism in land tenure has become widespread because the geography of rural and urban functions occur side-by-side.

In recent decades, hundreds of new cities, malls, complexes, and other forms of development have been built in peri-urban areas with a notable acceleration of this trend in the 2000s. Mosiane and Götz (2022) observe that there is often a mix of modern development, modern municipal land governance, informality and traditional authority governance. Thus, it is widely observed that political systems have been undergoing processes of hybridisation, through which democratic state institutions and precolonial, 'customary' ones are combined in a process of mutual adaptation and trans-formation (Beall et al. 2005; Beall and Ngonyama 2009; Burr and Kyed 2006; Logan 2009; Williams 2004). There is a new phenomenon that Mosiane and Götz (2022) calls displaced urbanism which generates momentum for change. There is change in land uses, ownership, and control of land in peri-urban areas that is characterized by intense competition engendering exclusion of access to land leading to customary land rights disputes (Akaateba 2008; Global Land Tool Network 2018). For instance, the conversion of agricultural lands has made local farmers in peri-urban areas become subjects of expropriation. Many farmlands are lost to residential and commercial developments without fair compensation to farmers (Peters 2010). Some local farmers, in their attempt to defend their rights, enter into a direct struggle with chiefs, while others physically attack new buyers from entering their lands or sometimes go to the extent of destroying newly built-structures (Kasanga and Kotey 2001; Ubink 2008a). Hence Woltjer (2014) observes that peri-urban areas have been characterized by their changing local economic and employment patterns from rising land values and dynamic land uses (Woltjer 2014). These developments will be discussed in the case of THLM with particular focus on the Ndebele Chieftaincy in Buhlebesizwe, Moloto and Magodongo villages.

13.6 Precolonial Era and the Origin of the Ndebele Chiefdom

The exact origin of the Ndebele is debated by numerous scholars but all converge around the precolonial character discussed below. Some scholars claim that the Ndebele originated from the border of Natal and Basotoland (Skhosana 2010). Others state that they came from the Thukela River. Some state that they originate from Ondini or Lundini near the Drakensberg mountains, and yet others state that they come from uMndeni in Kwazulu-Natal. Kosi Thugwana, the niece of Nkosi VW Mahlangu of the Ndzundza-Fene Traditional Council (interviewed in 2019) objected to the notion of the Ndebele as descendants of the Zulu nation. According to history passed on to him by his elders, the Ndebele were already in the Transvaal by the 1500s before the Zulu nation was formed by Shaka in the 1700s.

According to Lekgoathi (2003), KwaMmyanyana (northwest of Pretoria) is the most important location for the Ndebele because this is where the tribe split into two main groups, that is, into the Southern Ndebele and the Northern Ndebele. Musi, the last king to rule over a united Ndebele had two sons (from different mothers). According to Skhosana (2010), Musi died in 1630 and the succession battle continued until the Ndebele split into the Southern Ndebele and Northern Ndebele and smaller tribes. The Southern Ndebele are the Ndzundza and the Manala, the

Northern Ndebele are the Mtombeni. The focus of this chapter is on the Southern Ndebele.

13.7 The Formation of the Ndebele Homeland During Apartheid

Prior to the formulation of local municipalities in the 1990s, Thembisile Hani Municipality was part of the KwaNdebele homeland in what was called the Transvaal Province. It was not the plan of the apartheid government to establish a Ndebele homeland. The Ndebele were few and scattered throughout the country, thus the logic of the government was to allow them to merge with other nations within the various homelands (South African History Online 2016). The Ndebele were a homeless nation in a period where every ethnic group was designated a state of their own. This caused great anxiety amongst the Ndebele chiefs because this meant they were subjected to marginalisation and exclusion in non-Ndebele homelands (Lekgoathi 2003). By the 1960s, a group of elite Ndebele traditional authorities were actively promoting the unique identity of the Ndebele across the Transvaal. They put pressure on the apartheid government to establish a separate homeland for the Ndebele. KwaNdebele developed organically when Ndebele people were expelled from the neighbouring homeland of Bophuthatswana, leading the government to establish a massive resettlement program that saw the creation of 12 camps. The resettlement was also impacted by the forced removal of blacks from 'black spots' (ibid). Therefore, KwaNdebele originated from a series of trust farms and other homelands that were consolidated between Groblersdal and Marble Hall in the then Eastern Transvaal (Makanjee 1989). The Department of Bantu Administration released a draft plan of the KwaNdebele homeland in 1972 (Nielsen 1996). In 1973, the final plans of the homeland were located in Weltevreden. By 1986, KwaNdebele was set to be the country's fifth independent bantustan after Transkei, Bophuthatswana, Ciskei, and Venda, however, it was not granted independence but remained a self-governing state until 1994. After the democratic dispensation in 1994 and the creation of local municipalities, the KwaNdebele homeland was dissolved and demarcated into two municipalities, i.e., the THLM and the Dr. JS Moroka Municipality to the north.

13.8 The Case of Thembisile Hani Local Municipality

The THLM is situated in the Mpumalanga province. It is a category B municipality, meaning it shares executive and legislative authority with a category C municipality, that is, the Nkangala District Municipality. It is surrounded by the Dr. JS Moroka Municipality to the north, Elias Motsoaledi Municipality to the east, Steve Tshwete and Emalahleni Municipality in the south-east direction, Victor Khanye Municipality to the south. All these municipalities form part of the Nkangala District Municipality. It also shares provincial borders with the Tshwane Metropolitan Municipality in Gauteng to the south-west, and several municipalities with Limpopo to the north. The municipal area of Thembisile Hani is 2384 square kilometres (Thembisile Hani Local Municipality 2016) in extent and has the second-highest population of 333,331 in the district (Stats SA 2018) (Figs. 13.1 and 13.2).

The succession battle between the Ndzundza and the Manala continued even into the democratic era. The Nhlapo Commission report released in 2010 aimed to resolve the rightful heir of the Southern Ndebele Kingdom between the two paramountcy: Manala-Mbongo and Ndzundza-Mabhoko (Republic of South Africa 2003). Upon hearing all the oral accounts, the Commission concluded that the Manala-Mbongo were the rightful heir to the kingdom of AmaNdebele because Manala was the first son. However, despite this ruling which the Commission said was based on customary law, then President Jacob Zuma declared Mbusi Mahlangu of the Ndzundza-Mabhoko as king in a Government Gazette dated 5 November 2010. Thus, the kingdom of AmaNdebele has two paramountcy as depicted in the figure below. According to Sifile (2016), the Commission used an older version of the Traditional Law Act to make the ruling in 2008. This version did not give the president powers to change the Commission's decision. However, President Zuma used an amended version of the act (Traditional Law Act 2009) which gave him powers to change the decision in 2010 (Fig. 13.3).

This signifies a similar authority that the state has exercised in the past to determine the position of chiefs. Buthelezi and Vale (2019) highlight how the ANC has recognised traditional leaders as an important political statement in the post-apartheid era. They give an example of the January 2018 executive visit to three prominent traditional leaders (of KwaZulu Natal, Eastern Cape and Limpopo) by the ANC president Cyril Ramaphosa during his campaign as the next incoming president of South Africa. This is in line with the assumption that 'chiefs bring the

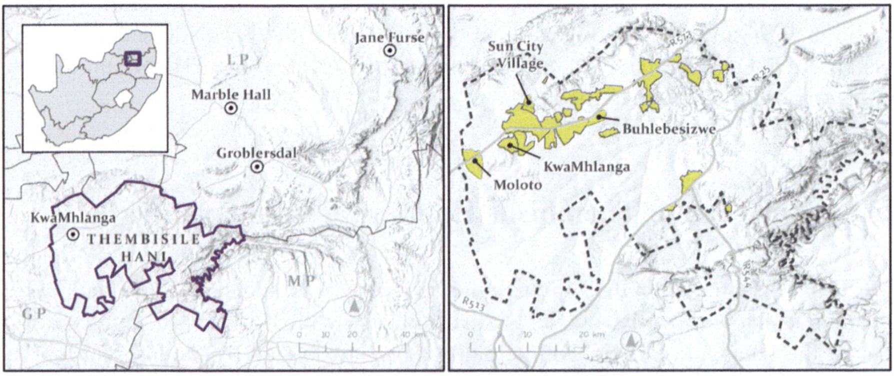

Fig. 13.1 Location map of Thembisile Hani local municipality (original to the author)

Fig. 13.2 Different housing typologies (original to the author)

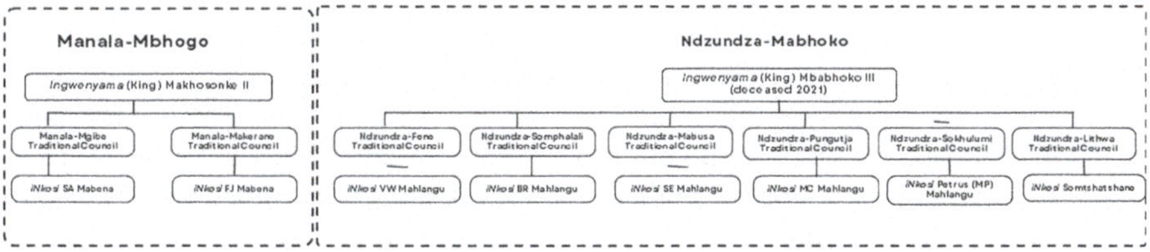

Fig. 13.3 The current Manala-Mbhoko and Ndzundza-Mabhoko chieftaincy (original to the author)

rural vote.' Therefore, the relationship between the state and traditional authorities is not stable but of constant opposition and bargaining.

13.9 Research Methods

In exploring the case of former KwaNdebele, secondary data was collected from organisational public documents of the THLM website, and grey literature was collected from the internet and other government policy documents. Official municipal reports that were not available on the internet were accessed physically from the municipality. National, regional, and local archival newspapers and magazines were used to better understand the social context of the case study. In addition, a field observation journal was recorded during fieldwork, to capture actual activities and the environment of the area.

Primary data was collected in the form of face-to-face interviews. This was done by carrying out door-to-door interviews using a set of semi-structured interview questions that were developed for each group of respondents, i.e., the traditional authorities, municipal officials, and community members. The first group of 35 households were interviewed from three villages, i.e. Buhlebesizwe, Moloto and Magodongo (See Fig. 13.4). Each village was chosen because of its unique traditional authority institution of practice. For instance, some sections of Moloto were governed by a headman, while some participants did not acknowledge him as legitimate. In Magodongo, the death of the headman left a void and several individuals claim to be headmen. In Buhlebesizwe, there is a legitimate headman.

Access to the community was obtained through the local ward councillor who selected a community member (who became the fieldwork assistant) to accompany the researchers to the households. The selection of households was determined by the availability of participants and by a snowball approach. In some cases, the fieldwork assistant influenced who to interview, for example, friends and family. Her role was to introduce the researcher and obtain permission to conduct the interview. As a well-known community member, the presence of the fieldwork assistant made access to households easier and mitigated community expectations. For example, a participant thought we were registering people for jobs, housing, or food parcels.

Therefore, it must be noted that research is an important socio-political signifier. As a researcher, one is perceived to belong to a higher social circle. Even the simplest items such as carrying a laptop, notebook, and pen, or driving a car may automatically indicates a higher social class. Therefore, researchers need to be aware of these aspects that change the perception of power dynamics in rural areas. The main challenge was the recordings. For instance, one community member was reluctant to an interview because she thought it was intrusive while another thought that it

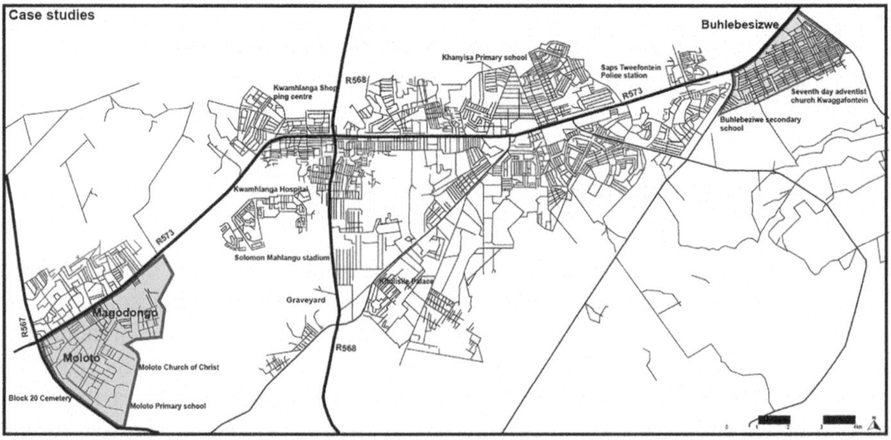

Fig. 13.4 The three villages: Moloto, Magodongo and Buhlebesizwe (original to the author)

would get her into trouble with the headman. Another participant wanted assurance that the information he provided would not reach the headmen, and proof that the interview was for research purposes. To mitigate this, assurance was given that the recordings were only for transcribing the conversation and identification cards confirming affiliation with the relevant tertiary institution was produced on request for evidence. Also, several community members saw the interview process as a way of bargaining for basic services. For example, because the two fieldwork assistants were known to the ward councillor, community members expressed their grievances during the interview in the hope that it would reach the ward councillor. In one extreme incident, a lady wept during the interview as she explained her struggles with poverty, unemployment, and a sick husband.

The second group of participants were the traditional leaders. A focus group interview was conducted with the Ndzundza-Fene Traditional Council members. This involved the senior traditional leader, his right-hand man and six headmen. The reason for choosing this council was that it had jurisdiction over two villages, i.e., Magodongo and Moloto. The council responsible for Buhlebesizwe could not be reached. In order to secure interviews, a letter of request was delivered to the secretary of the council who then delivered it to the council and secured a date and time for the interview. This protocol was crucial in securing the focus group discussion. It was not simply protocol but demonstration of respect for the council.

The last group of interviews was with two municipal officials and two ward councillors. Consent to interview officials was requested through email. Either there was no response, or the response was delayed. This meant that the researcher made random physical visits to the municipal offices. Additionally, there was no set procedure or protocol on how and who grants official permission for interviews. Another challenge was that the municipality had limited staff. In relation to ward councillors, permission had to be obtained through the speaker's office. Through telephone calls, the researcher secured appointments and met the ward councillors at their place of convenience. Coincidentally, the timing of the interview was in the same month that new councillors were appointed. Therefore, being interviewed while new on the job was exciting for the councillors.

13.10 Key Findings: Parallel Administration Between Municipal and Customary Land Management

From the field observations, it was clear that there was relative order among the two paramount chieftaincies concerning customary land governance, even though King Makhosonke II had taken former President Zuma to court over his decision to make King Mabhoko III Mahlangu sit on the throne as well (Republic of South Africa n.d.). The eastern side of the Moloto Road (R573) (see Fig. 13.4) is mainly Ndzundza-Mabhoko governed land while the west is mainly Manala-Mbhoko. The majority of the land in Thembisile Hani is owned by the state, and falls under the custodianship of the National Department of Agriculture, Land Reform and Rural Development and some parcels are owned by the provincial government of Mpumalanga. Only a few parcels of land dedicated to commercial farming are under private ownership, this also includes two formal townships (Thembisile Hani Local Municipality 2019). In an interview with the Ndzundza-Fene Traditional Council in 2021, the headmen indicated that since 1994, the municipality had not established any new village or township. Rather, it was the chieftaincy that was responsible for the development of new villages. This was confirmed by a municipal town planner in another interview. This suggests that regardless of the state-land ownership model, traditional authorities have a stronghold on the allocation of land, the spatial morphology and land use of Thembisile Hani. It also suggests that the municipality has a weak system of management and land use enforcement, as well as an inability to drive development in areas controlled by chiefs.

Each chieftaincy consists of several traditional councils headed by senior traditional leaders (as shown in Fig. 13.3). Under each senior traditional leader are headmen who are the leaders of the village wards. In each village ward, there is a tribal ward office which mainly houses a secretary and the headman. Individuals looking for land are directed to a tribal office (See Fig. 13.5) to enquire about its land availability. Should land be available, the headman accompanies the land seeker to the site for inspection immediately. If satisfied, they both return to the tribal ward office so that the *Lotsha* fee is paid (land allocation fee). Should land not be available, the land seeker is placed on a waiting list. Each council has a set value of the *Lotsha* fee based on the location, size of the land and its intended use. For example, the Ndzundza-Fene Traditional Council indicated that a site of 20 × 30 square meters of residential land in Moloto Village would cost R1200. The same size of land in Buhlebesizwe Village (under a different Traditional Council) costs R600. Once the *Lotsha* fee is paid, the secretary then writes a receipt which has the traditional authority stamp and emblem confirming payment. This receipt also gives the new land owner the right to occupy. This system of land allocation has attracted numerous people from further afield especially from urban areas.

13.11 Neo-Customary Practices and Competition Over Land Allocations

The attraction of people to these areas is motivated by the strategic location. Thembisile Hani is located between Gauteng and Limpopo and attracts those looking to access cheap land for housing and work opportunities. Pretoria serves as the main employment centre while those from areas such as Marble Hall, Jane Furse and Groblersdal in Limpopo migrate to the area to get closer to the city. The R573 Road (refer to Fig. 13.6), also known as the Moloto Road, connects the city of Pretoria in Gauteng and Thembisile Hani, then Marble Hall in Limpopo. The road stretches about 160 km from Gauteng, Mpumalanga, and Limpopo. It is the major

Fig. 13.5 The Ndzundza-Fene traditional council sub-office in Moloto Village (original to the author)

link between Tshwane, Thembisile Hani, Siyabuswa (Siyabuswa was the former capital of the KwaNdebele home and is now part of the DR JS Moroka Local Municipality.) and Marble Hall (Marble Hall is a small town that was formerly part of the Eastern Transvaal and is now part of Limpopo province.). It carries about 150,000 commuters daily to and from work, mainly Pretoria (Gwangwa 2018).

There is also an increase of migrants from Pretoria and surrounding townships to Thembisile Hani. Most migrants are individuals and families who cannot afford land and housing in the city. They take advantage of the lower costs of living, cheap and quick access to land and are willing to travel about 3–6 h between Pretoria and Thembisile Hani for work. For instance, a young single mother of two children relocated from Jane Furse to Magodongo Village because she believed that there were prospects of getting a job in Pretoria. An elderly woman moved from Soweto, a township in Johannesburg, to Sun City Village because she could build her rondavel hut where she could perform her rituals as a traditional healer. A young man from Atteridgeville, a township in Pretoria, was tired of the fast-paced life of the township and sharing his family home with his sisters. He wanted to have a place of his own. Interviews also revealed that there is a growing population of young adults who grew up in Thembisile Hani and want to build their own homes.

The increasing demand for land in Thembisile Hani has caused a shift in customary land allocations. While there is a systematic land governance system among traditional leaders as described earlier, there has been growing entrepreneurism. The pattern to some extent imitates a market-based approach that determines the price of land based on its demand and has led to its commodification. The gradual

Fig. 13.6 Formal and informal businesses along Moloto Road (original to the author)

shift from customary to neo-customary land practices has grown. On one hand, the chieftaincy has over the years gradually increased its allocation fees to take advantage of the demand for land. In other cases, fake headmen have sprung up in some villages. They allocate land without the authority of traditional leaders. An example of such a settlement is Magodongo village. Interviews with residents revealed that the village headman passed on in 2011 and because of his absence, random individuals, some even from his own family, pose as headmen and allocate land while charging higher and sometimes exorbitant prices for land as high as R5000 for a 20 × 30 square meter site. During interviews, one of the alleged illegitimate headmen questioned the presence of researchers in 'his section of the village' of Magodongo. He had established a permanent base in the village and had a 'site office' which is made up of an empty shack spray painted with his cell phone numbers. The shack is allocated along Moloto Road for visibility to enable those seeking land to find him.

Magodongo Village is one example where seemingly illegitimate headmen have brazenly taken over the land allocation system usually controlled by headmen. Interviews with community members have revealed that new land seekers often fall prey to these leaders through deceit, and those that are aware have somehow 'accepted' the existence of alleged illegitimate headmen. The main reason is that they give them what they need and that is, land. The entrepreneurial land allocation of both legitimate headmen appointed by traditional councils and the alleged illegitimate headmen leaves little room for new land seekers to determine who is legitimate and who is not. In a village such as Magodongo with no traditional leadership, residents identified different individuals as headmen. In Buhlebesizwe Village where there is an appointed headman, residents identified one individual as legitimate. In Moloto Village, some residents knew their headmen, some did not, while others questioned the legitimacy of his position. This suggests that the state of control over land distribution in each village may be different from the next. Some villages have a stable traditional authority leadership and therefore have control over land access. Others, particularly those that have a weak traditional leadership system have been subjected to alleged illegitimate headmen.

13.12 The Spatial Implication Resulting from Neo-Customary Practices

Consequently, the municipality battles to keep up with the sprawling residential areas. Electricity and road infrastructure are often delayed for years because of a lack of funds. In most cases, these new villages develop outside the Integrated Development Process (IDP) budget cycle meaning they are often left out of the municipal budgetary planning processes. Other municipal planning processes are also not aligned with what is happening on the ground. The municipal Land Use Scheme (LUS) is one example. Traditional authorities do not align their land allocations with the LUS, nor does the scheme resonate with the reality on the ground. Below is an extract from one of the municipal officials and their opinion on the municipal LUS.

> Eiish… I'm gonna bash the people (consultants) who did the scheme. It was supposed to be wall-to-wall "mara" (but), it looks like it was a desktop exercise. So, you will find where there is a school, they say it is residential. So, they just sat, and they (consultants) just gave colours, they just gave zonings [So, you are saying it does not represent the status quo] Yaa! [with affirmation] on the ground, yaa (Interview, 2021).

The shortage of technical and professional municipal staff and the over-reliance of consultants in the development of municipal policies has left a significant gap between the reality on the ground and the vision of the municipality. The consultant's approach is technical, i.e., they implement urban laws on a rural landscape. Even though there may be consultations between consultants and traditional leaders during projects, the participation of traditional leaders is too shallow to make significant changes to the technicalities of policies. On paper, the municipality disregards the powerful influence of traditional leaders on the daily land administration, management and spatial morphology of Thembisile Hani. Instead, it uses formal planning and legal approaches that are irrelevant to the status quo. Traditional authorities have never lost their presence in the villages. They are easily accessible, thus making them relevant to the daily needs of communities.

The blending of legitimate and fraudulent land allocation systems creates an uncoordinated administrative system. For example, each traditional council has its own form of receipt which bears the emblem and stamp. Alleged illegitimate headmen have somehow managed to intercept the traditional authority receipt system. One participant claimed that these 'fake' headmen colluded with legitimate headmen and made copies of the receipt system. This suggests a corrupted traditional authority system. With so many variations of the allocation receipt, new landholders have different versions, some get photocopied receipts, and others get a receipt from a simple receipt book.

The traditional authority annual levy system is not uniformly instituted in the villages. In Moloto Village, an interview with the secretary of the Nzundza-Fene Tribal Ward Office confirmed that residents are expected to pay an annual fee of R40 to the chieftaincy.

> There are annual fees. It's R40. [What is the payment for?] The money goes back to the community. Households are not the same. Some families have problems. You find that there is no maize, children are not at school, children have no shoes, and someone has died. We take from that money, and everything becomes alright (Interview, 2022).

In another part of the same village, an elderly woman said:

> We don't know of such fees. [She giggles nervously] Are you not going to get me arrested? [She laughs] Maybe they do pay. They know why they are paying. We don't know of such. A person must explain to you what the money is for. You see, they will know. We don't know what they are paying for (Interview, 2022).

Most participants did not pay the annual levies or simply did not know of them, others have boycotted them. A young man from Buhlebesizwe Village believed that the annual traditional authority levy system is not justifiable since traditional leaders have no role in community development. According to him, the chieftaincy is not accountable and transparent about what it uses the money for. In that case, he would rather pay for electricity because he can see what his money is doing.

The boycotting of paying the annual fees is not a surprising phenomenon for the chieftaincy. The senior traditional leader of the Ndzundza-Fene (in an interview in 2019) blamed the democratic government for creating a state-dependent society. He explained that when he was a young boy, the annual levy system was used to collect money to build roads, schools and clinics. In the past, the community took part in its own development, but according to him the democratic state provides free housing and expects him to be self-determined. According to the senior traditional leader, democracy has led to boycotting of the levy system.

Even though this is the case, state-led development projects are initiated by the municipality, provincial and national government. These projects have sometimes aligned with customary land allocations and sometimes have contested them. For instance, over the past three decades, Moloto Road has transformed into a corridor of dense occupation and activities. Informal businesses, small retailers, shopping centres and large commercial malls transformed the two-lane stretch of road into a business node. In interviews with traditional authorities and municipal officials, it was revealed that informal traders and small retailers access land through traditional leaders, while malls access land by first getting permission from traditional leaders and then approaching the municipality. While there may be some cooperation between the state and traditional leaders in some respects, there are also tensions. On the 18th of February 2023, SABC News broke a story that the South African National Roads Agency, responsible for the upgrades of Moloto Road, met with community members and traditional leaders about acquiring already built land on the road reserve to expand the road. The municipality has in its IDP expressed frustration over the irregular land allocation on road reserves, servitudes and environmentally sensitive areas by traditional authorities. For example, some residents of

Magodongo, a village along Moloto Road, have built their houses on the road reserve. Some parts of Magodongo and Moloto villages have underground water seeping from the earth onto their land. Like many other residents, a couple in Magodongo interviewed in 2022 choose to remain on the land and build a brick house despite 'living on top of water'.

> When we arrived here, we did not realize something about this stand. The first stand they allocated us, we realized that there was water on the stand, it is a watery area on that side. I think that about four stands were sitting on the water (Interview, 2022).

An elderly woman from Moloto Village has somehow made peace with the underground water and has made some investments in redirecting it into a self-made draining system that flows from inside the house to the street. Below is her explanation of how she managed to do that.

> Yes, the water comes out from underneath. I built on top of water actually. No one has built their home on top of the water like I have in this place. [And the house is still standing?] Yes. When we built this house, we lost a lot of money because we used the cement used to build bridges so that the cement can absorb water. We dug a drainage system. This drainage system comes from the house, and we built in a pipe so that the water drains out of the stand and goes outside. We also dug a drainage system behind the house, there it is. Even next to that wall, we dug a drainage system and we also dug here. And then we must find pipes and dig deep and close the pipe. Just like we did there, it was deep, we bought plastic, we put in concrete so that that water can drain to the outside. If you go into the house, you will see the water, but the water leaves the house and goes outside (Interview, 2022).

The woman's unwavering spirit towards her land plays a significant role in the attitude of newcomers. She has lived on her land since the 1980s, and even though she complains that the water has made her develop arthritis, her house is still standing. Many newcomers are encouraged to build on potentially dangerous land simply because they witness others building 'on top of water.' It is an incredible health and financial sacrifice to live in such an environment, but the desire to own land and build a home outweighs the potential danger of living on it.

A report from the National Department of Human Settlements (2015) indicated that between 2012 and 2015, there was a 37.2% growth in settlements. Most of the growth has occurred along the Moloto Road. An estimated, 7797 informal settlement structures were established in the 4 years, contributing to an average growth of 11% settlement physical growth per annum. Land allocations by headmen have driven much of the settlement growth. Some villages were established through legitimate headmen, others through alleged illegitimate headmen. In an interview with King Makhosonke II in 2023, he referred to the sprawling of new 'illegal' settlements that are brought on by alleged illegitimate headmen. Random individuals take advantage of the weak monitoring systems, chieftaincy disputes playing out in some villages, and the absence of legitimate headmen in others.

The report, like many of the municipal IDPs, indicates that headmen do not reserve land for community services such as schools, clinics, etc. There is a high demand for residential land. Therefore, new villages depend on the already existing community facilities and business nodes of neighbouring villages. Below is an extract from a municipal official and his opinion of the land allocation system.

> So yes, the allocation does affect the spatial pattern of the place. [How?] Most sites that the traditional authorities give are residential. So, when they do residential sites, they don't take into consideration issues like schools, churches, or public open spaces. No, they just go full-out residential. So, at the end of the day, where are the people that stay here supposed to go to school? Where are they supposed to play? It's gonna create a burden for already existing schools because they will have to move from where they are staying to existing schools and increase the number of pupils in that area (Interview, 2021).

Consequentially, the services that exist are not sufficient to cater to the growing residential areas. Mosiane and Götz (2022) state that even though the expansion of malls along Moloto Road offers some relief in accessing commercial and retail services, this is still not enough to cater to the entire population. For instance, queues for accessing banking and retail facilities are long, particularly at the end of the month. More complex services such as accessing higher education institutions are non-existent. Therefore, communities still rely on Pretoria to cater for local needs.

The concentration of the business node along the Moloto Road creates one way movement of people from the outskirts to the node adding pressure to the existing services. In addition, the outskirts have inadequate infrastructure development. Gravel roads in these areas mean that mini-bus taxis are not able to access these areas. Residents often have to pay extra to get transported or walk long distances from the main road while carrying heavy shopping bags to their homes. Such challenges cause vacant land along Moloto Road to be in high demand. Residents want to be close to services, traders, malls and shopping centres. Therefore, the price of land nearer to Moloto Road is much higher than land on the outskirts.

13.13 Concluding Remarks

Rapid urban development occurs in peri-urban areas and usually in areas that have been defined as rural in which governance of traditional chiefs prevails. While this sphere of governance was assigned to chieftaincy, there is a governance transition towards hybrid systems of governance. The global shifts in policy towards neoliberal/market principles have induced traditional leaders to be entrepreneurial in the land affairs. The land is now a subject of neo-customary practices involving commodification in the form of sale, lease, swops, and rent. Due to competition for power to lead these processes, there are conflicts among traditional leaders and between subjects. Traditional leaders are competing for positions, subjects are gunning for land grabs. Even corporates are scrambling for land sales and leases. This is compounded by tensions between traditional leaders and ward councillors.

Nevertheless, access to land in peri-urban areas is still relatively cheaper compared to urban areas. A significant consequence of this demographic change is the urban expansion occurring as a function of displaced urbanism. Small businesses and survivalist economic activities such as arts (woodwork), curios, game farm products, truck stops, farm stores, stores, spazas, tuck shops, welding workshops, and flower businesses, are emerging in peri-urban areas and along major arterial roads. Supermarket complexes and malls as well as dispersed tourist facilities especially along the main corridors are developing. There is an innovative co-existence of formal and informal land uses and activities. These are prolific acts of self-realisation by local residents trying to survive and pursue their aspirations; and, in turn, dynamic local economies from below—need to be taken much more seriously on their own terms. There are socio-economic webs and networks linking the city centres and peripheral economic centres as demonstrated in the flows of people, goods and services along the Moloto corridor. This marks the end of the binary urban–rural distinction to describe human settlement patterns in and around cities.

Despite neocolonial legacies that still prevail in the legislative and policy frameworks, chiefs persist in influencing the governance of the urban morphology. Their influence in land markets, systems of access to land, land allocation and how growth occurs in peri-urban areas heralds the development of new urban morphologies in the proverbial African city. Given the widespread control of land by chiefs in many African countries, this suggests that chiefs continue to influence and are drivers of urban change in the continent. However much of this growth is within the ambit of peri-urbanisation which promotes sprawl and informality. This trend is not sustainable and adverse to the achievement of the sustainable development goals.

Moving forward would require a seismic shift of mindset and worldview towards the institution of traditional authorities towards embracing the constructive holistic development between the so-called rural and urban. A huge undertaking of collecting geospatial data about peri-urban and rural areas is essential to facilitate the identification of assets both formal and informal. The data collection should involve local people who are familiar with the local environments. A set of policies should be developed in a decolonial approach so that community participation and documentation of land rights are also undertaken from the perspective of the grassroots. It would also be crucial to develop legislation and policy that affirm the rights of the locals rather than being a means of facilitating external incursions of people to engender elite capture. Lastly, functional implementation strategies should be designed and developed with locals to promote sustainable development and prefer a new imagination of the future African city.

References

Ainslie A, Kepe T (2016) Understanding the resurgence of traditional authorities in post-apartheid South Africa. J South Afr Stud 42(1):9–33

Akaateba MA (2008) Urban planning practice under neo-customary land tenure: the interface between government agencies and traditional authorities in peri urban Ghana. PhD Dissertation Thesis, TU Berlin

Alden WL, Hammond D (2001) Land security and the poor in Ghana. Is there a way forward? A land sector scoping study. DFID Ghana Rural Livelihoods Programme, Accra

Amanor KS (2010) Family values, land sales and agricultural commodification in South-Eastern Ghana. Afr South East 80:104–125

Amanor KS, Moyo S (2008) Land and sustainable development in Africa. Zed Books, London

Bagchi AK (2000) The past and the future of the developmental state. J World Res 2:398–442
Baldwin K (2015) The paradox of traditional chiefs in democratic Africa. Cambridge University Press, Cambridge
Bank L, Southall R (2014) 'Traditional leaders'; T. Gibbs, Mandela's Kinsmen: nationalist elites and Apartheid's first Bantustan (Woodbridge, James Currey)
Beall J, Ngonyama M (2009) Indigenous institutions, traditional leaders and elite coalitions for development: the case of Greater Durban, South Africa. Crisis States Research Centre Working Paper No. 55.2. London School of Economics and Political Science, London
Beall J, Mkhize S, Vawda S (2005) Emergent democracy and 'resurgent' tradition: institutions, chieftaincy and transition in KwaZulu-Natal. J South Afr Stud 31(4):755–771
Brenner N, Theodore N (2002) Cities and the geographies of 'actually existing neoliberalism'. Antipode 34(3):349–379
Bryceson DF (1996) De-agrarianization and rural employment in Sub-Saharan Africa: a sectoral perspective. World Dev 24(1):97–111
Bryceson DF (1997) De-agrarianisation in Sub-Saharan Africa: acknowledging the inevitable. In: Bryceson DF, Jamal V (eds) Farewell to farms: de-agrarianisation and employment in Africa. Ashgate, Aldershot, pp 3–20
Burr L, Kyed HM (eds) (2006) State recognition of traditional authority in Mozambique: the nexus of community representation and state assistance. Nordic Africa Institute, Uppsala
BusinessTech (2018) South Africa has a huge number of traditional leaders—here's how much they get paid. BusinessTech, 6 Aug 2018. https://businesstech.co.za/news/government/263191/south-africa-has-a-huge-number-of-traditional-leaders-heres-how-much-they-get-paid/. Accessed 10 May 2024
Buthelezi M, Vale B (2019) Collisions, collusions, and coalescences. New takes on traditional leadership in democratic South Africa- and introduction. In: Buthelezi M, Skosana D, Vale B (eds) Traditional leaders in a democracy. Resources, respect, and resistance. MISTRA, Johannesburg
Butler M (2002) Traditional authorities, land and policy: current debates and challenges. SARPN, Pietermaritzburg
Chauveau JP (2007) Changes in customary land tenure systems in Africa. IIED, London
Christopher AJ (1994) The atlas of apartheid. Routledge, London. City Press, 16 December 2001
Claassens A (2014) Denying ownership and equal citizenship: continuities in the state's use of law and "custom", 1913–2013. J South Afr Stud 40(4):761–779
Connolly C, Keil R, Harris Ali S (2021) Extended urbanisation and the spatialities of infectious disease: demographic change, infrastructure and governance. Urban Stud 58(2):245–263
Coplan DB, Quinlan T (1997) A chief by the people: nation versus state in Lesotho. J Int Afr Inst 67(1):27–60
Cotula L (2013) The great African land grab? Agricultural investments and the global food system. Zed Books Ltd, London/New York
Fair TJD, Murdoch G, Jones HM (1969) Development in Swaziland: a regional analysis
Global Land Tool Network (2018) Bridging the gap on land governance and tenure along the urban-rural continuum. UN-Habitat/Global Land Tool Network, Nairobi. Current debates and challenges. https://sarpn.org/CountryPovertyPapers/SouthAfrica/may2002/land/traditional_authorities_2.pdf. Accessed 9 Feb 2024
Gwangwa V (2018) 'Road of death' to be one of hope. Pretoria News, 31 July 2018
Kasanga K, Kotey NA (2001) Land management in Ghana: building on tradition and modernity. UK's Department for International Development (DFID), London
Keulder C (1998) Traditional leaders and local government in Africa: lessons for South Africa. Human Research Council, Pretoria
Knoetze E (2014) Legislative regulation of the development functions of traditional leadership: in conflict or cohesion with municipal councils? Speculum Juris 28(1):161–195
Le Roux P (1997) The Growth, Employment and Redistribution Strategy (GEAR): a critical discussion. Africanus 27(2):45–66

Lekgoathi SP (2003) Chiefs, migrants and North Ndebele ethnicity in the context of surrounding homeland politics, 1965–1978*. Afr Stud 62(1):53–77

Logan C (2009) Selected chiefs, elected councillors and hybrid democrats: popular perspectives on the co-existence of democracy and traditional authority. J Modern Afr Stud 47(1):101–128

Macagnano EV (2002) Peri-urban areas of South African cities: innovative technologies for sustainability. In: Brebbia CA, Martin-Duque JF, Wadhwa LC (eds) The sustainable city II, Southampton: WIT Press. pp 153–162

Makanjee V (1989) A unilateral declaration of independence. Indicator 5(1):62–65

Mamdani M (1996) Citizen and subject: contemporary Africa and the legacy of late colonialism. David Phillip Publishers, Cape Town

Martins M (2019) Greater co-operation between CoGTA and Gauteng Traditional Leadership. Mail and Guardian, 8 March 2019

Mengisteab K (2019) Traditional institutions of governance in Africa. Politics. https://oxfordre.com/politics/display/10.1093/acrefore/9780190228637.001.0001/acrefore-9780190228637-e-1347. Accessed 7 Jan 2023

Mosiane N, Götz G (2022) Displaced urbanisation or displaced urbanism? Rethinking development in the peripheries of the GCR. GCRO Provocation #08, Gauteng City-Region Observatory

National Department of Human Settlements (2015) Informal settlement: upgrading policy and strategy, assessment and categorisation report and upgrading programme. National Department of Human Settlements, Pretoria

Nielsen D (1996) Bringing together that which belongs together: The establishment of Kwandebele and the Incorporation of Mouse. Seminar Paper presented in the Richard Ward Building, seventh floor, Seminar Room 7003, University of Witwatersrand, Johannesburg, 11 March 1996

Ntsebeza L (2003) Traditional authorities, local government and land rights. In: Ray DI, Reddy PS (eds) Grassroots governance? Chiefs in Africa and the Afri-Caribbean. University of Calgary Press, Calgary

Ntsebeza L (2004) Democratic decentralisation and traditional authority: dilemmas of land administration in rural South Africa. Eur J Dev Res 16(1):71–89

O'Neill, A. (2024) Urbanization in South Africa 2023. 14 February, 2025 https://www.statista.com/statistics/455931/urbanization-in-southafrica/#:~:text=In%202023%2C%20over%2068.82%20percent%20of%20South%20Africa%27s,population%20from%20the%20total%20population%20of%20a%20country. (Accessed, 1 April. 2025)

Parliamentary Monitoring Group (n.d.) Urbanisation. https://pmg.org.za/page/Urbanisation. Accessed 6 Jan 2023

Peters E (2010) Contesting land and custom in Ghana. State, chief and the citizen. J Agrar Change 10:604–608

Pierre E (2002) Patterns and theories of traditional resurgence in tropical Africa. Monde en Dévelopement 30(118):51–64

Republic of South Africa (2003) Traditional Leadership and Governance Framework Act 2003. Pretoria

Republic of South Africa (n.d.) Determinations on the position of the paramount chiefs. Pretoria. https://www.gov.za/sites/default/files/gcis_document/201409/reports-paramountcies0.pdf. Accessed 12 Jan 2023

Rugege S (2003) Traditional leadership and its future role in local governance. Law Demo Dev 7(2):171–200

Rugege S (2009) Traditional leadership and its future role in local governance. Law Demo Dev 7(2):171–200

Sadiki P, Ramutsindela M (2002) Peri-urban transformation in South Africa: experiences from the Limpopo Province. GeoJournal 57(1/2):57–63

Scoones I, Amanor K, Favareto A, Qi G (2016) A new politics of development cooperation? Chinese and Brazilian engagements in African agriculture. World Dev 81:1–12

Siddle A, Koelble T (2012) The failure of decentralisation in South African local government. UCT Press, Cape Town

Sifile L (2016) Ndebele kings in court over turf wars. Sowetan Live, 28 May 2016
Sithole P, Mbele T (2008) Fifteen-year review on traditional leadership. Human Sciences Research Council
Skhosana BP (2010) The (ama)Ndebele of Africa and their name '(ama)Ndebele'. https://dspace5.zcu.cz/bitstream/11025/6424/1/Skhosana.pdf. Accessed 19 Aug 2021
Smith DM (ed) (1992) The Apartheid City and beyond: urbanization and social change in South Africa. Routledge, New York
South African History Online (2016) The influence of apartheid: the Ndebele and the apartheid state. https://www.sahistory.org.za/article/influence-apartheid. Accessed 3 Feb 2024
Southern African Regional Poverty Network (2002) Traditional authorities, land and policy: current debates and challenges. https://sarpn.org/CountryPovertyPapers/SouthAfrica/may2002/land/traditional_authorities_2.pdf. Accessed 9 Feb 2024
Stats SA (2018) Community survey 2016, provincial profile: Mpumalanga. Statistics South Africa, Pretoria
Thembisile Hani Local Municipality (2016) 2015/2016 reviewed integrated development plan. https://www.thembisilehanilm.gov.za/sites/default/files/FINAL%20IDP%2015_16%20VOL%201%20MAIN%20DOC.pdf. Accessed 10 Feb 2024
Thembisile Hani Local Municipality (2019) Thembisile Hani local municipality integrated development plan 2019–2022. https://www.thembisilehanilm.gov.za/sites/default/files/IDP%20REVIEW%202019%20%282%29%20%3D%3D%3D%3D%3D%3D.pdf. Accessed 8 Jan 2023
Ubink JM (2008a) Negotiated or negated? The rhetoric and reality of customary tenure in an Ashanti village in Ghana. Africa 78:264–287
Ubink J (2008b) Traditional authorities in Africa: resurgence in an era of democratization. Leiden University Press, Leiden
United Nations Economic Commission for Africa (2007) Relevance of African traditional institutions of governance. ECA, Addis Ababa
Van Kessel I, Oomen B (1997) 'One Chief, One Vote': the revival of traditional authorities in post-apartheid South Africa. Afr Aff 96(385):561–585
Walker C (2003) Piety in the sky? Gender policy and land reform in South Africa. J Agrar Chang 3(1 & 2):113–148
Wehrmann B (2008) The dynamics of peri-urban land markets in Sub-Saharan Africa: adherence to the virtue of common property vs. quest for individual gain. Erdkunde 62(1):75–88
Williams JM (2004) Leading from behind: democratic consolidation and the chieftaincy in South Africa. J Mod Afr Stud 42(1):113–136
Woltjer J (2014) A global review on Peri-urban development and planning. J Perencanaan Wilayah Dan Kota 25(1):1–16

Part III
Contemplating the Future

Chapter 14
Evaluating the Effectiveness of Spatial Development Frameworks as Strategic Plans

Catherine Stone

Abstract Spatial plans, or Spatial Development Frameworks (SDF), are key strategic planning tools within the territorial governance system in South Africa and a primary policy instrument for driving spatial transformation within this system. As frustration mounts with the slow pace of spatial transformation of our cities and towns, many conclude that spatial plans, and spatial planners, notwithstanding the policy commitments made in SDFs, have failed. The effectiveness of SDFs in the post-Apartheid era, and more specifically in the post-SPLUMA era of territorial governance, is evolving. There have been significant steps taken post-SPLUMA to more rigorously embed the role of the plan in governance systems and hold decision-making practices accountable to implementation of the spatial plan. Nevertheless, systemic challenges in the current territorial governance framework bring into focus the practical influence of a spatial plan in South Africa over significant levers for change in the built environment. In this context and considering that transformation of the built environment is inherently slow and incremental (Turok I, Planning for spatial transformation. In: Western Cape Government, land use planning and law reform in the Western Cape. Western Cape Government, pp 21–27, 2016), the role of SDFs in influencing decision-making is not always visible and perhaps underestimated. This chapter explores firstly, how the effectiveness of a spatial plan or SDF should be assessed, drawing on Faludi's (Plan Pract Res 15(4):299–318, https://doi.org/10.1080/713691907, 2000) suggestion that a strategic plan is a tool for supporting and influencing decision-making and cannot dictate outputs and outcomes. As such its effectiveness should be assessed based on its role as a 'learning framework' or decision support tool. Secondly, the chapter explores contextual and structural factors that sit between and SDF and its implementation. Scrutiny of these factors is important to more accurately understand the causes of the inertia that hinder significant change, despite the best laid out plans.

C. Stone (✉)
Independent Urban and Regional Planning Consultant, Cape Town, South Africa

D. du Plessis (ed.), *Territorial Governance and Spatial Transformation in Post-apartheid South Africa*, Local and Urban Governance,
https://doi.org/10.1007/978-3-031-85489-7_14

Keywords Spatial planning · Spatial plans · Spatial transformation · Spatial development frameworks · Territorial governance · Strategic planning

14.1 Introduction

Spatial plans are a central policy instrument in the territorial governance system in South Africa. The national Spatial Planning and Land Use Management Act (Act No. 16 of 2013) (SPLUMA) asserts an ambitious role for Spatial Development Frameworks (SDFs), to be prepared at national, provincial and municipal level. SDFs must inter alia represent a long- and medium-term vision; coordinate, integrate and align development policies and plans across government; guide planning and development decisions across all sectors of government and represent the trade-offs of these sectors. They must outline specific arrangements for prioritising, mobilising, sequencing and implementing public and private infrastructural and land development investment in priority areas (South Africa 2013). Based on the extensive public participation required in drafting of an SDF, and its approval by a Council of elected representatives, the SDF is a social compact with civil society on how government will develop and manage the built and natural environment. The Municipal Systems Act (Act 32 of 2000) (MSA) identifies municipal SDFs as a core component of the 5-year strategic plan for the development of a municipal area (the Integrated Development Plan).

SDFs have a particular responsibility to leverage this role to drive spatial transformation. The national Department: Planning, Monitoring and Evaluation's Revised Framework for Strategic Plans and Annual Performance Plans (DPME 2015)—plans prepared by national and provincial government departments—suggests that "spatial transformation and inclusive economic growth remain elusive" and describes the purpose of SDFs as to "radically change the rationale for and rules by which planning, budgeting and infrastructure investment and development spending in the country take place", in a context where public and private sector planning, investment and expenditure has "reproduced, entrenched and in some cases reinforced historic spatial patterns" (South Africa n.d.).

Spatial transformation is a fluid term in South African practice. As indicated in the introductory chapter definitions vary in their nuance and depending on the scale under consideration. Substantively, SPLUMA expects spatial planning and spatial plans to be transformative in such a way that they pursue spatial justice ("redress past spatial and other development imbalances"), promote principles of sustainability, efficiency, and resilience. The National Spatial Development Framework (2023) (NSDF) defines spatial transformation as "the carefully planned and well-managed process of placing infrastructure, social services and economic activities in settlements in such a way that (1) the segregated spatial patterns inherited from colonial and apartheid times are broken down, and (2) the inefficiencies, injustices and inequalities in access to opportunities resulting from these past patterns are corrected". The National Development Plan 2030 (2012) (NDP) proposes that a national focus on spatial transformation should result in "reduced travel distances and costs, especially for the poor households" and by 2030 "a larger proportion of the population should live closer to places of work, and the transport they use should be safe, reliable and energy efficient".

Yet, notwithstanding consistent commitments made in SDFs across the three scales and over the last 30 years,[1] there is frustration with the seeming lack, or slow pace of transformation of our cities and towns from the perspective of the definitions discussed. Well-located affordable housing opportunities remain inadequate, public transport connecting townships to the city centres has got worse, dormitory townships with low economic investment and long commutes to places of employment remain a stubborn pattern in South African cities (NPC 2023).

There is a substantial body of research that delves into the question of the performance of spatial planning in post-Apartheid South Africa (Denoon-Stevens et al. 2022; Moodley 2019; du Plessis 2013). This chapter seeks to contribute to this discussion by considering the question of spatial planning efficacy from the perspective of the role of the spatial plan as a decision-making tool in the South African territorial governance system and current and emergent practices in the Western Cape Province of South Africa. It does so by drawing on the knowledge and experience of the author as a public sector planning practitioner[2] and applying Faludi's (2000) framework to assess the effectiveness of a strategic plan, drawing from questions proposed by Mastop (1987) and situations identified by Wallagh (1988).

To start with, three foundational points need to be made:

(i) spatial planning in South Africa is effectively defined as a strategic planning function in the sense that the legislated tool is a spatial development ***framework***, not a master, or blueprint plan.
(ii) as such, the SDF is a policy—its role is to guide decision-making, while due attention is paid to the merits of the matter at hand.
(iii) as a tool for supporting and influencing decision-making, it cannot dictate outputs and outcomes. Its effectiveness should be assessed on this basis (Faludi 2000).

Underpinning Faludi's framework is an understanding of the strategic plan as a learning framework that should be evaluated as such. As a learning framework, the strategic plan assists decision-makers to understand a particular situation and the trade-offs implied by different responses to this situation. It recognizes that plans cannot always resolve all conflicts or address all interests equally, and that the

[1] The tradition of preparing SDFs precedes the advent of SPLUMA. For example, in metropolitan Cape Town, a metropolitan-wide, as well as sub-metropolitan scale SDFs guided development on an interim basis under the auspices of the transitional councils post-1994 and into the period of the newly unified metropolitan local government from 2000. The first SDF of the metropolitan government, as it is now, was adopted in 2012. A second iteration of this was adopted in 2018 and the third in 2023.

[2] A practitioner perspective brings to light processes (formal and informal) that are not necessarily visible to those outside of government.

context within which plans operate is dynamic (Faludi 2000). Plans can support decision-making with context, evidence, principles, and processes. This is particularly relevant in the South African context where the decision-making body within government, particularly local government, is dynamic and diverse.

14.2 An Appropriate Framework for Evaluating Strategic Plans in the South African Context

The inadequate theoretical and methodological underpinning of processes to evaluate the influence of spatial plans is not unique to South Africa but has long been recognised globally (Faludi 2000). Limited enquiry has been directed at the evaluation of implementation processes in the field of spatial planning and planners in general have inadequate knowledge about the implementation of spatial plans and their effects on spatial development (Talen 1996; Seasons 2003). This results in a limited understanding of decision makers regarding present and future problems they face (Faludi 2000; Laurian et al. 2004). This lack of evaluation of spatial planning implementation efforts has been attributed to factors such as the scarcity of methodologies for systematic analysis of plan implementation and the complex nature of quantified modelling of the effects of planning (Talen 1996; Laurian et al. 2004; Tian and Shen 2011). There are three main conceptual challenges underlying this situation: the issue of uncertainty in planning, the challenge of complexity and control, and the ability to establish causality between plans and outcomes (du Plessis 2019).

A critical point of departure in the process of spatial plan evaluation is that cities are complex systems and thus there is an implied requirement to evaluate both the whole and individual parts. The framework for spatial planning evaluation should thus be sufficiently flexible to make provision for the assessment of both individual processes and objectives, and the entire system (or both), while considering the interrelationships between the sub-systems and processes (du Plessis 2019). A flexible and modular approach to spatial planning evaluation is thus most suitable and should include elements of contextual evaluation, evaluation of the quality of the operational environment, and evaluation of various individual and combined stages of the spatial planning process and its outputs (spatial plans).

The evaluation of the success of a spatial plan consists of two distinct components: plan performance evaluation and plan conformance evaluation. This chapter is specifically focussing on plan performance evaluation. An important consideration is that operational decisions do not simply convert plans into reality, and that the success of spatial plans also entail deliberations about the consequences of alternative decisions in which spatial plans only form part of the information used by operational decision makers to reach their decisions (Mastop and Faludi 1997). This view of Mastop and Faludi (1997) that the usefulness of the plan in informing and guiding the decision-making process is more important than whether the final outcomes reflect the goals envisaged in the plan itself is regarded as directly relevant to the aim of this chapter and the context of territorial governance in South Africa.

14.3 Spatial Planning in Practice: A Dynamic and Challenging Context

There are several contextual factors that suggest that SDFs in South Africa are by necessity, decision-support tools.

14.3.1 SDFs Are Short- and Long-Term Plans

SDFs must set both a long-term direction and guide decisions made in government on a day-to-day basis. The planners and decision-makers must create plans that are responsive to immediate needs—anticipated and unanticipated. Plans must provide space for urbanisation, urban growth and meeting housing backlogs (formal and informal); open up opportunities for development that creates jobs; respond to shocks, such as COVID-19, or severe weather events as impacts of climate change and South Africa's energy crisis. At the same time, as spatial policy they must hold firm on what is important for long term transitions to tackle inefficient, costly urban infrastructure systems and massive inequalities, while securing land to protect resources, threatened species, and buffer risks associated with climate change impacts. On the one hand, the SDF must allow for flexibility and contextually appropriate interpretation of policy in decision-making, on the other, it must give certainty or predictability.

14.3.2 SDFs Depend on All-of Government for Implementation

National, provincial and municipal SDFs are integrative policy instruments within an inter-governmental system where no one state entity has authority over another—inter-governmental cooperation or rather, lobbying, persuasion, and negotiation, is the order of the day. Many different role-players shape, influence, and impact on urban territories in South Africa. The powers and resources of some are fundamental to possibilities for re-shaping South African cities.

Affordable housing, public transport and social facilities are potential catalysers of spatial transformation. For example, one of two sub-strategies supporting Cape Town SDF's first spatial strategy "to plan for inclusive economic growth and improve access to economic opportunities" is to "integrate land use, economic activities and transport planning that support the sustainable operation of the public transport network" (City of Cape Town 2023, b, c). Social rental housing is key to well-located affordable housing at appropriate densities. National government controls urban passenger rail services, housing delivery is a national and provincial mandate. Securing land for housing is the job of a national agency, the Housing Development Agency. Approval and subsidies for social rental housing are controlled by the national Social Housing Regulatory Authority. Provincial government build schools and healthcare facilities—space standards are set nationally.

SDFs' task of integrating, aligning, prioritising, sequencing and representing trade-offs with a parallel, largely nationally centralised system of sector planning, answerable to different political and economic imperatives, is challenging. Even when these plans are mutually supportive, the vulnerable state of public sector finances, capacity, capability, and stability has deepened the lack of predictability that sector plans will come to fruition (Palmer et al. 2017). In turn, testing the credibility of SDF policies and proposals that support or rely on these sector plans for implementation is imperative.

14.3.3 SDFs Mediate Sometimes Competing Interests

Widespread, persistent poverty and unemployment make the pursuit of economic growth (as a means to job creation) non-negotiable for leaders. SDFs must tread carefully in balancing this overriding political imperative with, for example, recommendations that might limit development to conserve rural and environmental assets and protect ecological infrastructure, or constrain land markets, such as suggesting placing obligations on real estate developers to provide a percentage of affordable housing. There are many, often competing interests at play in implementing an SDF. In practice, not everyone will take the necessary decisions to realise its vision, or at the pace and scale required. The plan can propose, motivate, set out a policy position but it does not control the decisions that do or don't implement the plan.

This begs the question—by what do we reasonably measure the success of a strategic plan like an SDF, a policy instrument that must "*guide* planning and development decisions", "*contribute* to a coherent, planned approach to spatial development" in the three spheres of government and "*promote* a rational and predictable land development environment" (South Africa 2013)? The logical desire (default) is to measure the performance of the plan by developments and projects (outputs) and, ultimately, spatial outcomes as experienced on the ground. Yet there are many processes that follow the approval of a plan, to the point of a decision to approve a development proposal, prioritise and budget for a certain infrastructure investment or allocate resources to further planning or land preparation. Decision-makers must interpret the meaning of the SDF principles, policies, guidelines, and proposals and navigate trade-offs in more immediate spaces 'closer to home'. In these processes contested interests lobby, negotiate, compromise, and patronise (Olver 2019).

Timing is important, budget cycles, spending deadlines and elections become decisive factors.

Cognisant of the messy realm of plan implementation, Faludi (2000) suggests that strategic plans *"must be evaluated, not primarily in light of their material outcomes, but for how they improve the understanding of decision-makers of future and present problems they face"*. The proposition is that the role of the plan, an SDF in this instance, is in sense-making and the plan-making process is one of "*mutual learning involving interaction between a multitude of actors*". Faludi recognises that in the process of plan-making and implementation, planners are constantly negotiating or mediating (guided by principles, policies and guidelines set out in the SDF), caught between competing interests, under pressure to approve development in pursuit of economic growth, tested by the complexity of day-to-day wicked problems and often not directly involved in processes of plan implementation. The question is rather whether they were able to clarify decision-makers' understanding of the situation within which decisions must be made, identify the potential consequences of different decisions and in so doing, assist these decisions?

To assist with answering this question, Faludi (2000) presents a method for evaluating plan performance, using conditions identified by Mastop (1987) and a checklist to evaluate the effectiveness of the plan set out by Wallagh (1988).

14.4 Assessing SDF Effectiveness as a Strategic Decision-Making Framework

Mastop (1987) identifies three conditions for an effective strategic plan:

(i) The plan must name the decisions for which it is intended as a framework (does the plan make clear the decisions it intends to support?)
(ii) The plan must be of continuing relevance to the situation as it evolves.
(iii) The plan must help in defining decision situations.

In the case of the latter, if not conforming to the plan, to what extent have implementors of the plan allowed themselves to be influenced by it in their decision-making? Linked to this question and linked to the second condition identified by Mastop, is whether, as Faludi asks, decision-makers deem the plan to be helpful in making sense of their situations, talking to the role of the plan in a learning process (Faludi 2000).

Wallagh (1988) suggests four situations that can assist the evaluator of plan performance:

(i) A decision explicitly conforms to the plan.
(ii) Departures from the plan are deliberate in the sense that the motivation draws on the plan itself.

(iii) The plan creates the basis for understanding the consequences of a decision that does not conform to the plan, bringing that decision into the ambit of the plan.
(iv) Future iterations of the plan use the original as the point of departure.

These situations are used to support the application of Mastop's conditions, framed as questions in the discussion to follow.

The decision-making system established by Municipal Systems Act (South Africa 2005) and SPLUMA assert the importance of plan-led governance, specifically the important role of the spatial plan, obliging government to be informed by and consult its own plans when making decisions. SDFs were used by local governments prior to the advent of SPLUMA, but SPLUMA has considerably strengthened their role in law as instruments for inter-governmental coordination, integration and making the necessary trade-offs in space, as well as better institutionalising their role in local government decision-making processes.

Current, unfolding practices in the use of SDFs in Cape Town and the Western Cape will be discussed using Mastop's conditions, to consider the effectiveness of spatial plans from Faludi's perspective of the plan as a tool for sense-making, learning, and influencing decisions.

14.5 Do SDFS Make Clear the Decisions They Intend to Support?

One might argue that SPLUMA has dealt with Mastop's first condition for the effectiveness of spatial plans, in that it makes clear the decisions that an SDF must guide, in relation to sectoral planning, investment (capital expenditure), land development and land use management decisions. It is useful however to unpack this further, to assess how SDFs are practically presenting in decision-making processes. There is evidence in the Western Cape and nationally of maturing systems that proactively assert the importance of spatial plans in decision-making processes central to plan implementation.

14.5.1 Implementation of SDFs Through Development Permission Decisions

SPLUMA ties more closely together historically parallel processes of spatial planning and land use management (the development permissions process) to support a policy-driven land use management system. It recognises that land use decisions are the day-to-day tool for implementing spatial plans, as the formal city gets built and re-built through individual development permissions. SPLUMA requires that municipal SDFs determine the purpose, desired impact and structure of the land use scheme. Land use schemes must give effect to MSDFs and include land use and development provisions to promote the effective implementation of the SDF and designate the areas to benefit from such provisions (South Africa 2013).

Municipalities have been slow to implement this requirement for tighter integration but there are a growing number of examples that in time can be evaluated for their impact on the transformation of the built environment. The City of Cape Town, for example, has identified transit-accessible precincts, aligned to its SDF sub-strategy of encouraging integrated settlement patterns, and has included these in its Development Management Scheme (zoning scheme) as an overlay zone.[3] The overlay zone for these precincts around public transport interchanges and routes (primarily train and bus rapid transit stations) allows for reduced parking ratios, in support of the SDF's transit-oriented densification policies (City of Cape Town 2023, b, c). A further example is a priority area for redevelopment identified in the Stellenbosch Municipality's SDF (2023). This area is referred to as the Adam Tas Corridor adjacent to the town centre running alongside the railway line. This corridor now benefits from an overlay zone in its Municipal Planning By-Law (Stellenbosch Municipality 2022), linking the allocation of additional or changed land use rights primarily to the provision of affordable housing supporting the SDF's objective of higher density, inclusive infill housing for this area (Stellenbosch Municipality 2019). The Mossel Bay Municipality also introduced an overlay zone to "*meet the spatial policies of mixed uses, densification, urban renewal, job creating and to increase access to healthcare …*" that permits additional uses to promote the development of a medical precinct (Mossel Bay Municipality 2022). The 2023 City of Cape Town SDF proposes a range of overlay zones to direct development in support of spatial transformation (City of Cape Town 2023, b). Such measures present the opportunity to incentivise developers to follow the plan by offering significant reductions in risk, cost, and time through regulatory streamlining. This is important because land use rights hold significant value as an instrument, within the control of municipalities, to proactively lever spatial change in urban development patterns in cities with strong land markets, to the mutual benefit of public good and private sector interests (National Treasury 2017).

SPLUMA further operationalises this integration in the case of land use management decisions, by providing that Municipal Planning Tribunals may only make decisions that depart from the plan on site specific circumstances, necessitating a motivation with reference to the plan. Thus, responding to Wallagh's second situation—departures from the plan are deliberate in relation to the plan. Otherwise, proponents of development that does not align to the SDF must engage with the process of reviewing, amending, or drafting a new SDF, to state their case for a change to be made to the plan in their favour. This case is thus put into the realm of public scrutiny and debate, forcing the municipality to consider the implications of

[3] Overlay zones are a mechanism to apply additional development management rules additional to a base zoning, to deal with specific concern policy objectives. Such rules can be restrictive or permissive aiding or incentivising the implementation of planning objectives identified in municipal plans (City of Cape Town 2015).

this proposed change to the plan, and related implementation tools such as transport and infrastructure plans, and the 10-year municipal capital expenditure programme (the Capital Expenditure Framework). This creates transparency in the system of agreeing to a plan and then diverging from it. This is important because the SDF sets out a long-term vision and maps out the plan to get there, linked to the IDP. If the SDF is a shared vision for the spatial future of the city, the implications of changing course or emphasis reverberate across sector planning and are a matter of public interest and accountability. A decision to approve a development proposal that diverges significantly from the SDF may no longer be made without transparent and thorough consideration of the consequences of doing so, hence responding to Wallagh's third condition—the decision to not conform with the plan is made within the ambit of the plan.

14.5.2 Implementation of the SDF Through Budget Allocation Decisions

SPLUMA also pursues a tighter relationship between spatial and capital expenditure planning decisions by requiring that a municipal SDF determine a Capital Expenditure Framework (CEF) for the municipality's development programmes, depicted spatially and to be developed within the context of a municipality's long term financial planning (Department: Cooperative Governance n.d.).[4] While the incorporation of CEFs into SDFs is presently an incremental effort,[5] both in terms of take up, the quality of content, and the use of the CEF to frame annual budgeting processes, the articulation of a prioritised 10-year programme of infrastructure investment required to implement the plan becomes an important tool for monitoring expenditure prioritisation against the plan. The National as well as Provincial Treasuries, in their oversight function, have asserted the importance of plan-led budgeting and specifically, the role of the SDF in expenditure planning and decision-making (National Treasury 2023).

It is useful to consider how influential departments such as the Treasuries are asserting the role of SDFs in relevant decision-making, supporting legislative intent. National Treasury prioritised support to metropolitan cities, to develop Built Environment Performance Plans (BEPPs) (in essence the same as a CEF), to

[4] While such a tool has existed in South African national guidelines in various guises including the MSA, limited attention was paid by municipalities until SPLUMA made this a legislative requirement of the SDF.

[5] To complement National Treasury's support to metropolitan city governments, the Department of Cooperative Governance & Traditional Affairs (COGTA) ran a programme to support the development of CEFs for intermediate cities. The Western Cape Government has been running an in-house support programme for other municipalities and more recently, COGTA and the Department of Rural Development & Land Reform have been looking to provide additional support to municipalities.

promote better spatial and budget planning alignment. This was incentivised by making BEPPs a prerequisite for certain conditional grants, administered by the national government. These included grants for infrastructure (including public transport and electricity), human settlement, land preparation and precinct development.[6] BEPPs had the explicit intention of more actively pursuing urban transformation by supporting municipalities to bridge the implementation gap between SDFs and budgets and improve leverage through better spatial targeting of grants. Their purpose was "to clarify [metropolitan municipal] spatial and development planning visions and assist them to initiate practical pipelining and preparation of investment programmes and regulatory reforms to progressively and measurably realise this vision" (National Treasury 2017). BEPPs required metropolitan municipalities to account for their spatial distribution of the budget in terms of defined spatial categories, as presented in the SDF. The City of Cape Town has advanced with its successor to the BEPP, its Infrastructure Plan and supporting Infrastructure Planning and Delivery Framework, "an administrative tool that enables the implementation and delivery of Council-approved plans and strategies, guided by the IDP and MSDF" (City of Cape Town 2021).

Municipal governance processes have gone a long way to institutionalise the spatial plan into decision-making processes. Similarly, national directives assert the role of all the spheres' SDFs in the broader national and provincial government planning system, including financial planning processes (South Africa n.d.). If 'on the ground' observations suggest to some that this has been of little effect, it has certainly prompted public sector land use, financial and sectoral planners to talk to one another, to understand one another's operating environment (the spatial informants to sectoral investment decision, for example), supporting a better quality, inter-disciplinary conversation about spatial coordination and integration, which may arguably lead to better plans—perhaps a step in the right direction. This is important when considering Mastop's next condition for an SDF to be considered effective—whether it is of continuing relevance to an evolving context?

14.6 Are SDFS of Continuing Relevance to the Situation as It Unfolds?

The processes discussed above that seek to close the gap between the plan and its implementation through tighter linkages to sectoral and long-term financial planning present further opportunity to 'reality-check' SDF proposals. Capital expenditure frameworks consider SDFs' proposals against a municipality's funding envelope, for example. There are a further two instances in the current spatial

[6] BEPPs were required of metropolitan municipalities from 2014/15 to 2019/20. Considered an interim measure to promote early adoption of Capital Expenditure Frameworks, and an ambitious version of such, they have since given way to the use of Capital Expenditure Frameworks, which in the case of Cape Town is presented as the Infrastructure Report (City of Cape Town 2022)

planning governance system that allow for testing the ongoing relevance of an SDF to decision-making systems.

14.6.1 Monitoring as a Test of the Relevance of an SDF

The link between the SDF and the IDP is important in South African legislation[7]—SDFs spatially locate future settlement growth and guide land use decision-making accordingly. This informs infrastructure planning and in turn, the municipality's capital and operating budget. Municipalities' performance in implementing IDPs must be reviewed annually and as such performance in implementing the SDF must also be considered. It is important to keep track of whether the day-to-day decisions that build the city are 'seeing through' the plan and at what pace. The provincial government of the Western Cape (Western Cape Government 2023) has asserted the importance of including the SDF in the annual IDP performance review process. It emphasises the value of keeping the implementation of the SDF front of mind for leading officials as annual corporate service delivery and budget planning processes commence and using this process to build evidence and reflect on how the day-to-day decision-making is in fact being guided by the SDF or not —whether it be in planning permissions associated with development applications, budget allocations or the implementation programmes of sectors. This data, and the insight it offers, creates public transparency and accountability to the plan. If the SDF is a social compact, how well is local government doing in its part to fulfil this compact? The 2023 Spatial Trends Report published by the City of Cape Town is an exemplar of emergent practice, presenting data on development patterns and interpreting these in relation to corresponding SDF policies (City of Cape Town 2023a, b, c). At the same time, this is also an opportunity to identify emerging changing circumstances that planners become aware of and if decisions are persistently diverting from the plan or seeking changes to be made to it, the implication will be that the plan, or aspects of it, may not be of relevance and requires review.

14.6.2 Review as a Test of Relevance of the SDF

As a component of an IDP, the provincial Minister responsible for local government oversight must assess the SDF for alignment with the plans of other spheres of government and whether the concerns of other spheres of government have been duly considered in the local government's finalisation of the plan. If not, the Minister may request an adjustment to the SDF, and by implication, the IDP. As an example,

---

[7] The MSA identifies the SDF as a component of the IDP. In the Western Cape, SDFs, and their amendments, are approved under the MSA and SPLUMA.

this provision has been invoked recently in the Bitou (Plettenberg Bay) Municipality's IDP approval process based on a disagreement between the provincial and local government on the delineation of the urban edge (boundary to limit urban growth) in the SDF approved by the Municipality. The provincial government was of the view that the revision to the urban edge to accommodate an additional 780 ha of land for development, over and above an existing 936 ha of land identified for new development within the original urban edge, was inconsistent with the SDFs own intent and that of the provincial SDF. The law provides for dispute resolution through an ad hoc Committee consisting of representatives of all three spheres of government. The process did not go so far as to require the Minister to appoint such a committee, as the Municipal Council agreed to adjust the SDF to return to the original, more limited urban edge. This followed substantive and political engagement on planning alignment between national, provincial, and municipal SDFs, the credibility of the planning rationale, the assumptions underlying the accommodation of such extensive additional land, and the gaps in the SDF itself that fall short of properly considering the implications on urban systems of the development of this additional land. Here, the plan itself (be it the municipal SDF or the provincial SDF, which advocates for urban compaction) created the basis for understanding the consequences of a decision that did not conform to the plan. The level of engagement to resolve the disagreement suggests a process that tests the relevance of the plan (and the acceptance of the plan)—either that of the provincial government of the local sphere. Either way, it tested the relevance of the planning argument for compaction.

Using Wallagh's situation as a basis for relevance or Faludi's "regenerative capacity of the plan", where the plan in need of review takes the original plan as a point of departure, the continuity of spatial strategies across several iterations of SDFs, in Cape Town for example, provides a case in point. There has been continuity across successive Cape Town SDF's, over the last 30 years, to promote a multi-nodal city, with economic investment in nodes closer to townships; enhanced public transport to support the integration of townships and residents into the wider city economy; and mixed land uses and densification to improve the viability of a mass-based public transport system and better reap the economic dividend of urban growth. Cape Town's SDFs have persistently advocated for the proactive development of opportunities for socio-spatial redress through infill development of affordable housing opportunities, identifying several large, well located state-owned land parcels that should be redeveloped. Through promoting infill, densification, and the use of an urban edge in the 2012 and 2023 SDFs and the identification of "discouraged growth areas" in the 2018 SDF, they promote compaction (City of Cape Town 2012a, 2018, 2023b).

However, it might also be argued that the persistent lack of visible transformation despite this continuity in planning strategies, suggests that rather than being regenerative, plans are stuck in traditions that may not be relevant. There are weaknesses in the plan-making which du Plessis (2013) usefully unpacks, associated with infrastructure, indicators, practical proposals, and other aspects. Perhaps the biggest threat to the relevance of SDFs is their struggle to grasp how to engage as the spatial plan with the reality of a divided city—the formal, less formal and increasing informality of growth and development in our urban centres. Plan-making is rooted in formal, western traditions of data gathering and interpretation, land use management, funding mechanisms and sector planning and delivery systems (Harrison et al. 2007). What an SDF can do to support decision-makers to engage with growing informality in settlement and the economy and with systems that govern informal practices requires a lot of deliberation, as well as different forms of gathering data and insights (du Plessis 2013). Insights that require plan-making processes that enable different forms of participation. Lack of attention to this may be the biggest threat to the relevance of SDFs going forward, but the problem sits within the broader systems of governance too. These are not designed or capacitated to work with more informal practices of engagement and organisation.

Processes of public scrutiny and political approval are the ultimate test of relevance for an SDF. This brings us to Mastop's third condition, whether the SDF is accepted as the spatial plan and is assisting decision-makers to understand the situation within which they must decide and the consequences of their decision, either way.

14.7 Is the Role of the SDF in Defining the Situation Within Which a Decision Must Be Made, Accepted?

Once approved, SPLUMA insists that decision-makers have little choice as to whether to accept the SDF as part of their decision-making. Those deciding on land use matters might be determined to make certain decisions regardless of the plan but in doing so must present an argument that motivates departure from the plan, for the decision to be rationale and limit legal challenge, bringing the decision into the ambit of the plan (Wallagh 1988). Alternatively, the plan must be amended, and due process followed, including public scrutiny, to secure such an alteration to the plan.

Wallagh's situation where the plan is providing the basis for understanding the consequence of a decision to depart from it (Wallagh 1988) can be seen in the record of decisions for two of perhaps the most significant departures from the spatial plan in Cape Town in the post-apartheid era, in scale and controversy. Namely,

(i) The Council's decision to change the SDF in November 2012, less than a year after its adoption, to pave the way for a proposed leapfrog development of 3101 ha, estimated to house a population of around 400,000 people, roughly equating, at the time, to the size of another small South African (capital) city, Bloemfontein—the proposed 'Wescape' development[8] (City of Cape Town 2012a, b, c); and

[8]Wescape was a proposed development along the West Coast of metropolitan Cape Town, 17 km beyond the existing built footprint of Cape Town and 1 km south of Atlantis, within the nuclear power station's evacuation zone. The proposal sought an amendment to the Cape Town SDF to

(ii) The second of two decisions to convert 281 ha of horticultural land to urban use in the Philippi Horticultural Area (PHA)[9] in the south-eastern plain of the city.

In both instances, assessments of these applications to depart from the SDF drew extensively on the evidence in the SDF, setting out the respective consequences of decisions that might go in their favour. In the case of the approval to amend the SDF to pave the way for the Wescape development, the decision's conditions drew heavily on these arguments to mitigate the risks identified—suggesting a level of acceptance of the SDF as a supporting framework for decision-making.

Ultimately, in the instance of the Wescape proposal, the alteration of the 2012 SDF was reversed in the 2018 Cape Town SDF (City of Cape Town 2012a, 2018). In the case of the PHA, neither of the proposals have materialised to date as intended. While the tussle between the protection of the PHA as a horticultural area (and its consequent importance to the city for food security and low-skilled employment) and the use of this land for housing and other real estate development has been ongoing, the SDF and supporting sub-metropolitan plans, approved by the Council, have consistently set out the unique attributes of this area, informing decision-makers of the opportunity cost of losing the PHA as a productive food growing area. This has prompted committees of Council, tasked with these deliberations, to request further input and independent expert opinion. The SDF, in this instance, has been serving its purpose as a learning framework and maintaining its relevance—bringing the decision into the "ambit of the plan" (Wallagh 1988), albeit decisions did not conform to the plan.

14.7.1 The Role of the SDF in Defining the Decisions of Others

The earlier example of inter-governmental engagement on plans for urban growth in Plettenberg Bay was an opportunity to test the relevance of both the provincial and local spatial planning policies. The persistent engagement, within defined processes, to resolve the matter suggests an acceptance of the SDF as the planning tool to mediate an agreement on the direction development must take between the two

---

include this area within the urban edge, paving the way for further land use and environmental authorisation applications.

[9] The Philippi Horticultural Area is a substantial farming area, established in the 1970s in the south-eastern part of Cape Town, surrounded by urban development, most of which is middle to low-income residential neighborhoods. It enjoys unique geological, hydrological and climatic conditions that are conducive to vegetable growing. In 2011 an application to amend the then plan to convert agriculturally designated land in the south-eastern corner for mixed use development was approved. In December 2012, the City of Cape Town's Mayoral Committee recommended a refusal, of a second application to transform 281 ha of land designated for agricultural use in the south-western corner of the PHA to urban development, through an amendment to its IDP (and the Cape Town SDF) (City of Cape Town 2012b 2012a, b, c).

spheres of government. These were two authorities with spatial planning responsibilities. Having noted the strategic importance of other sectoral arms of government to implement SDFs, a key question is whether SDFs assist them to define the situation within which they must make a decision? Notwithstanding SDFs' responsibility to reflect and reconcile all of government's programmes at the relevant national, provincial or municipal scale into a coherent spatial plan that guides planning and development decisions across all sectors of government (South Africa 2013), their role in influencing planning and investment decisions outside of where government must seek development permissions from municipalities is limited.

Much of the potential leverage to make a substantial difference in the spatial performance of cities and towns lies with national government entities responsible for transport systems, land acquisition and release, and housing subsidisation. Their decisions are made in board meetings, in which the local government plays no direct part. These include planning and investment in the port and port precinct (for coastal cities), freight rail and associated inland ports (Transnet), passenger rail (PRASA), major freeways (SANRAL) and the use of extensive tracts of well-located, underutilised state land in national government ownership (for example, Transnet, Department of Defence, Department of Public Works, Housing Development Agency). These decisions are more likely to be led by the real-politick of national resource distribution than bottom-up plans. In the author's experience of 15 years (2008–2023) of local and provincial government engagement with these entities, aimed at facilitating an adequate response to Cape Town's needs and partnering to realise potentially catalytic changes, in almost all instances, while only revealed later, these entities were amid serious maladministration (Stent 2022). At the time, frequent senior staff losses, changes, and organisational restructuring proved difficult to work with. This hampered their capability to support credible SDF proposals by effectively and honestly communicating their investment prioritisation criteria, their fiscal state and funding envelopes, their investment and asset management programmes and be influenced by SDFs in determining their implementation commitments. Two iterations of Cape Town SDFs have been betrayed by poor performance on the part of PRASA, making transit-oriented development policies supporting the rail as the mass transit backbone, difficult to realise. Aside from the very real challenge of opposition politics, other municipalities may face similar challenges to varying degrees, depending on their own context but also on their location.

Palmer et al. (2017) extensively unpack the complexity of the multi-scalar nature of the South African inter-governmental system including the contested terrain of the distribution of powers and functions related to urban infrastructure, state land and housing subsidies. To complicate matters further, regardless of the systems of spatial plans set up by SPLUMA, national government departments, outside of those with direct spatial planning mandates, have repeatedly imposed their own municipal scale spatial plans and spatial designations which municipalities must adopt if they want to benefit from conditional grants. These therefore define decisions around resource allocation, as opposed to the SDF itself. Perhaps this suggests that national government does not accept SDFs as the plan for a particular municipal area, although the linkages to SDFs, are sometimes, but not always, paid lip service to. Alternatively, it may suggest that national government is not convinced that the SDFs are adequately defining the situation for decision-making when it comes to pursuing a spatially targeted agenda.[10] This is clumsy terrain for spatial planners to navigate while maintaining a clear framework for supporting decision-making.

Examples of nationally imposed municipal-scale plans which seek to do what is already laid out in the municipal planning responsibilities, specifically the requirements of SDFs set out in SPLUMA, include:

- The Long Term One Plans, "a strategic framework to guide investment, service delivery and development in relation to district and metropolitan spaces" (not dissimilar from the role of the District Municipal IDP, SDF and CEF) expressing the inter-governmental work of the District Development Model, coordinated by the national Department of Cooperative Governance and Traditional Affairs (Department of Cooperative Governance 2020).
- The above-mentioned Built Environment Performance Plans[11] required by National Treasury of metropolitan local governments.
- Development plans commissioned by the Housing Development Agency for Priority Human Settlement Development Areas (PHSHDAs) gazetted by the National Department of Human Settlements. The purpose of designating these areas is to "ensure that the delivery of housing is used to restructure and revitalise towns and cities to overcome apartheid spatial patterns by fostering integrated urban forms" (Ministry of Human Settlement 2020). In many instances whole towns are designated as PHSHDAs. Development plans for them cannot help but to cover the same ground as SDFs.

Several of these plans impose their own spatial targeting criteria and spatial designations that must be delineated by municipalities. Without doing so, municipalities cannot apply for linked conditional grants (Table 14.1).

To assert their role as the spatial expression of all of government's plans and programmes within a municipality, SDFs must now incorporate and rationalise all these top-down spatial designations and reconcile them into a coherent spatial concept and strategy for the municipality, along with spatially articulating local needs and aspirations identified in the Integrated Development Plan and arising from the public participation process. These multiple layers of planning and myriads of spatial categories with overlapping objectives clearly create confusion and challenge the ability of the SDF to meet Mastop's three conditions for effective strategic plans. It also creates tension between national government and municipalities, as national government becomes the arbiter of what can be done and where it can be done in a

[10] Or it may be, as the recently published Ten Year Review of the National Development Plan 2012–2022 (National Planning Commission 2023) suggests, more a case of a lack of role clarity and discipline in national government in respect of spatial planning.

[11] Now discontinued in recognition of the overlap between BEPPs and SPLUMA's requirement for Capital Expenditure Frameworks required as part of MSDFs.

Table 14.1 Nationally imposed municipal-scale spatial planning designations and associated grants

Spatial designation	Spatial Planning mechanism	Grant	Responsible national government department
Restructuring Zones		Social Housing related Grants	National Department of Human Settlements
Priority Human Settlement Development Areas	Development Plans	Human Settlements Grants	National Department of Human Settlements
Integration Zones	Built Environmental Performance Plans (Metros only—discontinued)	Integrated Cities Development Grant	National Treasury
Urban Network Strategy components	Urban Network Strategy	Neighbourhood Development Programme rant	National Treasury

municipal area; and perverse consequences arise as spatial designations draw lines defining what is and isn't possible in and outside of the line to the detriment of overall policy objectives.

14.8 Conclusion

This chapter has proposed that assessing the role an SDF plays in the decision-making processes situated between the plan and its implementation is critical to evaluating SDF performance, supported by Faludi's argument that such plans cannot dictate the outcome of such processes. In the context of a dynamic, evolving and complex territorial governance system, and the incremental nature of spatial change enabled by many decisions in different spaces, are SDFs effectively serving decision-making processes? Are they adequately explaining the situation within which decisions need to be made and the consequences of decisions? Are they maintaining relevance in an evolving context, and fundamentally, are they accepted as decision-making support tools?

Current practices, primarily in the Western Cape, have been drawn on to cite examples of how SDFs are used in practice in terms of these key questions. The discussion suggests that there have been significant transitions post-SPLUMA to embed the role of the plan in governance systems and hold decision-making practices accountable to implementation of the spatial plan. SPLUMA positions SDFs centrally within the South African territorial governance framework as strategic plans that must coordinate, align, and mediate activities of spheres and sectors within and outside of government. It clearly sets out the nature of decisions an SDF is required to support. This is increasingly complemented by oversight processes putting emphasis on the role of SDFs in provincial and municipal budget planning processes. SDFs are increasingly playing a "tangible role in the choices of the actors to whom it is addressed" (Faludi 2000), even if the decisions are not always conforming to the plan, they are catalysing conversations needed to achieve alignment and identifying consequences such that divergent decisions set out mitigating conditions.

Critics of the spatial plans and planners might indeed question if the plans are of continuing relevance as a learning framework if the problems which it seeks to address remain stubbornly persistent. The role of SDFs is perhaps underplayed in that it is more effectively guiding incremental building processes and preventing undesirable development, than it is unlocking catalytic change. Which is indeed needed if urban spatial transformation it to be a reality. There are however various instances where SDFs are being taken forward through the development of further supporting decision-making tools, such as proactive changes to the rules governing land use rights in spatial priority areas.

The parallel decision-making environment of urban system functions key to local spatial transformation, yet centralised at the national level, is a challenge that links directly to where implementation falls short of the plan, and where the role of the plan in those decision-making spaces is weak. Drawing attention to these factors is important if we are going to target scrutiny towards turning around the inertia that hinders significant change, despite the best laid out plans—rather than play into the forces for whom it would be convenient to dismiss the credibility or point of spatial plans in and of themselves.

References

City of Cape Town (2012a) Cape Town Spatial Development Framework

City of Cape Town (2012b) Proposed amendment of the Cape Town Spatial Development Framework Applicable to Land Units in Philippi, Schaapkraal (in terms of the Municipal Systems Act No 32 of 2000 and the Land Use Planning Ordinance 15 of 1985)

City of Cape Town (2012c) Proposed amendment of the Cape Town Spatial Development Framework for Remainder Portion 1 (Klipvalley) of Cape Farm 41, Remainder Cape Farm (Lange Rug) 36, Cape Farm (Lange Rug) 37, Cape Farm (Brakkekuyl) 38 & 39; Cape Farm 1244, Cape Farm 1509, Cape Farm

City of Cape Town (2015) Municipal Planning By-Law, as amended

City of Cape Town (2018) Municipal Spatial Development Framework

City of Cape Town (2021) Infrastructure Planning and Delivery Framework

City of Cape Town (2022) Infrastructure report

City of Cape Town (2023a) Comprehensive Integrated Transport Plan

City of Cape Town (2023b) Municipal Spatial Development Framework

City of Cape Town (2023c) Spatial trends report

Denoon-Stevens SP, Andres L, Nel V, Jones P (2022) Unpacking planner's views of the success and failure of post-apartheid South Africa. Cities 130(103867)

Department of Cooperative Governance (2020) District development model: DDM implementation framework: annexure 3: one plan content guide

Department: Cooperative Governance (n.d.) Guidelines on capital expenditure framework for intermediate city municipalities

DPME (2015) Revised framework for strategic plans and annual performance plans
du Plessis D (2013) A critical reflection on urban spatial planning practices and outcomes in post-apartheid South Africa. Urban Forum 25(1):69–88. https://doi.org/10.1007/s12132-013-9201-5
Du Plessis, DJ 2019. The impact of spatial planning on the structure of South African cities since 1994. Doctoral dissertation. Stellenbosch: Stellenbosch University.
Faludi A (2000) The performance of spatial planning. Plan Pract Res 15(4):299–318. https://doi.org/10.1080/713691907
Harrison P, Todes A, Watson V (2007) Planning and transformation learning from the post-apartheid experience. Routledge
Laurian L, Day M, Berke P, Ericksen N, Backhurst M, Crawford J & Dixon J 2004. Evaluating Plan Implementation. A Conformance-Based Methodology. Journal of the American Planning Association 70(4):471-480.
Mastop H & Faludi A 1997. Evaluation of strategic plans: The performance principle. Environment and Planning B: Planning and Design 24(6):815-832.
Mastop JM (1987) Besluitvorming, handelen en normeren: Een emthodologische studie naaraanleiding van het streekplanwerk. Planologische studies 4, 2nd edn. Planologisch en Demografisch Instituut, Universitet van Amsterdam
Ministry of Human Settlement (2020) Government gazette volume No 43316, No 526: declaration of the priority human settlements and housing development areas (PHSHDAs)
Moodley S (2019) Why do planners think that planning has failed post-apartheid? The case of eThekwini Municipality, Durban, South Africa. Urban Forum 30:307–323. https://doi.org/10.1007/s12132-018-9357-0
Mossel Bay Municipality (2022) Mossel bay municipality: zoning scheme by-law
National Planning Commission (2023) Ten year review of the national development plan 2012–2022
National Treasury (2017) Supplementary guidance note for the built environment performance plan (BEPP) 2018/19–2020/21
National Treasury (2023) Addendum 5 to MFMA circular No. 88
Olver C (2019) A house divided. Jonathan Ball
Palmer I, Moodley N, Parnell S (2017) Building a capable state service deliver in post-apartheid South Africa, 1st edn. UCT Press, Cape Town
Talen E 1996. Do plans get implemented? A review of evaluation in planning. Journal of Planning Literature 10(3): 248-259.
Tian L & Shen T 2011. Evaluation of plan implementation in the transitional China: A case of Guangzhou city master plan. Cities 28:11-27.
Seasons M 2003. Monitoring and evaluation in municipal planning. Considering the realities. Journal of the American Planning Association, 69(4):430-440.
South Africa (2005) Act No. 32 of 2000: Municipal Systems Act, 2000 (as amended)
South Africa (2013) Act No.16 of 2013: Spatial Planning and Land Use Management Act, 2013
South Africa (n.d.) Revised Framework for Strategic Plans and Annual Performance Plans
Stellenbosch Municipality (2019) Stellenbosch Municipality Spatial Development Framework
Stellenbosch Municipality (2022) Adam Tas Corridor Overlay Zone. https://stellenbosch.gov.za/download/adam-tas-corridor-overlay-zone/. Accessed 3 Oct 2023
Stent J (2022) These graphics show PRASA's disastrous state. GroundUp
Turok I (2016) Planning for spatial transformation. In: Western Cape Government, land use planning and law reform in the Western Cape. Western Cape Government, pp 21–27
Wallagh G (1988) Tussen wens en werking. Masters thesis. University of Amsterdam, Amsterdam
Western Cape Government (2023) Practice note on the annual review of the implementation of municipal spatial development frameworks
National Treasury (2017) Land-based financing tools to support urban development in South Africa
National Treasury (n.d.) Cities support programme. https://csp.treasury.gov.za/ . Accessed 1 Jul 2024

Chapter 15
Smart Cities in South Africa—Can They Contribute to Spatial Transformation and Support Territorial Governance?

Engela Petzer, Tinus Kruger, Nosizo Sebake, Antony K Cooper, Matšubu Ragoasha, and Mark Napier

Abstract The undoing of the inherited segregated spatial settlement form has proved to be a major challenge, despite the implementation of numerous, deliberate interventions by the South African government over close to three decades. Municipalities, the territorial governance structures closest to citizens and mandated to lead local development, have employed various planning concepts with the aim of effecting spatial transformation. This chapter reflects on the concept of smart cities, including its international origins in the early 1990s and the lack of a collective understanding and varied definitions of the term. The chapter further reflects on what a smart approach to spatial transformation could look like. Lessons that can be learned from local South African smart initiatives are discussed. The chapter concludes that the impact of current South African municipal smart city initiatives on spatial transformation is not immediately apparent; however, when integrated into

---

E. Petzer (✉)
Council for Scientific and Industrial Research, Pretoria, South Africa

Department of Town and Regional Planning, University of Pretoria, Pretoria, South Africa
e-mail: epetzer@csir.co.za

T. Kruger · N. Sebake M. Ragoasha
Council for Scientific and Industrial Research, Pretoria, South Africa
e-mail: ssebake@csir.co.za; mragoasha@csir.co.za

A. K. Cooper
Council for Scientific and Industrial Research, Pretoria, South Africa

Department of Geography, Geoinformatics and Meteorology, University of Pretoria, Pretoria, South Africa
e-mail: acooper@csir.co.za

M. Napier
Council for Scientific and Industrial Research, Pretoria, South Africa

Faculty of Economic and Management Sciences, University of the Free State, Bloemfontein, South Africa
e-mail: mnapier@csir.co.za

D. du Plessis (ed.), *Territorial Governance and Spatial Transformation in Post-apartheid South Africa*, Local and Urban Governance,
https://doi.org/10.1007/978-3-031-85489-7_15

the system of territorial governance—which aims to unify a place-based approach and multi-level governance—the smart city concept may contribute to spatial transformation over the long term, given an appropriate implementation approach.

Keywords Smart cities · Spatial transformation · Territorial governance · Place-based approach · Multi-level governance

15.1 Introduction

While there are many examples of smart cities internationally, the implementation of the concept of a smart city is relatively new in South Africa. However, more and more municipalities are exploring ways of becoming smarter, or are implementing smart initiatives. Many stakeholders, including officials, politicians, communities, technology companies and other service providers, put pressure on municipalities to introduce smart technologies to improve efficiency and address service delivery challenges.

In addition to the need to improve service delivery, South Africa must also contend with the inequitable, inefficient, and unsustainable settlement forms stemming from the country's past (Turok 2013). In response, various plans, frameworks, and strategies, including the National Development Plan 2030 (South Africa 2013a), the Integrated Urban Development Framework (CoGTA 2016), and the National Spatial Development Framework (NSDF) (DALRRD 2023) place a high priority on spatial transformation. The NSDF focuses specifically on spatial transformation and inclusive growth as essential for realising the National Spatial Development Vision of all people "…living in shared and transformed places in an integrated, sustainable and competitive national space economy" (DALRRD 2023: 86).

There seems to be general agreement that progress in moving away from the key characteristics of spatial and societal segregation of cities, towns and neighbourhoods has been insufficient (Joseph 2013; Turok 2013, 2018; Parnell and Crankshaw 2013; Hall 2014; Ndifuna Ukwazi 2018). Despite some improvements, the impact of post-94 spatial planning on the sprawl, separation and fragmentation of the South African settlement has been limited (Phuhlisani NPC 2017). The purpose of this chapter is to explore the possible contribution of smart cities to bringing about spatial transformation in South African settlements. Potential contributions will be viewed against the backdrop of territorial governance.

The chapter commences with a brief background to the smart cities' phenomenon, including a summary of definitions and interpretations. This is followed by a discussion of a South African interpretation of smart cities given the local context. In the next section the potential linkages between smart city initiatives, spatial transformation, and territorial governance are addressed. The chapter concludes with an attempt to answer the question whether smart city initiatives could contribute to spatial transformation and to supporting territorial governance.

15.2 Background to Smart Cities

15.2.1 An International Perspective

More than half of the world's population lives in urban areas (Chourabi et al. 2012). By 2050, 2.5 billion more people will live in cities and almost 90% of those people will be in either Africa or Asia (Henreckson 2018). Urbanisation is a necessary catalyst for economic prosperity (Turok and McGranahan 2013; Turok and Borel-Saladin 2014). However, urbanisation also has negative consequences due to its far-reaching social, economic, and environmental implications (Turok and Borel-Saladin 2014). Mosannenzadeh and Vettorato (2014) suggest that the implementation of the concept of smart cities not only responds to urban challenges like rapid urbanisation and climate change, but that it has also exploited new opportunities, such as the growing advancements in information and communication technologies (ICT). Angelidou (2014) concurs, adding that smart cities characterise a conceptual urban development model that is centred (amongst others) on the application of human and technological capital to enhance the development and prosperity of urban settlements.

In the early 1990s, municipalities started to use ICT infrastructure and e-governance to appeal to high-tech industries and promote economic growth. These cities referred to themselves as "smart" (Petzer and Kruger 2020). The improved development practices advocated by the North American "smart growth" theories of the 1990s have also been linked to the origins of the smart city concept (Hollands 2008). The smart growth movement was specifically focused on a "new paradigm of intelligent urban development" (Supangkat et al. 2018).

The smart cities concept also spread to other parts of the world during this time. Examples include a proposal for a "City of the Future" in the Australian city of Adelaide (Winchester and Chalkley 1990), the 1997 re-planning of the two cities of Cyberjaya and Putrajaya in Malaysia as intelligent garden cities (Söderström et al. 2014) and Singapore's vision of becoming an intelligent island (Choo 1997).

In the early 2000s, major ICT companies started to enter the smart cities field (Cocchia 2014). Companies selling so-called "urban technologies" identified governments, and specifically municipalities, as an untapped market (Townsend 2013). Corporate market growth strategies were often supported by extensive marketing campaigns, the publication of guidelines, and the establishment of councils and coalitions aimed at promoting the smart city agenda (Petzer and Kruger 2020). Private sector involvement in smart cities include initiatives by IBM, Deloitte, CISCO, Huawei, the Smart Cities Council, and the World Economic Forum's G20 Global Smart Cities Alliance (Petzer and Kruger 2020).

15.2.2 Smart Cities Definitions and Characteristics

There are many descriptions and definitions of smart cities, but there does not seem to be a collective understanding of the concept. Hollands (2008) accused the smart cities movement of "definitional impreciseness", having "numerous unspoken assumptions" and being "self-congratulatory". However, Karuri-Sebina and Masango (2022) make a compelling argument for using "good enough" working definitions, as the smart city is not a static destination, and its application should always be context-specific. Aurigi and Odendaal (2021) warn against the "prevalently pre-packaged, product-like version of the smart city" and Balkaran (2019) argues that "each city is unique, has different problems and should address them with specific solutions."

While ICT is a recurring theme in most descriptions of smart cities, recent definitions reflect the growing realisation that "smart" technologies should not be prioritised at the expense of issues such as social inclusion and sustainability (Petzer and Kruger 2020). There are limits to what technology can accomplish. Poverty, inequality, weak institutions, and crumbling infrastructure cannot be solved by dashboards, sensors, and apps (Aggarwala et al. 2018). Salter (2022) makes a strong case for smart cities to be people-focused and argues that the smartness of our cities comes from the diverse human bodies of different genders, cultures, and classes whose rich, complex, and even fragile identities ultimately make the city what it is.

The United Nations specialised agency for information and communication technologies (ITU) presented a useful smart cities definition:

> "A smart sustainable city is an innovative city that uses information and communication technologies (ICTs) and other means to improve quality of life, efficiency of urban operation and services, and competitiveness, while ensuring that it meets the needs of present and future generations with respect to economic, social, environmental as well as cultural aspects." (ITU 2022)

As mentioned in "A Guide to Establishing a Municipal Smart City Initiative" (CSIR 2022), cities and towns involve a complex system of systems, and a smart city should not be viewed as an end product in a state of completion. A smart city is constantly evolving and is (hopefully) on the way to becoming ever smarter, rather than reaching the point of perfect smartness. Smart cities involve a combination of various smart components related to, for instance, infrastructure elements, or they could be different operational areas that are central to a city's functioning, such as the economy, people, governance, mobility, and the environment (Petzer and Kruger 2020).

A South African Smart Cities Framework, published in 2021 by the Department of Cooperative Governance (DCoG) and the Council for Scientific and Industrial Research (CSIR), suggests that a useful point of departure in understanding smart cities is to focus on the two terms, namely "smart" and "city". Both terms should not be defined in a narrow sense. The terms are described as follows in the framework document (DCoG, CSIR 2021):

"The term "smart" is generally associated with a range of technological and digital concepts and interventions, especially ICT. There also seems to be a particular focus on 4IR (Fourth Industrial Revolution) technologies. However, in addition to this technology-intensive interpretation, smart could also mean "intelligent", or "knowledge-intensive". The understanding of the term "technology" could be expanded to also include innovative approaches, techniques and processes, as well as non-conventional interventions and scientific innovation.

The word "city" has multiple meanings in the smart city conversation. It is a catch-all phrase that includes various types of settlements, or parts of settlements. It could refer to any of the following:

- Cities, towns and villages of any size, including those in rural locations.
- Municipalities (metropolitan, district, local).
- A custom-built greenfield development ("city") which may or may not be linked to an existing city.
- Large new precinct developments linked to an existing city (e.g. business parks).
- Upgrading or retrofitting aspects (e.g. transportation or connectivity) or parts (e.g. an educational precinct) of an existing city or town.
- New residential, commercial or mixed-use developments, such as privately developed gated communities."

15.3 Smart Cities in the South African Context

15.3.1 South African Realities

The Neighbourhood Planning and Design Guide (*Red Book*) (CSIR, DHS 2019: 17) states that "South African human settlements are inextricably linked to the country's socio-economic context. This means that poverty, unemployment, inequality, crime and violence and other challenges have an impact on the sustainability of cities and towns." Concerns about smart cities being a misplaced priority becomes pertinent as the country battles with these challenges (Musakwa and Mokoena 2017). The development of smart city initiatives in a time of financial austerity and weak local government therefore becomes a case of competing priorities and trade-offs. The reality is, smart cities initiatives are developed and implemented in a context where they are confronted with critical spatial, socio-economic and governance challenges, and if smart cities initiatives are to make a meaningful impact, they need to respond appropriately to this context ((Ragoasha 2020). Moreover, capacity and resource constraints of establishing smart cities, especially within local government, need to be well understood (Odendaal 2006).

According to the World Bank's Systematic Diagnostic Report of 2018, many South Africans are still excluded from the productive economy and remain locked in persistent poverty with extremely low chances of being fruitfully integrated into the labour market (Gardner 2018; Zizzamia et al. 2019). The argument as posited by the CSIR is that "certain technologies may not appropriately address the priority needs of the poor, particularly the chronic poor, due to the additional financial burden they may place on them" (Petzer and Kruger 2020).

Various challenges negatively affect municipal service delivery in South Africa, including capacity and resource constraints, corruption, and mismanagement due to "...political instability or interference, corruption and incompetence" (Brand 2018). As noted in the *Red Book* (CSIR, DHS 2019), many cities, towns and neighbourhoods struggle as a result of service level deficiencies, inadequate amenities, facilities, infrastructure and public spaces etc. In addition, the country also faces challenges related to informality, especially regarding housing and economic activities (CSIR, DHS 2019).

15.3.2 The Regulatory Environment

Smart city initiatives should be considered within the context of the South African regulatory environment. Considering the multifaceted challenges that exist in South Africa's cities, numerous efforts have been made to strengthen municipal planning and reform the regulatory planning system. A host of acts, policies, frameworks, and strategies emerged with a specific focus on guiding spatial planning and developing sustainable and inclusive human settlements. This objective is guided by a national vision of transforming human settlements as set out in the National Development Plan 2030 (NDP) (South Africa 2013a).

The NDP is complemented by, for instance, the National Spatial Development Framework (NSDF) (DALRRD 2023), the Integrated Urban Development Framework (IUDF) (CoGTA 2016), the Spatial Planning and Land Use Management Act (SPLUMA) (South Africa 2013b), Integrated Development Plans (IDPs), and a range of guidelines and toolkits that guide decision-making processes concerning urban and settlement development such as the *Red Book*, that guides the planning and design of integrated settlements and neighbourhoods (CSIR, DHS 2019).

While the NDP does not refer to the concept of smart cities directly, it does identify ICT as a critical enabler of economic activity and envisage it playing an increasingly significant role. Policy documents that support this notion include the Public Service Corporate Governance of ICT Policy (2012), Electronic Government: The digital future—a public service IT policy framework (2001), National Broadband Policy (2013), Cyber-security Policy Framework (2015), and the National Integrated ICT Policy White Paper (2016).

15.3.3 A South African Approach to Smart Cities

In the 2019 State of the Nation Address (SONA), President Cyril Ramaphosa (2019, 2020), expressed his dream of building a South African smart city. This was followed, in the 2020 SONA, with an announcement of more concrete plans to make the South African smart city dream a reality (Ramaphosa 2020). In their response to the 2019 SONA, Horber, for the Young Urbanists (2019), highlighted the risks of smart urban development, emphasising the potential to further entrench urban spatial segregation and inequality. Horber (2019) further cautioned that "the [President's] proposal to build a new city [took] away from where our energy and focus should be – on consolidating and optimising existing infrastructure and urban centres". Serfontein (2020) also questioned the viability of the South African smart city, given the country's need for "genuine solutions" to "very real problems". In line with Horber (2019), he suggests, that "the real Smart City can be found through Traditional Urbanism, with its liveable and sustainable places designed for the people".

Different national sector departments have addressed the topic of smart cities. For instance, The Department of Telecommunications and Postal Services (DTPS) published a *National Smart Communities Framework* in 2020. The DTPS framework aims to prepare municipalities to become data-driven for efficient and effective service delivery.

A South African Smart Cities Framework was published in 2021 by DCoG and the CSIR, to "guide decision-making and provide all role players with a structured approach to identifying, planning and implementing smart city initiatives that are appropriate to the local context" (DCoG, CSIR 2021: 5).

The implementation of smart city initiatives should be driven by both business and local government (SACN 2020) and Fataar (2020) suggests that a city may in its IDP partially articulate its vision of a smart city, then potentially support this in a separate strategy with more specific objectives and initiatives. Smart city interventions could involve various initiatives, including web services for citizens, businesses and other stakeholders (such as directing and tracking queries, managing municipal accounts, uploading meter readings, and paying municipal accounts, traffic fines and other charges), smart electricity and water meters, and internet connectivity for residents and organisations, particularly in the form of free Wi-Fi access points or hotspots and publicly accessible computers in libraries.

15.4 Spatial Transformation and Territorial Governance in South Africa

While smart cities have been touted as the answer to many ills, what contribution might the smart cities movement make to South Africa's spatial transformation, away from the inequalities and spatial patterns of exclusion of the past? The term "spatial transformation" is used in plans and policies in various parts of the world. It is mostly used to describe "major urban change or restructuring" (Turok 2014) over time. When used in these contexts, the phrase usually does not communicate the form that the change has taken or whether the change is regarded as generally positive or negative (Petzer et al. 2020). However, the term has a more specific meaning when used to describe South African cities and towns.

Transformation in South Africa generally refers to a process that will result in a "fairer balance of outcomes", where this balance pertains to fairness not only in the economic features of ownership, production, distribution and consumption, but also fairness in how these outcomes each have "racial, income and gender dimensions" (Gasa 2016). This transformation will ultimately facilitate access to a range of opportunities, leading to a more inclusive, participative and cohesive society (Gasa 2016).

A shared understanding of spatial transformation will have to acknowledge the broader transformation agenda (Petzer et al. 2020) as described above, but should also refer specifically to the undoing of the segregated spatial patterns inherited from colonial and Apartheid times (DALRRD 2023). The breakdown of Apartheid spatial patterns requires an in-depth understanding of the built environment and its role players and systems, acknowledging that attitudes and complex practices are deeply rooted (Joseph 2013). Furthermore, Maritz et al. (2016: 19) argue that "To contribute towards achieving spatial transformation objectives, and tracking progress in this regard, the need for the development, extension and use of spatially and temporally aligned sub-city level indicators is evident."

Although there is a large body of knowledge on what spatial transformation means in South Africa, it remains a complex phenomenon and a challenge to capture in a single definition. The NSDF (DALRRD 2023: 19–20) definition of spatial transformation focuses on the role of planning and infrastructure investment in achieving spatial transformation. The NSDF definition highlights aspects that should be "broken down" or "corrected" and does not pertinently propose what a transformed settlement should look like:

> "The carefully-planned and well-managed process of placing infrastructure, social services and economic activities in settlements in such a way that (1) the segregated spatial patterns inherited from colonial and Apartheid times are broken down, and (2) the inefficiencies, injustices and inequalities in access to opportunities resulting from these past patterns are corrected."

Another key concept that is influencing thinking about effective spatial interventions to achieve spatial transformation is the idea of territorial governance. Territorial governance is defined in many ways and its application depend on the context. It can be described as an approach to governance that involves place-based policy making. This means that policy making should be able to adapt to changing (territorial) contexts and should take local realities into account. It has a particular focus on cooperation and participation between a wide range of stakeholders.

Local government is arguably best placed to drive real change in the built environment that leads to spatial transformation (SACN 2016). As the government closest to the people, municipalities should be able to respond to residents' needs through local interventions. Municipalities are also able to observe changes that happen at a local level as a result of city-, province-, or country-wide interventions and respond to these in an appropriate manner—working with communities to harness or mitigate these changes.

15.5 A Smart Approach to Spatial Transformation

A South African Smart Cities Framework (SCF) (DCoG, CSIR 2021) promotes the development of smart cities that are inclusive, benefit all residents, and contribute to the well-being of the city as a whole. Six interdependent principles are identified that support inclusivity and provide guidance regarding the identification, planning and implementation of smart initiatives and technologies. These guiding principles provide an indication of the potential linkages between smart city initiatives, spatial transformation, and territorial governance. The principles can be summarised as follows (DCoG, CSIR 2021):

(i) **Inclusive smart cities are smart for all.**
Cities should implement a range of smart technologies and initiatives that would benefit the city as a whole and will not negatively affect some parts of the city or certain sectors of society. These technologies and initiatives should not exclude those with limited financial and other resources and with a limited understanding of technology.

(ii) **Inclusive smart cities use technology as an enabler rather than a driver.**
Technological innovation and adoption might improve service delivery and contribute to economic growth, but technology, specifically ICT, should not be seen as a panacea that could solve any problem. Technology can enable or facilitate improvements in efficiency, but it should not be the only factor that guide decisions.

(iii) **Inclusive smart cities are shaped by, and respond to, the local context.**
Local realities and conditions should guide decisions regarding the selection of smart technologies and initiatives. Appropriate interventions would be cognisant of factors such as poverty, inequality, and unemployment, informality, crime, municipal capacity, the availability of resources etc.

(iv) **Inclusive smart cities are informed by the real needs of the community.**
Communities should be involved in the process to identify, develop, implement and maintain smart city initiatives. The active participation of the beneficiaries and other stakeholders, as opposed to mere consultation, plays a critical role in the success of interventions. The needs of communities should not be assumed, and therefore participatory decision-making should be integral to the entire process.

(v) **Inclusive smart cities embrace innovation, partnerships, and collaboration.**
Both the public and private sector, as well as members of the community, could make valuable contributions to the identification of innovative technologies and initiatives. Encouraging collaboration and establishing formal partnerships could result in new and creative ideas and improve the quality of smart interventions.

(vi) **Inclusive smart cities are sustainable, resilient, and safe.**
Smart technologies and interventions could contribute to improving sustainability, building resilience by enhancing the city's ability to deal with and

prevent chronic stresses and acute shocks, and creating safer communities by assisting the police and other law enforcement agencies with crime prevention and detection.

Some of the principles listed above could be linked to spatial transformation and territorial governance. The following linkages are evident:

- The need for smart cities to be smart for all relates to spatial transformation in that both are aimed at supporting inclusive growth to allow all people to live in shared and transformed places in an integrated, sustainable, and competitive national space economy, as promoted in the NSDF (DALRRD 2023).
- If used thoughtfully, smart technologies could contribute to spatial transformation by breaking down the segregated spatial patterns of settlements. Smart technologies (e.g. e-governance) could also enhance territorial governance by reducing inefficiencies and inequalities, and by improving access to opportunities for all residents, as outlined in the NSDF (DALRRD 2023).
- Smart interventions that are informed by the real needs of the community and a vision of a more compact city will address spatial fragmentation and will inform territorial governance.
- Smart initiatives that involve partnerships and collaboration support the objectives of territorial governance and could result in improved service delivery.
- By incorporating the principles of sustainability, resilience and safety into smart city initiatives, these initiatives could contribute to the equitable redistribution of resources within cities, as well as the building of systems that facilitate affordable and safe access to the full ambit of resources that makes urban living productive. This is critical to spatial transformation as described by the Cities Support Programme (National Treasury 2017).

These potential positive outcomes of smart city interventions will require deliberate effort if they are to result in more inclusive cities and economic growth. A technology-dominated push towards smartness without carefully targeted regulatory requirements or incentives to achieve inclusion and transformation may, by default, result in greater inequality and exclusion, whether social, economic, or spatial.

15.6 Learning from Some Examples

The Department of Public Works and Infrastructure (DPWI) highlighted five planned smart cities in their progress update on selected smart cities towards the end of 2023: the Eastern Seaboard New Smart City, Nkosi City, the Durban Aerotropolis, Mooikloof Mega City, and the Lanseria Smart City (DPWI 2023). These five high profile smart city initiatives are all in different phases of planning and implementation.

- The Eastern Seaboard New Smart City has been identified as a South African flagship by the Technical Inter-Ministerial Committee in 2021 (DPWI 2023). This initiative is framed as being part of the State President's vision to develop "new post-apartheid smart cities" (MISA 2023). CoGTA, through the Municipal Infrastructure Support Agent (MISA) has embarked on a process to develop the Eastern Seaboard between Port St Johns and Margate, which will culminate in "one or more African coastal smart cities in the region" (MISA 2023). Spanning multiple administrative boundaries, the process started with the development of an Eastern Seaboard Regional Spatial Development Framework (ESRSDF) published in 2023. The ESRSDF specifically emphasises the importance of creating a digitally connected region and it plans to pilot areas for smart city development, referring to smart villages where "both traditional and modern networks and services are improved using advanced digital technology, telecommunications, innovation, and knowledge utilisation" (MISA 2023).
- Nkosi City borders the Kruger National Park. This smart city initiative intends to develop small-scale farming land parcels next to state-subsidised affordable housing, creating so-called 'urban farms' where intensive and sophisticated agriculture will be established and products will be processed on site, and packaged and refrigerated locally. The smart development will ultimately cover around 5000 hectares with housing offered in a range of typologies, as well as schools, tertiary institutions, health facilities and retail. A township establishment process for Nkosi City has been completed and the City of Mbombela Municipality has reportedly approached National Treasury for bulk engineering funding through the Budget Facility for Infrastructure (BFI) (DPWI 2023).
- The Durban Aerotropolis is another smart city development that entails a purpose-built city. Led by the KwaZulu-Natal Department of Economic Development Tourism and Environmental Affairs, a 50-year Durban Aerotropolis Master Plan was completed in 2018. The Dube TradePort Special Economic Zone (SEZ) and the King Shaka International Airport are central to the plans to create a smart manufacturing, trade, and logistics hub for Southern Africa, ultimately covering 32,000 hectares of integrated development. Of the five smart city initiatives reported on by DPWI, the Durban Aerotropolis seems to be the most advanced. However, implementation has been significantly disrupted by the COVID-19 pandemic (Luthuli 2020) and initial assessments of the socio-economic benefits of the initiative remain inconclusive (Ngwenya et al. 2022).
- The Mooikloof Mega City in Pretoria is a public-private collaboration. Recognised as one of the government's 62 Strategic Integrated Projects (SIPs) gazetted in 2020, the smart city initiative is geared towards addressing the housing needs of the so-called "missing middle" i.e. catering for first time homeowners that do not qualify for a full government subsidy but cannot afford to buy through a long-term mortgage loan. A total of 50,000 sectional title residential units are planned. Mooikloof Mega City's smartness will mostly be focused on the promotion of energy efficiency by implementing green building principles and introducing solar energy solutions, water harvesting, and others. According to results for the 6 months ending on 31 August 2023, published by the Balwin

Group, the Mooikloof Mega City private sector partner, 2500 apartments of the Greenkloof development (part of the Mooikloof Mega City) are being constructed with 124 apartments already sold. Construction for the further 10,234 apartments planned for this phase have a status of 'Inactive' (Balwin Properties 2023). The DPWI (2023) reports that infrastructure upgrades are being completed to support Phase 1 of the development. However, there are reports of challenges with the remainder of the infrastructure investment committed to by government (see DPWI 2023; Moneyweb 2023).
- The Lanseria Smart City development is led by the Investment and Infrastructure Office in The Presidency and promises to be home to between 350,000 and 500,000 people by 2030. President Ramaphosa promised in 2020 that "It will not only be smart and 5G-ready but will be a leading benchmark for green infrastructure continental and internationally" (Ramaphosa 2020). Consultants assisting with the Greater Lanseria Master Plan view their involvement as an opportunity to shift smart city paradigms, promising that Lanseria "will not reflect an extravagant tech-heavy city, but instead, a true post-apartheid city in which innovative technologies are used to provide smarter, more sustainable solutions and services that will change the apartheid spatial architecture and provide social and economic opportunity for all South Africans" (Du Plessis 2024). According to DPWI (2023) the initiative is dependent on "the collective energies of all three spheres of government, a range of relevant parastatals and various layers of private sector actors to realise". During a parliamentary question and answer session in March 2024 the Minister of Human Settlements, Mmamoloko Kubayi, acknowledged that certain approvals were outstanding "due to the lack of institutional and government arrangements", but gave the assurance that these have been resolved (Maqhina 2024).

In contrast to the above often favoured and more popular notion of new smart city development, many municipalities—struggling with failing infrastructure—are focusing "on transforming the urban places and spaces where most of South Africa's population already lives" (Horber 2019 referring to the IUDF). Specifically, metropolitan municipalities like (but not limited to) the cities of Tshwane, Ekurhuleni, eThekwini and Joburg have implemented numerous smart initiatives, including smart meters (for water and electricity), public Wi-Fi (provided in public spaces and facilities), and digital platforms (for managing municipal accounts and pre-paid electricity purchases, accessing job opportunities, reporting faults, and accessing maps and municipal documents). Luke Boyle published four parts in a series exploring Cape Town's journey from the early 2000s to position itself to become Africa's "first truly digital city" (Boyle and Staines 2019; Boyle 2019a, b, c). In addition to these examples, several municipalities provide computer training and other scarce skills, often in collaboration with private companies or non-profit organisations, such as Cape Town's SmartCape initiative.

What is evident from the smart cities' interventions implemented thus far, is that a smart city is a "smart" way of doing things rather than a fixed state (DBSA, World Bank Group 2021). It is also worth noting that successfully implementing such initiatives requires different partnerships between government, the private sector, civic society, and academia (DBSA, World Bank Group 2022: 10).

15.7 Conclusion

Based on the discussions in this chapter, it is evident that smart cities could contribute to spatial transformation and support territorial governance if certain conditions are adhered to. The guiding principles described in the SCF (see Sect. 15.5) should support decision-making for the identification of suitable smart initiatives and technologies. Appropriate smart interventions could result in more effective, data-driven decision-making, new economic development opportunities, improved quality of life, safer communities, enhanced engagement between municipalities and residents, and cost savings.

When developing and implementing smart initiatives unintended negative consequences must be avoided. For instance, poorly considered smart technologies and interventions could exacerbate spatial fragmentation, they could limit economic opportunities for poor and vulnerable residents and widen the digital divide.

Against the backdrop of South Africa's socio-economic realities, it could be argued that, rather than developing smart cities, funding should be prioritised for the provision of basic services such as sanitation, electricity, water, waste management, healthcare and housing (DCoG, CSIR 2021).

While more in-depth research will hopefully bring conclusive evidence on the alignment of initiatives to the six SCF guiding principles, the smart city examples discussed in Sect. 15.6 provide an indication of possible challenges and concerns. Smart city ideas are often implemented without comprehensive consideration of local contexts and needs. Piloting a smart city innovation in one locality does not ensure successful roll-out in others. A thorough understanding of a place and its specific requirements remain essential (Petzer and Kruger 2020). Partnerships are central to smart city developments and setting up public-private partnerships and inter-governmental partnerships have proved to be time consuming and complex. Added to that, the ethical issues associated with acknowledging and managing vested corporate interests and sometimes allowing (maybe unintentionally) the private sector to set agendas for the development debate comes to the fore.

Smart initiatives should respond to local challenges if they were to contribute to improving the quality of life of all South Africans. They should be appropriate to the South African context and should not be based on generic models that may not be suitable for local conditions.

To increase the potential of smart cities to contribute to spatial transformation and to support territorial governance, it is important to consider putting in place proper metrics to measure the social, economic, environmental, and spatial results of smart city interventions. These metrics would assist with monitoring the impacts of smart initiatives and with addressing unintended consequences. Such a smart cities impact framework would align well with other endeavours to measure spatial transformation of South African cities.

References

Aggarwala RT, Hill K, Muggah R (2018) Smart city experts should be looking to emerging markets. Here's why. In: World Economic Forum. https://www.weforum.org/agenda/2018/10/how-the-developing-world-can-kickstart-the-smart-cities-revolution/. Accessed 13 Nov 2019

Angelidou M (2014) Smart city policies: a spatial approach. Cities 41:S3–S11. https://doi.org/10.1016/j.cities.2014.06.007

Aurigi A, Odendaal N (2021) From "Smart in the Box" to "Smart in the City": rethinking the socially sustainable smart city in context. J Urban Technol 28:55–70. https://doi.org/10.1080/10630732.2019.1704203

Balkaran S (2019) Smart cities as misplaced priorities in South Africa: a complex balance of conflicting societal needs. J Manag Admin 2019:1–30

Balwin Properties (2023) Unaudited interim financial results for the six months ended 31 August 2023. Balwin Properties, Johannesburg. https://balwin.co.za/investor-relations/2023-Interim-Report.pdf

Boyle L (2019a) URERU smart city report series, part 4: the way forward for the City of Cape Town and what it means to be "Smart" in Africa. Urban Real Estate Research Unit, University of Cape Town, Cape Town

Boyle L (2019b) URERU smart city report series, part 3: identifying the opportunities and challenges that exist for Cape Town's smart city development. Urban Real Estate Research Unit, University of Cape Town, Cape Town

Boyle L (2019c) URERU smart city report series, part 2: the current state and characteristics of Cape Town's smart city development. Urban Real Estate Research Unit, University of Cape Town, Cape Town

Boyle L, Staines I (2019) URERU smart city report series, part 1: overview and analysis of Cape Town's digital city strategy. Urban Real Estate Research Unit, University of Cape Town, Cape Town

Brand D (2018) The heavy price SA is paying for dysfunctional local government. News24. https://www.news24.com/news24/opinions/analysis/the-heavy-price-sa-is-paying-for-dysfunctional-local-government-20180904

Choo CW (1997) IT2000: Singapore's vision of an intelligent Island. In: Droege P (ed) Intelligent environments: spatial aspects of the information revolution. North-Holland, Amsterdam, pp 49–65

Chourabi H, Nam T, Walker S, Gil-Garcia JR, Mellouli S, Nahon K, Pardo TA, Scholl HJ (2012) Understanding smart cities: an integrative framework. In: 2012 45th Hawaii international conference on system sciences, pp 2289–2297

Cocchia A (2014) Smart and digital city: a systematic literature review. In: Dameri RP, Rosenthal-Sabroux C (eds) Smart city: how to create public and economic value with high technology in urban space. Springer International Publishing, Cham, pp 13–43

CoGTA (2016) Integrated Urban Development Framework: a new deal for South African cities and towns. Department of Cooperative Governance and Traditional Affairs, Pretoria

CSIR (2022) A guide to establishing a municipal smart city initiative. CSIR, Pretoria. https://www.ukesa.info/library/view/a-guide-to-establishing-a-municipal-smart-city-initiative

CSIR, DHS (2019) The neighbourhood planning and design guide: creating sustainable human settlements (Red book). Department of Human Settlements, Pretoria

DALRRD (2023) National Spatial Development Framework, March 2022. Department of Agriculture, Land Reform and Rural Development, Pretoria

DBSA, World Bank Group (2021) SmartCity.za Bulletin #1: preparing common ground. https://www.dbsa.org/sites/default/files/media/documents/2023-09/SmartCity.za%20Bulletin%20Issue%201%20-%202021.pdf
DBSA, World Bank Group (2022) SmartCity.za Bulletin #2: the case for smart city partnering. https://www.dbsa.org/sites/default/files/media/documents/2023-09/SmartCity.za%20Bulletin%20Issue%202%20-%202022.pdf
DCoG, CSIR (2021) A South African Smart Cities Framework: a decision-making framework to guide the development of smart cities in South Africa. Department of Cooperative Governance and Traditional Affairs, Pretoria
DPWI (2023) Presentation to portfolio committee on Public Works and Infrastructure: progress on selected smart cities as at 24 August 2023. Portfolio Committee, National Assembly, 5 September. https://pmg.org.za/files/230905DPWI_Presentation___Portfolio_Committee_on_PWI__Smart_Cities_DRM_05_Se....pptx
Du Plessis D (2024) Lanseria smart city: designing tomorrow's sustainable cities. SMEC, Johannesburg. https://www.smec.com/project/lanseria-smart-city/. Accessed 30 May 2024
Fataar R (2020) Unpacking smart city development in Cape Town and Johannesburg. South African Cities Network, Johannesburg
Gardner D (2018) South African urbanization review: analysis of the human settlement programme and subsidy instruments. The World Bank, Washington, DC
Gasa B (2016) In pursuit of a comprehensive economic transformation agenda. In: Vision 2030: Our future our plan. Topco Media, Cape Town, pp 247–251. https://issuu.com/topcomedia/docs/vision_2030_publication_2015
Hall R (2014) The legacies of the Natives Land Act of 1913. Scriptura: J Context Hermeneut Southern Africa 113:1–13. https://doi.org/10.10520/EJC159820
Henreckson J (2018) Benefits of smart cities in Africa. In: The Borgen Project. https://borgenproject.org/benefits-of-smart-cities-in-africa/. Accessed 3 Oct 2023
Hollands R (2008) Will the real smart city stand up? Creative, progressive, or just entrepreneurial? City 12:302–320. https://doi.org/10.1080/13604810802479126
Horber J (2019) A new city is not a smart idea for South Africa. Daily Maverick. https://www.dailymaverick.co.za/article/2019-07-29-a-new-city-is-not-a-smart-idea-for-south-africa/
ITU (2022) Digital transformation for people-centered cities. International Telecommunications Union. https://www.itu.int/cities/about/. Accessed 4 Oct 2023
Joseph S-L (2013) Directing urban land towards spatial transformation. In: The urban land series, vol 1. South African Cities Network (SACN), Johannesburg, p 9
Karuri-Sebina G, Masango L (2022) Preface from the editors. In: Smartcity.Za bulletin—issue 2: smart cities need smart partnerships. Development Bank of Southern Africa (DBSA), World Bank Group, Johannesburg, p 3. https://www.ukesa.info/library/view/smart-city-za-bulletin-issue-2
Luthuli N (2020) Airlines, airports… and the Durban Aerotropolis on the Skids! College of Law and Management Studies. University of KwaZulu-Natal. https://ww1.clms.ukzn.ac.za/airlines-airports-and-the-durban-aerotropolis-on-the-skids/
Maqhina M (2024) Master plan developed for the Lanseria smart city – Kubayi. Cape Times, 11 March. https://www.iol.co.za/capetimes/news/master-plan-developed-for-thelanseria-smart-city-kubayi-b669b337-b640-4536-902a-607f9096216e
Maritz J, Van Huyssteen E, Le Roux A, Pieterse A, Ndaba D, Mans GG, Ngidi M (2016) Are we achieving spatial transformation in South Africa? Can sub-city spatial indicators make a contribution? South African Planning Institute, Johannesburg, p 19
MISA (2023) Eastern Seaboard Regional Spatial Development Framework: Spatial Proposals and Implementation Framework. Municipal Infrastructure Support Agent, Pretoria, p 118. https://www.misa.gov.za/wp-content/uploads/2023/02/DRAFT-EASTERN-SEABOARD-RSDF-2023.pdf
Moneyweb (2023) Promises, promises … Mooikloof Mega City developer left out of pocket. Moneyweb, 31 October. https://www.moneyweb.co.za/investing/property/promises-promises-mooikloof-mega-city-developer-left-out-of-pocket/

Mosannenzadeh F, Vettorato D (2014) Defining smart city. A conceptual framework based on keyword analysis. TeMA J Land Use Mob Environ. https://doi.org/10.6092/1970-9870/2523
Musakwa W, Mokoena BT (2017) Smart cities in South Africa! A case of misplaced priorities? University of Johannesburg. https://hdl.handle.net/10210/244608
National Treasury (2017) MFMA Circular No. 88: Municipal Circular on Rationalisation Planning and Reporting Requirements for the 2018/19 MTREF
Ndifuna Ukwazi (2018) Inclusionary housing should mitigate spatial apartheid. Ndifuna Ukwazi, Cape Town
Ngwenya K, Naude MJ, Wissink HF (2022) The Durban Aerotropolis strategy for sustainable socio-economic development. Africa's Pub Serv Deliv Perform Rev 10(1):10. https://apsdpr.org/index.php/apsdpr/article/view/618
Odendaal N (2006) Towards the digital city in South Africa: issues and constraints. J Urban Technol 13:29–48. https://doi.org/10.1080/10630730601145997
Parnell S, Crankshaw O (2013) The politics of 'race' and the transformation of the post-apartheid space economy. J Housing Built Environ 28:589–603. https://doi.org/10.1007/s10901-013-9345-6
Petzer E, Kruger T (2020) Inclusive smart cities appropriate to the South African context: key findings of a CSIR study. CSIR, Pretoria
Petzer E, Napier M, Molefe N, Le Roux A (2020) Developing a tool to monitor neighbourhood spatial transformation. South African Local Government Association, Pretoria
Phuhlisani NPC (2017) The role of land tenure and governance in reproducing and transforming spatial inequality. Commissioned report for High Level Panel on the assessment of key legislation and the acceleration of fundamental change, an initiative of the Parliament of South Africa. Phuhlisani NPC, Cape Town, South Africa. https://www.parliament.gov.za/storage/app/media/Pages/2017/october/High_Level_Panel/Commissioned_Report_land/Commissioned_Report_on_Spatial_Inequality.pdf
Ramaphosa C (2019) President Cyril Ramaphosa: 2019 State of the Nation Address.
Ramaphosa C (2020) President Cyril Ramaphosa: 2020 State of the Nation Address
Ragoasha MV (2020) How can "smart cities" thinking benefit urban planning and development in South Africa? Learning from international smart city approaches. MRTP mini dissertation, University of Pretoria.
SACN (2016) State of South African Cities Report 2016. South African Cities Network (SACN), Johannesburg. 416pp. https://www.sacities.net/publication/state-of-south-african-cities-report-2016/
SACN (2020) Smart cities paper series: smart governance in South African cities. South African Cities Network
Salter C (2022) The smart city is a perpetually unrealized utopia. MIT Technology Review. https://www.technologyreview.com/2022/06/24/1053969/smart-city-unrealized-utopia/. Accessed 3 Oct 2023
Serfontein B (2020) Smart cities in the time of physical distancing. https://www.buildingand-decor.co.za/smart-cities-in-the-time-of-physical-distancing-by-bouwer-serfontein/, https://www.buildinganddecor.co.za/smart-cities-in-the-time-of-physical-distancing-by-bouwer-serfontein/. Accessed 3 Oct 2023
Söderström O, Paasche T, Klauser F (2014) Smart cities as corporate storytelling. City 18(3):307–320. https://doi.org/10.1080/13604813.2014.906716
South Africa (2013a) National Development Plan 2030: our future—make it work. National Planning Commission, Pretoria
South Africa (2013b) Spatial Planning and Land Use Management Act
Supangkat SH, Arman AA, Nugraha RA, Fatimah YA (2018) The implementation of Garuda Smart City Framework for smart city readiness mapping in Indonesia. J Asia-Pacific Stud 32:169–176. https://doi.org/10.57278/wiapstokyu.32.0_169
Townsend AM (2013) Smart cities: big data, civic hackers, and the quest for a new utopia. WW Norton, Company, New York

Turok I (2013) Transforming South Africa's divided cities: can devolution help? Int Plan Stud 18:168–187. https://doi.org/10.1080/13563475.2013.774146

Turok I (2014) Settlement planning and urban transformation. South African Cities Network (SACN), Johannesburg, pp 74–99

Turok I (2018) Worlds apart: spatial inequalities in South Africa. In: Smith MN (ed) Confronting inequality: the South African crisis. Jacana Media, Johannesburg, pp 129–151

Turok I, Borel-Saladin J (2014) Is urbanisation in South Africa on a sustainable trajectory? Dev South Afr 31:675–691. https://doi.org/10.1080/0376835X.2014.937524

Turok I, McGranahan G (2013) Urbanization and economic growth: the arguments and evidence for Africa and Asia. Environ Urban 25:465–482. https://doi.org/10.1177/0956247813490908

Winchester HPM, Chalkley BS (1990) The Japanese-Australian multifunction polis: context and issues. Urban Stud 27:273–281. https://doi.org/10.1080/00420989020080221

Zizzama R, Schotte S, Leibbrandt M (2019) Snakes and ladders and loaded dice - Poverty dynamics and inequality in South Africa, 2008–2017 (Working Paper No. 25/2019). United Nations University World Institute for Development Economics Research (UNU-WIDER), Helsinki, Finland. https://www.wider.unu.edu/publication/snakes-and-ladders-and-loaded-dice

Chapter 16
Thirty Years of Magical Planning and the Reluctant Pragmatic Turn

Claus Rabe

Abstract This chapter explores the evolution of economic development planning discourse in South African cities from 1994 to the present day. Drawing on the life and death of three urban policy concepts, 'Local Economic Development', 'Transit-oriented Development' and 'Smart Cities', this paper contrasts the situational merits of normative planning in the post-conflict/nation-building period (1994–2008) with its role in the subsequent economic malaise gripping South African cities. Out of necessity rather than Damascene conversion, the South African body politic is implicitly reassessing the role of the State in the realisation of its economic objectives. Heralding an incipient pragmatic turn, policymakers are reluctantly looking to private initiative to deliver public goods in the vacuum left by State failure.

Amid this interregnum, urban practitioners face a choice either eschewing or embracing pragmatism in practice and discourse. The first path—the continued elevation of abstract planning ideals over the messy reality on the ground—leads to the deepening irrelevance of our profession in broader policy debates and co-option by short-term political interests. In contrast, the pragmatic planner, freed from grand narratives, is absorbed by identifying and understanding interventions *that work*—that impact citizens' material wellbeing in an immediate and measurable way. To distil, adapt and replicate interventions that work, the pragmatic planner appreciates the role of functional theory—how cities *actually work*—as an essential precondition for effective policymaking. By augmenting each stage of the policy process—from problem definition and design to implementation and post hoc evaluation—with empirical evidence and applied theories on proximate causes and effects, the pragmatic planner is better positioned to regain a foothold at the policy table.

Keywords Local economic development · Transit-oriented development · Smart cities · Magical planning · Pragmatic planning · South African cities

C. Rabe (✉)
Palmer Development Group, Cape Town, South Africa
e-mail: claus@pdg.co.za

D. du Plessis (ed.), *Territorial Governance and Spatial Transformation in Post-apartheid South Africa*, Local and Urban Governance,
https://doi.org/10.1007/978-3-031-85489-7_16

16.1 Introduction

In happy contrast to the grim prognostications of my economics professors, I had the privilege of witnessing South Africa's cities thrive and millions of Black South African households ascend into the middle class. I vividly recall a moment in 2007 when it felt that we—the windswept, hollow-eyed South African people—had achieved the unimaginable: emerging from a place of darkness, transcending the rules of history, and arriving on the world stage as a beacon of hope for troubled post-conflict nations worldwide.

Bristling with impatience to play my part as an urban planner, little did I know that this moment marked the turning point in our economic and urban trajectory. 2008 was the year when nearly all our indicators of material wellbeing flatlined, and, perhaps irreversibly, commenced a slow and stubborn descent.

Most analysts agree on the macro causes behind the downturn and the subsequent malaise blighting our cities: the Global Financial Crisis, Polokwane, commodity cycles, COVID. But where does that leave local government, mandated by the Constitution to be the coalface of the developmental State? What role has planners played in shaping local economic outcomes?

In this retrospective spanning 30 years of local economic development planning in South Africa, between 1994 and the present, I offer a frank account of my observations and reflections as a researcher, government official and public sector consultant. Drawing on the life and death of three urban policy bandwagons, 'Local Economic Development', 'Transit-oriented Development' and 'Smart Cities', I evaluate South Africa's legacy of economic development planning not through the lens of high-minded principles but rather, in terms of its cumulative material impact on the everyday South African citizen.

Aimed at fellow practitioners who share a working knowledge of our collective track record on local economic development planning and implementation, this subjective account spares the reader from yet another recitation of the statistics of malaise afflicting South Africa since 2007. It is a synthesis of 20 years of local research, innumerable meetings, conversations, published accounts, direct project experience, and my own published work on place-based economic policy (Rabe et al. 2015; Rabe and Selod 2023).

16.2 A Reality Check

Before we embark on our history, let us try and circumvent our language traps and approach the topic from beyond the echo chamber: First, from the outside-in (as a visitor from Outer Space), and then, from the bottom-up—what South Africans say they want. Given the macro context and the needs, we then situate the role of local economic policy within South Africa's constitutional framework of powers and functions, distilling the *de facto* mechanisms available to local government to influence local economic outcomes.

16.2.1 An Economy in the World

First, an outside-in, matter-of-fact perspective of the South African economic context is distilled into the following five essential, distinguishing characteristics:

1. *A small, open, consumption-led economy.* South Africa is a small, remote economy stuck in a middle-income trap, where income levels stagnate rather than converge with developed economies. Our export potential is bounded by a small regional market and remoteness to global markets. Similar to other resource-based economies, recent urbanisation is accompanied by consumption rather than industrialisation (Henderson 2020; Hausmann 2022).
2. *A decade of economic stagnation.* South Africa has missed almost all the targets set in its National Development Plan. Over the last 10 years, GDP per capita has declined in real terms by 7%, alongside fixed capital investment by private and public sectors. Narrow unemployment has risen sharply to 32% as of December 2023.
3. *Declining state capability.* The rapid deterioration of state-owned enterprises, local governments' inability to provide basic services, the spread of extortion rackets, and the build-up of debt all signal a dramatic decline in State capability since 2007. Consequently, South Africa's fiscal multiplier has dropped below one, implying that every additional Rand taxed and spent by government has a negative impact on national income (Hausmann 2022).
4. *Weak entrepreneurial culture.* South Africa's prospects for realising inclusive economic growth are hampered by a historically weak entrepreneurial culture in South Africa. Whereas the expected rate of 'own-account work' for a country with South Africa's income profile is around 20%, we can only muster 4% (Shah 2022). Furthermore, the vast majority of informal workers are engaged in survivalist rather than growth-oriented activity. Current conditions, such as low urban densities (Hausmann 2023) and prevailing stigma against success in poor communities (Hare and Walwyn 2019), perpetuate the legacy of Apartheid-era corporatism, whose features - such as market concentration and the predominance of salaried employment - undermined the development of entrepreneurial culture.[1]
5. *Regional hysteresis.* The income differential in South Africa between rural and urban households, and between regions, is exceptionally high.[2] Regional economic prospects are tightly bound to their natural advantage and are path dependent on political design and economic processes over preceding centuries (Loewald et al. 2021). Historical unevenness in economic prospects is perpetu-

[1] See Glaeser's discussion on intergenerational effects on entrepreneurial acumen.

[2] Average household incomes per district municipality range from 39% below (OR Tambo) to 34% higher (City of Cape Town) than average income at national level (R272k/month). Comparable figures for the top 2.5% of households range from 74% below national average (Alfred Nzo) to 86% above national average (City of Johannesburg) (Source: Quantec (2022) RSA Standardised Regional Household Current Income Deciles).

ated by, inter alia, structural unemployment,[3] precarious land tenure in traditional areas, and the concentration of footloose skilled workers in cities.

16.2.2 What Do Citizens Actually Want?

Second, let's consider our cities from the bottom up. Our plans and policies bristle with assumptions about the needs and aspirations of our citizens. The language used—inclusion, spatial transformation, sustainability—is often imported from the international conference circuit rather than expressed by citizens themselves. What then do we in fact know about what citizens actually want?

First, participation in general elections provides a binary indication as to level of trust and hope in democratic processes. According to the IRR, voter turnout has declined from 87% in 1994 to a meagre 46% in the 2021 local government elections, when less than 20% of the population aged between 18 and 35 were registered to vote.[4] And in 2020, the Bureau for Economic Research found that confidence in local government was at 34%.[5]

Second, the Afrobarometer's survey of citizen's top five priorities[6] has been consistent: (1) Unemployment, (2) Crime, (3) Electricity, (4) Corruption and (5) Water Supply. Nothing about gentrification, bike lanes or other urbanist bandwagons. Nothing about access to land, nationalisation or other such populist tropes either. For the remainder of this paper, we will refer back to these 'Big Five', since citizens' perceptions on the State's progress on these bread-and-butter issues will be its final arbiter.

16.2.3 The Two Levers of Local Government

Finally, let's cut through the policy jargon and speak plainly: what are the levers available to local government to improve the material wellbeing of households at scale? I would argue that municipalities essentially have two: (1) providing and maintaining basic services for households, and (2) improving the operational environment for formal and informal businesses.

---

[3] Structural unemployment describes a situation where large sections of the labour force disengage permanently from the economy due to their income earnings potential being less than the value of social grants.

[4] https://irr.org.za/fan/media/south-africas-low-voter-turnout-is-no-laughing-matter-5-crucial-takeaways

[5] NDC 2022 Report on Monitoring NDP Indicators and Targets.

[6] Afrobarometer 2022.

The first lever is typically redistributive: the municipality draws on national grants and surplus from paying municipal customers to provide and maintain basic services to households, like electricity and water.

The second lever is explicitly aimed at achieving economic objectives: the municipality provides and maintains the networked services and regulatory environment necessary for businesses to survive, thrive and create jobs.

In this section, I have provided a straightforward description of local economic conditions in South Africa, beginning with an outside-in perspective and then shifting to a bottom-up view. I conclude by identifying the de facto levers available to municipalities to improve the material well-being of citizens. Now, having taken our proverbial ginger pills, let's embark on a retrospective journey of 30 years of local economic planning.

16.3 A History of Local Economic Planning in Three Acts

In South Africa, local economic planning has manifested in three successive policy waves: 'Local Economic Development', 'Transit-Oriented Development' and 'Smart Cities'. In this section, these policy waves are briefly discussed chronologically, including their conceptual origins, implicit theories-of-change, applications and operationalisation within government machinery.

16.3.1 Act I: Local Economic Development (1994–2007)

Local Economic Development ('LED') is a slippery concept open to interpretation and manipulation, often resulting in policy confusion. Section 152 of the Constitution obliges municipalities to promote the economic development of local communities, what "economic development" entails however remains open to interpretation.

Whereas international definitions characterise LED as a locally-driven but growth-oriented activity, sympathetic local authors view LED as responding to both growth- and equity imperatives (Rogerson 2014). In my experience, however, Ababio and Meyer's definition of LED as "a mechanism for poverty alleviation" (2012: 6) is most reflective of its application in practice in South Africa—that is, a tool to ameliorate the effects of extreme poverty through spatially targeted job creation projects.

Although LED is a statutory function, most municipal LED units lack the capacity and resources to implement anything beyond small-scale projects such as food gardens, pedestrian walkways, tourism kiosks, and basic infrastructure for informal traders. Despite modest capability, municipal economic visions and strategies tend to be ambitious, disconnected from reality and lacking in economic rationale (SACN 2019). While LED teams in better resourced municipalities may include a qualified development economist, the prevailing practice is more socially oriented. The unit is tasked with preparing 'doorstop' reports, supporting IDP processes, engaging ward councillors and other community representatives, and provide arms-length support to the implementation of one or two local projects. Project selection is a product of political contestation rather than technical analysis, a negotiation between political principals and ward councillors. Programmatically, LED uses the 'watering can' approach[7] (South Africa. The Presidency, 2006), with a litany of small, unrelated projects that share vaguely articulated rationales, untethered to a unifying theory-of-change, and lacking measurable performance indicators or outcomes.

Despite initial fanfare, LED projects rarely survive beyond one political term. The municipal landscape is littered with the stripped husks of incubators, propellers, hives, markets, cooperatives, cultural and tourism centres. In his 2018 paper, Rogerson's conclusion that 'LED does not promise to radically alter the status quo' is, from my perspective, a noble understatement (2018). I am not aware of any evidence in South Africa of LED projects having any observable impact on the economic massing of municipalities or towns, or indeed, any measurable benefit to the material wellbeing of households, whether aggregated to ward or municipal level.

This has called into question the relevance of LED within national development strategies. Tellingly, neither the NDP nor the IUDF—the two premier policy frameworks guiding urban policy—mention LED. Within metros, local intervention became subsumed by programmatic responses aligned to network considerations (e.g., Urban Network Strategies) and the emerging practice of Township Economic Development. While LED units continue to exist, they are marginal entities engaged with isolated, ward-based projects.

16.3.2 Act II: Transit-Oriented Development (2008–2019)

> "Today we are taking transit-oriented development another step forward, with the introduction of a project that will forever change the urban structure of Johannesburg and eradicate the legacy of Apartheid spatial planning." (Mayor of Johannesburg, Parks Tau, 2013)

It all started with a trip to Curitiba, a city in Southern Brazil. African transport officials were treated to a demonstration of the awesome possibility of TOD, with Bus Rapid Transit (or 'BRT') as its backbone. BRT was touted as "something close to a miracle for cash-strapped cities, capable of easing congestion and improving public transportation without breaking the bank. By separating buses from the irregularities of traffic, BRT can approach the same capacities and speed as costlier rail service. Buses can travel faster, maintain reliable schedules, and potentially turn a profit." (Bloomberg.Com 2013).

In the run up to the 2010 World Cup, National Treasury made significant capital funding available to metros able to present their proposals for BRT systems. Faced

---

[7] The watering can approach is implicitly premised on deconcentrating or diffusing economic activity from existing nodes, towards economically marginalised areas.

with limited time and even more limited experience in systems modelling, grant funding decisions were made based on a so-called 'beauty pageant approach'—that is, evaluating technical proposals based solely on attractive bus and station designs as rendered in 3D computer simulations. The sudden release of free money for a high visibility capital spend programme triggered a feeding frenzy amongst ambitious politicians.

Almost immediately, implementation faltered at planning stage in a number of cities. In Cape Town, George and Johannesburg, however, the buses started running. Political alarm bells rang early on—the buses were running empty. Cities were only recovering a fraction of their ballooning operating costs from ticket sales. Despite consuming 15% of the metro's capital budget (Harrison et al. 2019), Rea Vaya in Johannesburg served roughly 3000 commuters per working day in 2016/2017, recovering 35% of its direct vehicle operating costs from fare revenue between 2017/2018 and 2020/2021 (City of Johannesburg 2017). Before COVID, MyCiti was estimated to serve 2% of Cape Town's commuters (unchanged from 2013 to 2020), equivalent to approximately 70,000 passenger trips compared to 830,000 provided by minibus taxis and 255,000 trips by the conventional scheduled bus service (City of Cape Town 2023).

The reasons for poor financial sustainability are complex and beyond the scope of this chapter. Nuance did not stop BRT's political champions from finding a scapegoat: urban planners. According to this narrative, the problem with BRT wasn't a misalignment between system design and effective demand, but rather that urban planners, complicit in the legacy of Apartheid spatial planning and conspiring with faceless NIMBYists, were deliberately frustrating the realisation of the land use densities required to fill buses. They were not on the 'transformation-through-TOD' bus. Urban planners, and the property development fraternity in general, needed a 'mindset change'.

What BRT champions refused to appreciate, however, is the simple truth that developers don't base their investment decisions on beautifully rendered station designs or vision statements alone. The provision of an attractive BRT bus stop is simply not incentive enough to induce densification at the scale and rate required to meaningfully impact on ridership levels within 10, even 15 years. BRT does not (in itself) supplant the fundamental determinants of development viability: macroeconomic conditions, effective demand and building costs. As Harrison found, while TOD "depended on inducing the private real estate sector into the Corridors"… "BRT, with its mainly low-income ridership, held less attraction" (Harrison et al. 2019: 458).

Change *is* happening, albeit slowly and selectively, as governed by local market dynamics. Despite ongoing challenges, some networks are expanding, efficiencies are being gained, high potential corridors are densifying, and ridership is growing (Engineering News 2023). Where pragmatism prevailed over fantasy, BRT systems have even demonstrated agility and innovation: in George, the system is comparatively affordable owing to contextually appropriate network specifications: dedicated lanes were deemed unnecessary given a population of 160,000. In Cape Town, services are being rationalised and the viability of future phases closely scrutinised:

while remaining committed to completing the next phase of the MyCiti roll-out, the City is "moving away from big legacy transport projects for at least the next five years" (Kretzmann 2022).

At its heart, the experience of 'BRTOD' has exposed a disconnect between, on one hand, grand ambition—specifically, the idea that public transport-enabled density would naturally lead to social mixing across class and race—and, on the other hand, the reality that our city governments possess limited influence over locational decisions by the private sector. With grant funding drying up and belts being tightened, it is safe to say that TOD, once the go-to planning instrument for economic transformation, is no longer *de rigueur*.

16.3.3 Act III: Smart Cities (2019–)

With LED having lost credibility and TOD being swept in the fiscal undertow, Ramaphosa had to dig deep to evoke a hopeful tone during his inaugural speech in 2019. He duly announced his dream of three new 'Smart Cities': Nkosi City, bordering the Kruger National Park, the African Coastal Smart City in the Eastern Cape, and the Lanseria Smart City in Gauteng. Unlike LED and TOD, local economic planning would be driven by the Presidency itself.

Where does this concept of 'Smart Cities' come from? While the term first appeared in the early 1990s, it was popularised in international urban discourse in 2008 with the launch of IBM's Smarter Planet project. While its founding advocates saw the intention of smart cities is to use information technology to make a city more intelligent and efficient (Yin et al. 2015), it quickly became a 'slippery' concept (Hollands 2008). In Africa the term is more commonly applied to new cities, whereas in South Africa advocates attempt to de-emphasise the role of technology and efficiency, and elevate social inclusion, liveability and environmental sustainability. Despite different emphases in marketing, they have the following in common: (1) they are large scale, (2) they override formal city plans, and (3) they consist of 3D visions of future cities rather than detailed land use plans (Watson 2014).

By way of example, Lanseria—billed by the President as South Africa's 'first post-Apartheid city'—will cover 430 square kilometres and straddle three municipalities. The plan envisages a massive greenfield development—home to half a million people within 10 years—adjacent to existing urban fabric, largely using conventional building and infrastructure technology, and urban design principles. The only real departure from conventional urban development is that the planning concept is being used as a vehicle to raise funding for the massive (conventional) bulk infrastructure investment required to unlock the proposed development. The implicit strategy appears to be that political support and the sheer scale of proposed development would be sufficient to overcome the funding and regulatory constraints that has hitherto prevented development in this region.

At the time of writing, no ground has yet been broken. As with previous attempts at developing new cities and not unlike the TOD experience, the Smart Cities agenda—driven by politicians and urban designers—saw utopian visions scuppered by basic economics. As with TOD, little consideration was given to the interplay between macro-economic conditions, effective demand and development costs. For example, the vision of a post-Apartheid, socially inclusive city in Lanseria has foundered on the realities of the South African housing market, where unmet demand is concentrated at the bottom end of the housing ladder. This backlog is translated into effective demand through government's capital housing subsidy, which, in turn, only affords RDP-style housing on the urban periphery, a typology antithetical to social inclusion and liveability (Saba 2023). And, ironic given its environmental goals, the only economic activity that has been attracted to date is a plastic production plant.

16.4 Magical Planning in Urban Policy

> "Vision without action is a daydream. Action without vision is a nightmare." (Japanese Proverb)

16.4.1 Defining Magical Planning

In this section, I introduce the psychological concept of magical thinking and explain how 'magical planning' is a useful diagnostic for the divergence between the grand narratives shaping our policy and the lived reality.

Before introducing the concept of magical thinking, I first outline a simple theoretical framework. In his seminal work on cities, Kevin Lynch (1984) distinguishes between two types of theory informing urban policy: normative and functional. Normative theory deals with the connection between human values and economic outcomes, while functional theory attempts to explain *how* local economies function.

The Continental tradition, rooted in Kantian idealism, essentialism, and the inherent rightness or wrongness of actions (i.e., deontological ethics), tends to prioritise normative questions over functional considerations. In contrast, the pragmatic tradition, rejecting grand narratives given the limits of our cognition and contingency of our beliefs, prioritises practical outcomes and experiential insights. Whereas the essentialist judges actions by intentions, the pragmatist evaluates actions by their consequences.

In South Africa, our planning approach, rooted in the Continental tradition, has blended with a proclivity for magical thinking, borne of collective trauma, shortsightedness and exceptionalism, forming a unique, creolised praxis I call 'magical planning': the belief that policy intentions, unmediated by a functional understanding of systems, will lead to broader change.

I borrow the concept of 'magical thinking' from the field of psychology, defined by Pronin as 'the belief in the ability to influence events at a distance with no known physical explanation' (2006). It offers the possibility of positive outcomes through non-logical means.

The implicit purpose of magical thinking is to create magical hope, override group differences and defend against frustrated expectations. Conditions which give rise to magical thinking include stress and anxiety, cultural and social influences (e.g., superstitions and religious practices), trauma and grief, a lack of scientific understanding, and a desire for hope and optimism.

For newly minted planning professionals seeking to craft their identity, magical planning offers a route towards unbounded creative expression. For academics it provides a safe medium through which to burnish their moral credentials. For politicians long on intentions but short on time, much is to be gained from perpetuating magical hope. For a society under stress and fearing loss of control, magical hope helps create a sense of order.

However, for experienced practitioners who have seen one project fail after the other once planning ideals make contact with reality, Hegel's argument that our cities are merely products of human thought limited only by our imagination and progressive ideals, may ring hollow.

16.4.2 Manifestations of Magical Planning

Magical thinking shares three themes: (1) the search for causality, (2) the tendency to essentialise, and (3) culturally endorsed elevation of belief over reason (Rosengren and French 2013). In this section I use these characteristics to structure a discussion on how magical planning manifests in the policy development process and ultimately gave rise to the three policy bandwagons introduced earlier.

I contrast the prevailing approach with a pragmatic policy development framework that bridges normative ('what do we value?') and functional theory ('how does it work?') and emphasises experiential insight over grand narrative.

16.4.2.1 The Search for Causality

The first step in urban policy is answering the question, 'what is to be done?'. Answering the question typically involves articulating a desired future state, problematising the status quo, and drawing a plausible path between the two based on a suite of interventions. Ideally, the policy development process iterates between normative and functional knowledge domains (see Fig. 16.1).

Reflecting on the three case studies, the bridge between what we value and how things work is missing. The causal assumptions between the desired future state, the status quo, and the path from one to the other, is untethered from a functional understanding of social processes and economic systems.

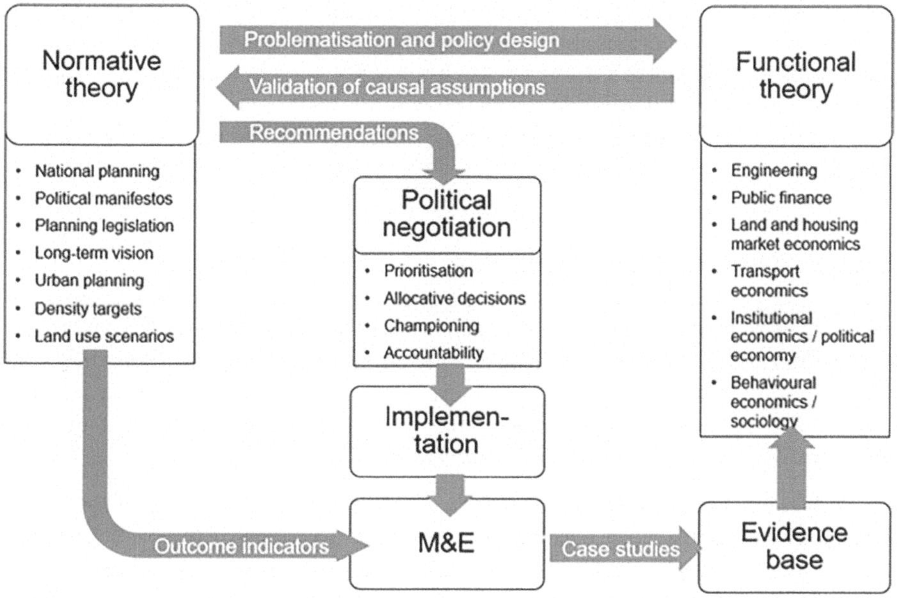

Fig. 16.1 A pragmatic framework for policy development (original to the author)

Elevating the Spatial Over the Structural Implicit in all three case studies is the belief that economic malaise and poverty alleviation can be overcome through localised, microeconomic interventions alone rather than through structural reform. This elevation of the spatial over the structural is, in my experience, less a product of solid deductive reasoning than a reflection of the limited tools available to planners, evoking the expression "to a hammer, everything looks like a nail". In reality, however, relying on ad hoc, local interventions in a society where poverty is generalised doesn't cure the patient but temporarily and selectively provides relief while facilitating continued decline.

Conflating Need and Opportunity One of the key characteristics of the language of local economic development planning in South Africa is the conflation of social development and economic growth mandates by local government. While social assistance is critical given the disparities within South Africa, these programmes in themselves will never 'grow the economy'. At best, it will temporarily provide a 'leg up' for a lucky few. The mounting impact of 30 years of chronic underinvestment in economic infrastructure by South African municipalities directly affect the lived experience of most citizens.

Local Government at the Centre of Economic Development Whereas there is a prevailing tendency in policy discourse to place the developmental local government at the centre of local economic development, the reality is that more than two-thirds of our local economies are made up of private transactions between individuals, households and businesses. These private transactions are influenced by a range of factors over which local government has no control, such as macro-economic stability, migration, proximity to major markets, education and health, policing, courts and SOEs.

16.4.2.2 The Tendency to Essentialise

The concept of psychological essentialism is the notion that individuals treat objects and entities in the world as if they have an underlying core nature or 'essence' that determines their properties and characteristics. The problematisation underlying LED, TOD and, in particular, Smart Cities, essentialises South African cities and communities at the cost of effective intervention:

South African Cities as Husks of the Past The prevailing narrative in the South African planning discourse essentialises our cities as husks of the past. Issues of segregation, housing affordability, inequality, congestion, car-dependence, informality, environmental degradation, poverty and unemployment are readily ascribed to Apartheid spatial planning. While South African cities are indeed outliers on many of these issues, we by no means have a monopoly. Indeed, these pathologies are evident to varying degrees in cities worldwide where Apartheid as a system of spatial organisation was never implemented. It is convenient for us to monocausally attribute poor outcomes today to the past rather than hold a mirror to ourselves as citizens, as policymakers, as leaders.[8] Our cities are not husks but living, breathing organisms. The reproduction of these pathologies lies in the decisions we make today, and how they impact on both ongoing and emergent social, economic and technological processes that shape our cities.[9]

The Reification of Community The merits of harnessing existing social capital within groups and leveraging community leaders as "nodes of influence and power" within cooperative networks, is emphasised in development studies. During typical project implementation, the interests of the 'Community' are ostensibly articulated by local leaders. Having consulted "the Community" and the legitimacy of community leaders affirmed, implementation can commence. While the role of local leaders in building trust and modelling cooperative behaviour is acknowledged as pivotal conditions for local project success, I would argue that the reification of

[8] Loss finds that South Africa's education system—weakened by under-resourcing, school mismanagement, poor teacher and student discipline—reproduces patterns of inequality along the lines of race, location and socio-economic status.

[9] The NDP's problematization of economic concentration as a legacy of Apartheid underscores this point. It ignores ongoing and universally observed agglomerative processes such as urbanisation and localization economies.

'Community' as a local economic system is a significant analytical barrier to effective economic intervention. Local economic policy is unable to identify opportunity within given settlements if its population is treated as a singular, undifferentiated mass, homogenous in terms of material circumstances, values, aspirations, and productive potential.

16.4.2.3 Cultural Endorsement of Magical Belief

The development and maintenance of magical thinking involves cultural support for an alternative belief system that transcends reasoning (Rosengren and French 2013). For the purpose of our discussion, cultural support is equated with the prevailing political culture within which policy discourse occurs.

The Politicisation of Policy Design In many municipalities, the policy development process is subverted by the demands of party-political patronage. With exceptionally low levels of literacy amongst councillors, local politics becomes the only viable route to economic mobility. In social contexts where material needs are pressing and common purpose beyond kin is weak, discrete, ad hoc and bricks-and-mortar interventions serve as perfect vehicles for councillors to signal their influence over allocative decisions. Seemingly random ad hoc projects are justified based on the direct benefits arising from construction rather than their operation. In contrast, interventions that are hidden, networked in design, and preventative in intent—such as refurbishing bulk infrastructure or regularising shacks to accommodate emergency vehicle access—are maligned, neglected and subject to extortion.

Requisite Technical Capacity Is Missing Even under ideal circumstances, municipalities are complex organisations that require experienced and suitably qualified leadership with requisite technical experience of financial, institutional and programmatic management. At the same time, there is no greater threat to political patronage than a senior municipal official with a functional understanding of the built environment. The less experienced a senior municipal official, the more beholden they are to the beneficence of their patron. Politicians need pliant officials to legitimise the subversion of the policy development process by drafting municipal plans that are rich in pathos but starved of reason. Consequently, many municipal departments are run by political deployees with limited experience, based on the fiction that they will 'grow into' their position faster than the department falls into irreversible dysfunction. The consequent technical capacity gap in municipalities is the main reason for the chronic delays in infrastructure roll-out, leading Alec Moemi, the head of Infrastructure South Africa, to lament: '[municipalities] have struggled to get out of the starting blocks.'

Neglect and Ignorance of Markets Economic literacy is another victim of the political culture enveloping local economic planning. While social grants have become the main form of income in a growing tally of failed local economies, the decisions of non-State economic actors—households, businesses, entrepreneurs—lie at the centre of regional and urban economies. In project design, however, economic actors that operate outside of local patronage networks are typically ignored, sidelined or treated with suspicion.

Lack of Measurable Targets A pragmatist would argue that a policy is only as good as its outcome. As no systematic empirical analysis has been conducted on the impact of any of the three policy bandwagons, there is no evidence that any of the policy bandwagons have positively impacted positively on the aggregate material wellbeing of households, whether at ward, municipal or national level. When policy processes are subverted by the demands of political patronage, there is a clear incentive to undermine any attempt at defining and committing to measuring outcome.[10]

There may be critical points in a country's political history when there is good cause for planned ideals to be elevated beyond our functional understanding of human processes. During a political interregnum, for example, the risk of instability may be deemed greater than the cost of non-delivery. Under these circumstances, the power of magical planning lies in its ability to build a tenuous consensus across interest groups, rather than its ability to execute vision. When left unreconstituted for too long, planning discourse becomes turbid, and practitioners' ability to intervene decisively and effectively, diminished.

An ineffective State is an inherent consequence of magical planning, which withers under technical, theoretical and empirical scrutiny. Scarce resources are misdirected, opportunities for broad-based material impact are lost, and structural reforms deferred. It results in South Africans being a little bit poorer and less engaged this year than they were the year before. As the gap between the planned ideal and reality widens, magical planning is complicit in eroding trust in the Constitution and the legitimacy of the democratic process. Left unchecked, magical planning creates the conditions for State failure and the commensurate rise of private initiative.

16.4.3 The Legacy of Magical Planning

> "Policies since 1994 have unintentionally perpetuated many aspects of spatial exclusion." (Hausmann 2023)[11]

In South Africa, magical planning may have served its purpose in the 1990s. I would argue, however, that it is no longer fit for purpose. There is no evidence that local

---

[10] This reluctance is often enabled by postmodernist academics who argue that we cannot compare and evaluate knowledge except in very limited contexts. While pragmatists and critical realists would agree that our knowledge is always partial and fallible, they would argue that all knowledge is not equally fallible (Danermark et al. 2019).

[11] Hausmann, 2023. "Growth through inclusion in South Africa." No. 434. CID Faculty Working Paper. Growth Lab, Boston, MA. https://growthlab.hks.harvard.edu/sites/projects.iq.harvard.edu/files/growthlab/files/2023-11-cid-wp-434-south-africa-growth-through-inclusion.pdf

economic planning has yielded broader economic returns over the last 30 years. Despite the inherent political preference for small, ad hoc capital projects, these projects, at best, reach a small number of beneficiaries for a few years. More often than not, projects tend to fail within one political term due to inadequate long-term planning, budget constraints, poor coordination, vandalism, extortion or corruption. More recently, transit-oriented development ('TOD') offered a more programmatic approach to economic intervention, harnessing the possibilities created by public transport grants in the run-up to the 2010 World Cup. With the benefit of hindsight, however, it is evident that these investments alone were not sufficient to compensate for unevenness in the underlying urban land market and unlock the expected land use responses along public transport corridors. The process of spatial transformation will be measured in decades, not years. And change will not comply with our urbanist fantasies. With Smart Cities, the urban agenda has lurched further away from reality and into the realm of the absurd.

The legacy of magical planning in South Africa cautions that unless practitioners and policymakers are able to transition from relying on magical hope towards crafting *realistic hope*, visions for the future will remain just that—dreams.

16.4.4 The Reluctant Pragmatic Turn

In his 2023 address to the SAAPAM conference, Thabo Mbeki cited a speech by Dr. Endres from the Institute of Race Relations to characterise seismic shifts in the delivery apparatus of public goods:

> "Here we have this receding power of the state, its loss of authority and credibility, its inability to translate plans into action, and the growing disconnect between the ruling elite and those they govern, and this is where South Africa's greatest opportunity for the future is to be found, in its innovative and resilient private sector and civil society, which are solving problems in the growing absence of the State and doing so successfully…"

What Mbeki labels as a 'counter-revolution' I describe as a 'reluctant pragmatic turn'. The pragmatist puts ideology aside and accepts that the delivery of public goods through private initiative is the second-best alternative to a capable State. What do I mean by pragmatism? It comes from the Greek word *pragma*, meaning action, from where 'practice' and 'practical' comes from. William James—an American philosopher—describes the pragmatist as one who 'turns away from abstraction, from verbal solutions, from bad a priori reasons, from fixed principles, closed systems, pretended absolutes and origins'. The pragmatist rejects 'dogma, artificiality and the pretence in the finality of truth', and in doing so, she 'stands for no particular result.' The pragmatist turns instead towards 'concreteness, adequacy, towards facts and towards action'. He concludes by describing pragmatism as 'looking away from first things, principles, categories and supposed necessities', and 'looking towards last things, fruits, consequences and facts' (James 1906).

16.4.5 Barbarians at the Gate

The tentative bridging of our planned ideals with reality may signal a shift in our discourse towards pragmatism, a concept anathema to the South African policy discourse since 1994. Four ways in which this reluctant pragmatic turn is manifesting in our body politic are (1) evidence-based planning, (2) the market, (3) the entrepreneur and (4) the citizen:

Enter Evidence-Based Planning The last few years have seen marked improvement in the availability and use of evidence in policy decision-making. In terms of availability, third-party geocoded environmental, settlement and economic data has proliferated exponentially in recent years. The spatialisation of SARS tax data spearheaded by the CSP, and the application of machine learning techniques to satellite imagery by third party service providers are but two ways in which the evidence base for understanding social processes at a spatially granular level has expanded dramatically. National Treasury has—as part of its reporting reforms process—invested significant resources into improving the mechanisms by which local government reports local performance outcomes. The City of Cape Town is developing data-driven economic appraisal tools and embedding these into the allocative decision-making process. By harnessing the growing availability of administrative, official and third-party data, local government is better positioned than ever before to gain insight into the functioning of the city economy, and on that basis identify and prioritise interventions that are more likely to yield meaningful developmental returns.

Enter the Market Over the last 40 years, private service providers have supplanted the State in the provision of health services, security, and, more recently education. The precipitous decline in state-owned logistics and energy sectors have saw the beating heart and arteries of our economy falter. Collapsing water infrastructure has reached into the daily lives of households. After years of lip service about PPPs, the Presidency launched Operation Vulindlela which aims to drawing the market into sectors previously dominated by the State, including energy, ports, freight, visas and telecommunications. The involvement of the private sector isn't limited to major network industries, but increasingly extends to core municipal functions: with only 10 working firefighting vehicles and the procurement of 17 additional engines mired in corruption, the City of Johannesburg has admitted its inability to service communities. Gauteng is now served by three private firefighting companies (News24 2023).

Enter the Entrepreneur The focus on private initiative is not limited to better resourced, established industry players. In recent years, the development of small-scale rental units by micro-developers in former townships has gathered pace and geographic spread. Some metros are revisiting regulatory impediments to backyarding, microdevelopment and informal entrepreneurship to accelerate housing supply and economic growth.

Enter the Citizen The final, and most consequential, barbarian confounding our planned ideals but charting our future is the citizen. The roll-out of rooftop solar panels is emblematic of the power and resilience of the individual, creating the energy generation capacity of 4500 MW in 1 year (it took the State 20 years and R200bln to build Medupi). Neighbourhood Watch, Gift of the Givers, City Improvement Districts, Tidy Towns, 'Ignite and Revive Kroonstad' and Kimberley's 'Bring Back the Sparkle' are but some of a myriad new civil society formations which have emerged in recent years in the wake of deteriorating urban management. There is also a dark side to self-help: households invading strategically located parcels of public land, connecting illegally to electricity grids, and exacting vigilante justice.

Where magical thinking helped the country navigate through a period of transition and uncertainty, pragmatic policy may help us cope better with complex problems. Not because it resolves longstanding ideological disputes, but by shifting attention away from the normative as the final arbiter of theoretical respectability. To paraphrase Hoch (2017), pragmatism makes the pursuit for some abstract notion of a better society appear tangential and inconsequential to the business of coping with the pressing problems of the day.

16.5 A Manifesto for the Pragmatic Planner

What is the role of the planner in this brave new world? The following points-of-departure guide my practice as a built environment professional:

16.5.1 Check Yourself

The first three guidelines aim to provide an intellectual orientation towards our practice, by drawing on useful disciplines, tools and ways of thinking about social processes across geographic scale:

The Small Stuff Is the Big Stuff The spatial and economic outcomes we see today are perpetuated not by the persistence of grand narrative but rather the culmination of a myriad daily decisions taken by a wide variety of economic actors, from workers, consumers and households to entrepreneurs and investors. These micro-decisions, based on personal preferences, incentives and limited information, can lead to spatial and economic outcomes far removed from our vision statements (Schelling 1978).[12] The discipline of economics offers a range of empirical methods to shed light on the nature of these decisions and its relationship to broader neigh-

[12] See Nobel Laureate Thomas Schelling's seminal Micro-motives and Macro behavior where he accounts for segregation in US cities.

bourhood and citywide outcomes. Within economics there are particular fields which focus specifically on identifying and analysing regularities in social processes across spatial scale: above all, urban economics and behavioural economics derive from both microeconomics and social psychology, and provide useful frameworks for us to understand cities from the bottom up.

Be a Model Thinker In contrast to normative planning, economics is a logical-positivist social science, and its theory building is deductive: the appropriate relationship between each economic actor will depend on local context, macro conditions, and the capacity of each actor. Unlike normative theory, an economist *aspires* to be value-free by describing reality from the bottom up, starting with the 'small stuff', and only *then* attempts to intervene (Garza and Garza 2024). We do so by hypothesising the causal relationships between phenomena, building models to represent those hypothesised relationships, and then validating the model through real-world data. While formal economic modelling is a specialised skill beyond the ability of your author, fortunately one doesn't have to *be* a modeler to *think* like a modeller.

What do I mean by 'model thinking'? Models are simplifications of reality which help us planners think about and intervene in complex phenomena like cities more effectively (Page 2018). Model thinking helps the pragmatic planner in three ways: (1) since all humans—even planners—are subject to personal biases and face limits of cognition, models allow us to interrogate our own assumptions, embrace uncertainty and not fixate on specific outcomes—in short, it helps us to reason coherently;[13] (2) by requiring precise language, explicit causal mechanisms and measurable objectives, models help us measure systematically and communicate clearly; (3) last but not least, models stimulate our creative minds by allowing exploration of possibilities and hypotheticals. These are some hallmarks of model thinking.

Municipalities as Political Economies It is naïve however to think that urban development *only* happens from the bottom up. Non-market forces such as politics, institutions and social norms exert a powerful influence over spatial and economic outcomes. Indeed, every city since the bronze age is to a large extent a product of collective government decisions (Garza and Garza 2024).[14] The pragmatist views these organisations not formalistically as powers and functions but rather *networks of self-interested individuals* operating within a set of implicit organisational norms, making rational decisions based on limited information and negotiating perverse incentives and risks. It is through this lens, for example, that we can identify the conditions under which endemic corruption take root and Constitutional impera-

[13] See Edmund Burke's exposition on 'epistemological modesty'.

[14] Negating the role of the State in spatial outcomes would lend credence to the dismissal of economics as the 'science of neoliberalism' (Zuidhof 2014) bringing deregulation, privatisation policies and the rollback of the welfare state in its wake. This, in my view, mischaracterises economics as an ideology rather than a practical tool, since the pragmatic planner shouldn't hold vigil to any specific dogma about how society *should* function.

tives like subsidiarity founder. By drawing on institutional economics, the pragmatic planner is able to identify whether the requisite political and organisational conditions exist to support a proposed intervention.

16.5.2 Map Out Your Assumptions

Next, I share three points-of-departure which assist me in developing robust and useful theories-of-change to make explicit the causal assumptions underlying the policy design procs:

Focus on the Here and Now The challenge for the pragmatic planner is to move from the abstract to the concrete, from long-term visions to intermediate outcomes. While this is no easy task for a student of humanities such as myself, the following techniques have proved enormously useful in my practice. First of all, *ditch the jargon*. When initiating a new project, apply a scalpel to vague terms with no settled meaning, like 'spatial transformation', 'resilience' or 'spatial justice'. Second, *keep things separate and distinct*. Resist the temptation to conflate social and economic development objectives, or to conflate your objectives with constraints. Third, get into the habit of mapping out your causal assumptions during project inception. A conceptual framework consisting of measurable phenomena is the intellectual work horse of any complex project, there to pull you out of the morass of circular reasoning and ambiguity.

What Matters Is What Works The pragmatic planner should always keep one foot planted in reality (Lynch 1984). Let go of ideological purity—worry about results. As goes the saying 'It doesn't matter if a cat is black or white, as long as it catches mice', a policy is only as good as its practical consequences. A solution that meets certain definitional qualifications, but fails to solve the problem, is meaningless. A pragmatic planner identifies and adapts tried and tested solutions that have been shown to work elsewhere. In this sense, the pragmatic planner is more like a plumber, absorbed with 'making things happen'. However, being policy agnostic does not imply being myopic: if a particular problem is shown to lack precedent, the pragmatist must revisit the chains of causation that underlie social processes and the regularities by which they articulate over space and time.

Fewer Objectives, More Constraints Understanding the difference between what we are trying to achieve, and what might limit our options, is an essential (yet sorely lacking) ingredient for effective policymaking. As our experience with the NDP shows, one of the main reasons why we fail to achieve our visions is that we set too many objectives. In trying to cover all possible grounds, from economic to social and environmental, we achieve none of our goals. One reason why we set too many objectives is because planners tend to conflate goals with constraints (Bertaud 2018). By differentiating between our goals and constraints, not only are we better able to focus on a limited set of objectives, but our strategies are better rooted in reality.[15] Clear separation helps in prioritizing actionable plans.

16.5.3 People Then Places

The next four guidelines help the pragmatic planner avoid common mistakes in defining the problem and identifying solutions:

Invest in People, Then Places Whereas essentialists are wedded to Place, pragmatists see the citizen as the measure of all things. Informality, urbanisation and migration reminds us that people are not rooted to place—they move to opportunity. An increasing number of authors caution against the use of place-based policies except in exceptional circumstances, arguing that development policies based on spatially blind strategies are more effective at enhancing efficiencies and generating positive distributional impacts (Rabe and Selod 2023). What is the implication for local government, whose levers are inherently place-based? In a context where two-thirds of our local municipalities are classified as failures (Ramaphosa 2022), the municipality which focuses all its energy on providing core municipal services affordably, effectively and universally, while maintaining core infrastructure like roads, street lighting and reservoirs, is excellently positioned to retain existing and attract new economic activity from beyond its borders.

Inhabit the Mind of the Notional Investor The pragmatic planner asks 'who are the actors whose decisions shape cities?' How are development and funding decisions made? Which factors influence these decisions? When is development considered viable? As private actors central to urban development, investors and property developers play a crucial role in shaping urban development outcomes yet are often expected by policymakers to be compliant partners that share the planner's concerns and objectives. Do you understand the motives and behaviours of property developers or the structure and the organisation of the development industry? Keeping the processes by which urban development happens front-of-mind is a pre-requisite to respond effectively to the Big Five concerns at scale. By engaging with these critical decision-making processes, planners are better able to influence aggregate outcomes.

No Place Is an Island Economic outcomes and the causal processes that govern them cannot be understood unless spatial structure is accounted for. Problems that are narrowly defined and interventions narrowly designed are blind to the trade-offs and consequences of localised interventions on the urban system, and vice versa. Similarly, the prospects of intervention success are significantly affected by market dynamics which operate beyond the boundaries of the precinct and suburb. We may consider, for example, the impact of an unyielding urban edge on the periphery of

[15] For example, aiming to reduce traffic congestion (a goal) should be distinct from the limited budget for infrastructure improvements (a constraint).

the city, on housing affordability at its centre (Pfeiffer et al. 2019). Or the impact of deteriorating operating environments in industrial areas in Johannesburg South on job accessibility for residents of Soweto. In all these cases, the prospect for achieving a place-based objective depends on what happens elsewhere in the region.

Target Economic Potential Whereas social strategy is about targeting vulnerability, economic strategy is about harnessing opportunity.[16] Every region in South Africa harbours unmet development potential. However, not all places offer the same potential for economic intervention. A policy may succeed in one location and fail completely in another. Since location potential will rarely align geographically with political and equity priorities, a key challenge for policymakers is to avoid reductive and decontextualised assumptions about what makes a location 'strategic'. The inherent unevenness of location potential suggests that local economic development planning should seek to develop each region or city to its maximum potential rather than pursue 'administered convergence' at all costs.[17] Maximising local potential requires understanding locational comparative advantage from the perspective of investors rather than government officials. It is the responsibility of local government to define pathways to investment that speak to each region's specific attributes.

16.5.4 Speak up

The pragmatic planner challenges magical thinking at every turn, in politics, in discourse, on social media, in project meetings. When urban planners hide behind abstract terms—undefinable, mutable, unmeasurable—they cede terrain to magical thinking and become complicit in the subversion of the policy development Whereas politics is about the next 5 years, pragmatic planning is about results, about execution. It is about forming a bridge between normative planning (*imagining* a better society) with functional theory (*how* we can change our *existing* society for the better) using evidence-based, causal reasoning.

16.6 Quo Vadis?

> "There's a lesson in all of this for those who govern SA today: they should find a balance between their ideals and economic realities. The world is what it is. The real — not ideal — world, is complex, and it is a world where the only success for countries lies in making hard

---

[16] Economic strategy that only responds to need rarely generates benefits beyond the once-off social transfer between State and Citizen.

[17] The National Spatial Development Perspective—which provided a clear roadmap in directing resources based on an evidence-based comparative assessment of location potential and social need—was sadly ahead of its time.

> choices and painful trade-offs." (Jabulani Sikhakhane Business Day 2023 "Balancing idealism and pragmatism")

In this retrospective spanning 30 years of local economic development planning in South Africa, between 1994 and the present, I offer a frank account of my observations and reflections as a researcher, government official and public sector consultant. Drawing on the life and death of three urban policy bandwagons, 'Local Economic Development', 'Transit-oriented Development' and 'Smart Cities', I evaluate South Africa's legacy of economic development planning not through the lens of high-minded principles but rather, in terms of its cumulative material impact on the everyday South African citizen. Tracing the seeds of our current economic and urban malaise to the embrace of normative planning in the heady years of nation-building, the inescapable conclusion is that a generation of built environment practitioners have pursued grand spatial narratives beyond our appreciation of how cities actually work—that is, the myriad interconnected micro-decisions made by individuals, entrepreneurs, households, developers and investors, that shape and reproduce the living organism that is the city.

Whether the elevation of vision over execution from the mid-1990s onwards was symptomatic of excessive zeal and exceptionalism, or a necessary compromise in our fragile democracy-building process, every metric tracking material outcomes serves as a warning that the prevailing approach to planning is no longer fit for purpose. Whether through intention or oversight, we have engaged in 'magical planning', the conceit that we can manifest progress towards a desired future state solely by articulating and promoting that vision, regardless of the State's capacity to implement or our understanding of the social processes governing urban development. The yawning gap between our vision statements and the lived experience of the average South African citizen has never been greater.

Yet, recent developments suggest that the South African body politic is resigning itself to the accommodation of private initiative as a delivery agent for public goods as a supplement to deteriorating State capability. As the rest of South Africa lurches reluctantly towards pragmatism, planning discourse risks getting left behind.

16.7 Afterword

The reader will excuse the bruises and biases evident in my account, written on the slow train from Samarkand to Khiva in September 2023. At best, it will encourage a more eloquent, considered soul to tackle a sympathetic but honest history of our practice spanning the last three decades, a history that rises above the language games constricting our discourse; moreover, a history that neither condemns nor spares those of us, living or dead, illustrious or middling, that tried—each in our particular and flawed way—to help the country we love.

References

Ababio EP, Meyer DF (2012) Local economic development. Admin Publica 20(4) https://journals.co.za/doi/pdf/10.10520/ejc-adminpub-v20-n4-a2

Bertaud A (2018) Order without design: how markets shape cities. The MIT Press, Cambridge, MA

Bloomberg.Com (2013) The limits of bus rapid transit: a Cape Town case study, March 14, 2013. https://www.bloomberg.com/news/articles/2013-03-14/the-limits-of-bus-rapid-transit-a-cape-town-case-study

City of Cape Town (2023) Comprehensive Integrated Transport Plan 2023–2028. Cape Town

City of Johannesburg (2017) Rea Vaya Bus Rapid Transit System. Report to National Parliament. https://static.pmg.org.za/1/170912_COJ.pdf

Danermark B, Ekstrom M, Karlsson, JC (2019) Explaining society critical realism in the social sciences. Routledge Second edition. Abingdon/New York: Routledge. | Series: Routledge studies in critical realism | Translation of the author's book Att fèorklara samhèallet

Engineering News (2023) Cape Town's MyCiTi bus service logs 18.32-million trips. Engineering News. https://www.engineeringnews.co.za/article/cape-towns-myciti-bus-service-logs-68-growth-since-2021-2023-09-18

Garza N, Garza J (2024) Urban strategy in an era of public policy assessment: beyond the methodological divide. Rev Dev Econ 28(4):2105–2120. https://doi.org/10.1111/rode.13017

Hare C, Walwyn D (2019) A qualitative study of the attitudes of South African Spaza shop owners to coopetitive relationships. S Afr J Bus Manage 50(1):1–12. https://doi.org/10.4102/sajbm.v50i1.1295

Harrison P, Rubin M, Appelbaum A, Dittgen R (2019) Corridors of freedom: analyzing Johannesburg's ambitious inclusionary transit-oriented development. J Plan Educ Res 39(4):456–468. https://doi.org/10.1177/0739456X19870312

Hausmann R (2022) South Africa's macroeconomic risks after a decade of microeconomic turbulence. Working Paper Series 404. Centre for International Development, Cambridge, MA. https://growthlab.cid.harvard.edu/publications/macroeconomic-risks-after-decade-microeconomic-turbulence-south-africa-2007

Henderson JV (2020) Cities in developing countries. Lectures on urban economics. London School of Economics, London

Hoch C (2017) Neo-pragmatist planning theory. In: The Routledge handbook of planning theory. Routledge

Hollands RG (2008) Will the real smart city please stand up? Intelligent, progressive or entrepreneurial? City 12(3):303–320. https://doi.org/10.1080/13604810802479126

James W (1906) Pragmatism. In: Pragmatism. Routledge

Kretzmann S (2022) City of Cape Town's new transport plan ditches big legacy projects. GroundUp News, 26 October 2022. https://www.groundup.org.za/article/city-of-cape-towns-new-transport-plan-ditches-big-legacy-projects/

Loewald C, Makrelov K, Worgotter A (2021) Addressing low labour utilisation in South Africa. WP/21/09. Working Paper Series. South African Reserve Bank

Lynch K (1984) Good city form. MIT Press

News24 (2023) Gauteng now has 3 private firefighting companies. One of them dealt with 14 fires on its first day. News24, 2023, sec. News24. https://www.news24.com/news24/southafrica/news/gauteng-now-has-3-private-firefighting-companies-one-of-them-dealt-with-14-fires-on-its-first-day-20230822

Page SE (2018) The model thinker: what you need to know to make data work for you, 1st edn. Basic Books, New York

Pfeiffer BF, Rabe C, Selod H, Viguie V (2019) Assessing urban policies using a simulation model with formal and informal housing: application to Cape Town, South Africa. SSRN Scholarly Paper, Rochester. https://papers.ssrn.com/abstract=3430533

Pronin E, Wegner DM, McCarthy K, Rodriguez S (2006) Everyday magical powers: The role of apparent mental causation in the overestimation of personal influence. J Pers Soc Psychol 91(2):218–231. https://doi.org/10.1037/0022-3514.91.2.218
Rabe C, Selod H (2023) Place-based economic policies: international lessons for South Africa. Economic Research Southern Africa. https://econrsa.org/publications/place-based-economic-policies-international-lessons-for-south-africa/
Rabe C, McGaffin R, Crankshaw O (2015) A diagnostic approach to intra-metropolitan spatial targeting: evidence from Cape Town, South Africa. Dev South Afr 32(6):726–744
Ramaphosa C (2022) Key note address by President Cyril Ramaphosa. South African Human Rights Commission, Johannesburg. https://www.sahrc.org.za/index.php/sahrc-media/news/item/3276-this-is-a-crisis-ramaphosa-bemoans-state-of-municipalities
Rogerson CM (2014) Reframing place-based economic development in South Africa: the example of local economic development. Bull Geogr Socio-Econ Series 24:203–218
Rogerson C (2018) Local economic development in the changing world: the experience of Southern Africa. Routledge
Rosengren K, French J (2013) The Oxford handbook of the development of imagination. Oxford University Press, Oxford
Saba A (2023) Smart city remains but a dream. The Mail & Guardian (blog), 6 February 2023. https://mg.co.za/news/2023-02-06-smart-city-remains-but-a-dream/
SACN (2019) Rethinking LED in intermediate cities, Johannesburg
Schelling TC (1978) Micromotives and macrobehavior. WW Norton & Company, New York
Shah K (2022) Diagnosing South Africa's high unemployment and low informality. Working paper series 2022.138. Harvard University, Cambridge, MA. https://growthlab.cid.harvard.edu/publications/diagnosing-south-africa-high-unemployment-and-low-informality
South Africa. The Presidency (2006) The National Spatial Development Perspective. Pretoria.
Watson V (2014) African urban fantasies: dreams or nightmares? Environ Urban 26(1):215–231. https://doi.org/10.1177/0956247813513705
Yin C, Xiong Z, Chen H, Wang J, Cooper D, David BT (2015) A literature survey on smart cities. Sci China Inf Sci 58(10):1–18. https://doi.org/10.1007/s11432-015-5397-4
Zuidhof P (2014) Thinking like an economist: The Neoliberal Politics of the Economics Textbook. Rev Soc Econ 72(2):157–185. https://doi.org/10.1080/00346764.2013.872952

Index

D. du Plessis (ed.), *Territorial Governance and Spatial Transformation in Post-apartheid South Africa*, Local and Urban Governance,
https://doi.org/10.1007/978-3-031-85489-7

The manufacturer's authorised representative in the EU is Springer Nature Customer Service Centre GmbH, Europaplatz 3, 69115 Heidelberg, Germany. If you have any concerns regarding our products, please contact ProductSafety@springernature.com

Printed and bound by CPI Group (UK) Ltd, Croydon, CR0 4YY
01/10/2026
02225960-0015